Springer Complexity

Springer Complexity is an interdisciplinary program publishing the best research and academic-level teaching on both fundamental and applied aspects of complex systems – cutting across all traditional disciplines of the natural and life sciences, engineering, economics, medicine, neuroscience, social and computer science.

Complex Systems are systems that comprise many interacting parts with the ability to generate a new quality of macroscopic collective behavior the manifestations of which are the spontaneous formation of distinctive temporal, spatial or functional structures. Models of such systems can be successfully mapped onto quite diverse "real-life" situations like the climate, the coherent emission of light from lasers, chemical reaction-diffusion systems, biological cellular networks, the dynamics of stock markets and of the internet, earthquake statistics and prediction, freeway traffic, the human brain, or the formation of opinions in social systems, to name just some of the popular applications.

Although their scope and methodologies overlap somewhat, one can distinguish the following main concepts and tools: self-organization, nonlinear dynamics, synergetics, turbulence, dynamical systems, catastrophes, instabilities, stochastic processes, chaos, graphs and networks, cellular automata, adaptive systems, genetic algorithms and computational intelligence.

The two major book publication platforms of the Springer Complexity program are the monograph series "Understanding Complex Systems" focusing on the various applications of complexity, and the "Springer Series in Synergetics", which is devoted to the quantitative theoretical and methodological foundations. In addition to the books in these two core series, the program also incorporates individual titles ranging from textbooks to major reference works.

Editorial and Programme Advisory Board

For further volumes:
http://www.springer.com/series/712

Springer Series in Synergetics

Founding Editor: H. Haken

The Springer Series in Synergetics was founded by Herman Haken in 1977. Since then, the series has evolved into a substantial reference library for the quantitative, theoretical and methodological foundations of the science of complex systems.

Through many enduring classic texts, such as Haken's *Synergetics* and *Information and Self-Organization*, Gardiner's *Handbook of Stochastic Methods*, Risken's *The Fokker Planck-Equation* or Haake's *Quantum Signatures of Chaos,* the series has made, and continues to make, important contributions to shaping the foundations of the field.

The series publishes monographs and graduate-level textbooks of broad and general interest, with a pronounced emphasis on the physico-mathematical approach.

Philippe Blanchard • Dimitri Volchenkov

Random Walks and Diffusions on Graphs and Databases

An Introduction

 Springer

Philippe Blanchard
Dimitri Volchenkov
Universität Bielefeld
Fakultät für Physik
Universitätsstr. 25
33615 Bielefeld
Germany
blanchard@physik.uni-bielefeld.de
dima427@yahoo.com

ISSN 0172-7389
ISBN 978-3-642-26842-7   ISBN 978-3-642-19592-1 (eBook)
DOI 10.1007/978-3-642-19592-1
Springer Heidelberg Dordrecht London New York

Cover design: SPi Publisher Services

Printed on acid-free paper

Springer is part of Springer Science+Business Media (www.springer.com)

*Dedicated to our parents with thanks for
their care and love.*

Preface

Classical graph theory has been developed in regard to its applications in urban planning, transport, energetics, and many other fields. The general optimization mindset dominating these researches has addressed to graph theory the questions which were often related to finding the *shortest path* between nodes, as being of the minimum time delay for information transmission and of the minimum cost for connection maintenance. Not surprisingly, the very definition of distance between two vertices in a graph is given as the geodesic distance, i.e., the shortest path connecting them. With respect to the graph metric, a complex network of weighted edges is rather considered as a minimum weight spanning tree of the underlying graph, i.e., a subset of paths that has no cycles but still connects to every vertices at the lowest total cost. However, in many problems of practical interest found in everyday life the existence of many paths of different lengths as well as a nexus of cycles traversing the nodes in many complex networks do matter!

In contrast to classical graph theory paying attention to the shortest paths of least cost, in the approach that we discuss in our book, *all possible paths* between the two vertices in a connected graph are taken into account, although some paths shall be more preferable than others. Such a formulation of graph theory can be called as of a *"path integral"*, since "integral" means "to include all." Random walks respecting all graph symmetries assign a probability to each path in the graph to be traversed by a random walker. Then, in order to find the expected first-passage distance between the two vertices, one integrates over all possible paths of the system in between them. Consequently, each vertex is characterized with respect to the entire structure of the graph by its own "path integral" vector accounting for the sum of the probability amplitudes for every possible path leading to that from a randomly chosen vertex. Perhaps, the most interesting fact about such a "path integral" approach to graphs is that the probabilistic distance naturally induces a Euclidean metric on a graph (sometimes called the 'diffusion metric', or the 'effective resistance metric') allowing for a geometric representation of the relationships between vertices in a graph, in terms of distances and angles, as in Euclidean geometry of everyday intuition. Vertexes of graphs and units of data bases that cast in the same mold with respect to the individual data features are revealed by geometric proximity in Euclidean space that might be either exploited visually, or accounted analytically. High-dimensional Euclidean representations of graphs and databases are characterized by the rank-ordering of data traits providing us with the natural geometric framework for dimensionality reduction facilitating the data analysis and further interpretation of results.

Perhaps, Lagrange was the first scientist who investigated a simple dynamical process (diffusion) in order to study the properties of a graph (Lagrange 1867). He calculated the spectrum of the Laplace operator defined on a chain (a linear graph) of N nodes in order to study the discretization of the acoustic equations. Nowadays it is well known that random walks could be used in order to investigate and characterize how effectively the nodes and edges of large networks can be covered by different strategies (see Tadic 2002; Yang 2005; Costa and Travieso 2007 and many others).

In this book, we follow the interdisciplinary lecture course on the stochastic analysis of complex networks and databases delivered by us at the University of Bielefeld (Germany) during the Fall semester 2008 and the Spring semester 2009 and targeted to bring about a more interdisciplinary approach across diverse fields of research including complex network theory and data analysis, as well as sociology, bio-informatics, urban planning and linguistics. The book contains a wealth of material generously equipped with suggestions for further reading and the glossary of term and concepts in graph theory that is helpful for those at the beginning of their acquaintance with the subject.

In the subsequent ten chapters of this book, we describe a fascinating journey from the elementary discrete mathematics (Chaps. 1, 2) to the elements of algebraic graph theory (Chap. 3), to a detailed analysis of complex multicomponent systems and databases (Chaps. 4–7), to the applications of random walk methods for the components analysis of complex networks and databases (Chap. 8). In the Chap. 9, we discuss the dynamical processes in models containing a large number of positive and negative feedbacks such as epidemic spreading, synchronization, and self-regulation in complex genetic networks. Finally, in the Chap. 10, we consider strongly non-linear transport phenomena in large complex networks containing regular subgraphs.

Many colleagues helped over the years to clarify many points throughout the book. Our thanks go to Sven Banisch, Bruno Cessac, Pierre Collet, Jean René Dawin, Sergey Dorogovtsev, Jürgen Jost, Dmitri Krioukov, Andreas Krüger, Tyll Krüger, Ricardo Lima, Zhi-Ming Ma, Rui Vilela Mendes, Walter Pauls, Filippo Petroni, Helge Ritter, Gabriel Ruget, Maurizio Serva, Ludwig Streit, Sören Wichmann.

We are further indebted to Dr. Christian Caron for competent advices and assistance in the completion of the final manuscript. His assistance is gratefully acknowledged.

Bielefeld   *Philippe Blanchard*
                                                                             Dimitri Volchenkov

Contents

Contents   xiii

Chapter 1
Introduction to Permutations, Markov Chains, and Partitions

A foliage garden of orchard apple trees can be fun and rewarding, no matter what apple variety is grafted onto root stocks. Depending on variety and climate you can expect to harvest your apples at different times. Once the base skin color of the apples and a pleasing taste convince you that the fruits are harvested mature they require quick ripen off the plant to ensure excellent quality. Then you must decide the order of picking apples from the trees. A major question is now to enumerate trees in the garden by saying which tree "precedes" another. How many different orders are possible?

In the present chapter, we review this and related questions.

1.1   Permutations and Their Matrix Representations

Certainly, in the garden $\mathcal{G}$ consisting of N apple trees, you have precisely N choices for the first tree you pick apples from. Then you have $N - 1$ choices for the tree which goes second. Proceeding in a similar manner we have $N - 2$ choices, for the plant to be served third and so on, until the only choice, for the remaining tree harvested last (Fig. 1.1).

The number of orders in which the trees can be harvested equals

$$N! = N \cdot (N - 1) \cdot (N - 2) \cdot \ldots \cdot 2 \cdot 1, \tag{1.1}$$

the rather large number even for small N. Any arrangement of orchard trees into a linear order by harvesting each of them once, and only once is called a *permutation*. A permutation $\Pi : \mathcal{G} \to \mathcal{G}$ over a finite set $\mathcal{G}$ is a one-to-one correspondence (*bijection*) that is nothing else but a certain enumerating rule for its elements,

$$t_k = \Pi(t_i), \quad t_i, t_k \in \mathcal{G}. \tag{1.2}$$

P. Blanchard and D. Volchenkov, *Random Walks and Diffusions on Graphs and Databases*, Springer Series in Synergetics 10, DOI 10.1007/978-3-642-19592-1_1, © Springer-Verlag Berlin Heidelberg 2011

Fig. 1.1 The number of all harvesting orders in this garden of five apple trees is $5! = 120$. The picture was drawn by Andrew Volchenkov

Each equation of (1.2) constitutes an elementary transposition that swaps two elements of the set $\mathcal{G}$.

It is convenient to write the complete set of (1.2) in a matrix form. The set of vectors having 1 at the ith-position,

$$\mathbf{e}_i = (0, 0, \ldots, 1_i, \ldots, 0), \quad i = 1, \ldots N, \tag{1.3}$$

forms an orthonormal set, as all vectors $\mathbf{e}_i$ in that are of unit length and are mutually orthogonal with respect to the inner product

$$(\mathbf{e}_i, \mathbf{e}_j) = \delta_{ij}$$

where

$$\delta_{ij} = \begin{cases} 1, i = j \\ 0, i \neq j \end{cases}$$

is the Kronecker delta symbol. The orthonormal set (1.3) forms an orthonormal basis $\{\mathbf{e}_i\}_{i=1}^{N}$ of a vector space over $\mathcal{G}$. Any permutation Π over $\mathcal{G}$ can be *uniquely* represented with respect to the basis $\{\mathbf{e}_i\}_{i=1}^{N}$ by a specific square binary *permutation matrix* (which we denote therefore by the same symbol $\mathbf{\Pi}$ but bold faced) that has exactly one entry 1 in each row and each column and 0's elsewhere.

For instance, the swapping over a set of five elements, in which 1 holds its place, 2 changes to 4, 3 to 2, 4 to 5, and 5 to 3,

$$\Pi = \begin{pmatrix} 1 & 2 & 3 & 4 & 5 \\ \downarrow & \downarrow & \downarrow & \downarrow & \downarrow \\ 1 & 4 & 2 & 5 & 3 \end{pmatrix}, \tag{1.4}$$

is described with respect to the basis $\{\mathbf{e}_i\}_{i=1}^{N}$, by the permutation matrix

$$\mathbf{\Pi} = \begin{pmatrix} 1\,0\,0\,0\,0 \\ 0\,0\,0\,1\,0 \\ 0\,1\,0\,0\,0 \\ 0\,0\,0\,0\,1 \\ 0\,0\,1\,0\,0 \end{pmatrix}. \tag{1.5}$$

The identity permutation which fixes each element of an ordered set on its own place is the $N \times N$ identity matrix,

$$\mathbf{\Pi}_{\mathrm{id}} = \mathbf{1},$$

under which each element of the set holds its order.

It is easy to check that the transpose of a permutation matrix $\mathbf{\Pi}^{\top}$, which is computed by swapping columns for rows in the matrix $\mathbf{\Pi}$, describes the inverse permutation

$$\mathbf{\Pi}^{\top} = \mathbf{\Pi}^{-1}$$

that swaps elements back to their initial positions. Permutation matrices are orthogonal matrices, as

$$\mathbf{\Pi}\,\mathbf{\Pi}^{\top} = \mathbf{1}. \tag{1.6}$$

Given the two different permutations, $\mathbf{\Pi}_1$ and $\mathbf{\Pi}_2$, the composition of both,

$$\Pi_1 \circ \Pi_2 : \mathcal{G} \to \mathcal{G},$$

is also a permutation, which is naturally defined by the following rule,

$$\Pi_1 \circ \Pi_2\,(\mathfrak{t}_k) = \Pi_1\,(\Pi_2\,(\mathfrak{t}_k)). \tag{1.7}$$

The permutation matrix of the composition (1.7) is the product of the permutation matrices $\mathbf{\Pi}_1$ and $\mathbf{\Pi}_2$,

$$(\Pi_1 \circ \Pi_2)_{ij} = (\mathbf{\Pi}_1\,\mathbf{\Pi}_2)_{ij}. \tag{1.8}$$

We conclude that all permutations of a finite set of elements forms a group under matrix multiplication with the unit matrix $\mathbf{1}$ as the identity element. The group of all permutations is called the *symmetric group* $\mathbb{S}_N$. The matrix representation of the symmetric group $\mathcal{S}_N$ consists of $N!$ permutation matrices. Matrix representation of a group is important because it allows many group-theoretic problems to be reduced to problems in linear algebra.

1.2 Orbits and Fixed Points

Any permutation $\Pi \in \mathbb{S}_N$ over a finite set $\mathcal{G}$ determines an *equivalence relation*

$$\mathfrak{t} \sim \mathfrak{t}', \quad \text{for} \quad \mathfrak{t}, \mathfrak{t}' \in \mathcal{G}$$

if there is an integer number $n > 0$ such that

$$t' = \Pi^n(t). \tag{1.9}$$

The equivalence relation (1.9) partitions $\mathcal{G}$ into a set of *equivalence classes* $[t]$,

$$\mathcal{G}/\sim = \{[t] : t \in \mathcal{G}\}, \quad [t] = \{t' \in \mathcal{G} : t' \sim t\}, \tag{1.10}$$

called *disjoint cycles* (or *orbits*),

$$\mathcal{G} = [t_1] \cup [t_2] \cup \ldots \cup [t_k], \quad [t_i] \cap [t_j] = \emptyset, \text{ iff } i \neq j. \tag{1.11}$$

For example, the permutation (1.4) partitions the set of five elements into two orbits,

$$\mathcal{G} = [1] \cup [3],$$

in which $[1] = 1$ consists of the only element that holds its place as the permutation (1.4) advances, and

$$[3] = (3 \to 2 \to 4 \to 5 \to 3)$$

is a cycle. The elements of an ordered set that hold their places under a permutation are called the *fixed points* of the permutation. The permutation (1.4) has the only fixed point, $1 = \Pi(1)$.

The trace of a permutation matrix, the sum of its diagonal elements, equals the number of fixed points of the permutation,

$$\text{Tr } \mathbf{\Pi} = \text{card}\{t = \Pi(t) : t \in \mathfrak{G}\}. \tag{1.12}$$

The characteristic equation of a permutation matrix $\mathbf{\Pi}$ is the equation in one variable λ,

$$\det(\mathbf{\Pi} - \lambda \cdot \mathbf{1}) = 0. \tag{1.13}$$

The solutions of the characteristic (1.13) are the eigenvalues of the matrix $\mathbf{\Pi}$. The set of eigenvalues of a permutation matrix always consists of the two real points, $+1$ and -1, of some multiplicity and a number of complex conjugated pairs of eigenvalues. For example, the eigenvalues of the permutation matrix (1.5) are

$$\lambda = \left[1, 1, \left\{ \begin{matrix} i \\ -i \end{matrix} \right\}, -1 \right], \quad i = \sqrt{-1}.$$

The multiplicity of the maximal eigenvalue $\lambda_{\max} = 1$ of a permutation matrix equals the number of orbits in the permutation Π. The eigensubspace belonging to the maximal eigenvalue $\lambda_{\max} = 1$ of the permutation matrix (1.5) is spanned by the orthogonal eigenvectors,

$$\mathbf{X} = \begin{array}{l} \langle\, 1\ 0\ 0\ 0\ 0\, \rangle \to [1] = (1) \\ \langle\, 0\ 1\ 1\ 1\ 1\, \rangle \to [3] = (3 \to 2 \to 4 \to 5 \to 3), \end{array} \tag{1.14}$$

representing the set of disjoint cycles in the permutation Π.

The counting of different orbits of a permutation can be performed automatically by considering the matrix $\mathbf{X}$ with the columns formed by the eigenvectors belonging to the maximal eigenvalue $\lambda_{\max} = 1$ of $\mathbf{\Pi}$ as a rectangular matrix. The product $\mathbf{XX}^\top$ is a square diagonal matrix, the rank of which equals the number of disjoint cycles in the permutation, and the diagonal elements of which are the lengths of those cycles,

$$\mathbf{XX}^\top = \begin{pmatrix} 1 & 0 \\ 0 & 4 \end{pmatrix}. \tag{1.15}$$

Permutation $\Pi \in \mathcal{S}_N$ is said to–belong to a *cycle class*

$$\{1^{\alpha_1}\, 2^{\alpha_2} \ldots N^{\alpha_N}\}, \quad \text{where} \quad 1\alpha_1 + 2\alpha_2 + \ldots N\alpha_N = N, \tag{1.16}$$

if it contains precisely α_l orbits of length $l = 1, 2, \ldots N$. For example, the permutation Π defined by (1.4) belongs to the cycle class $\{1^1 4^1\}$.

Different cycle classes include different numbers of permutations. Although the number of all possible permutations over a finite set of N elements equals $N!$, many of them might be of the same cycle class. By using the *decomposition of unity*,

$$1 = \sum_{1\alpha_1 + 2\alpha_2 + \ldots N\alpha_N = N} \frac{1}{1^{\alpha_1} 2^{\alpha_2} \ldots N^{\alpha_N} \alpha_1! \alpha_2! \ldots \alpha_N!}, \tag{1.17}$$

in which the summation is taken over all possible solutions of the (1.16) in integer numbers, we can show that the total number of equivalent permutations within the cycle class $\{1^{\alpha_1} 2^{\alpha_2} \ldots N^{\alpha_N}\}$ is equal to

$$C\,(\alpha_1, \alpha_2, \ldots \alpha_N) = \frac{N!}{1^{\alpha_1} 2^{\alpha_2} \ldots N^{\alpha_N} \alpha_1! \alpha_2! \ldots \alpha_N!}. \tag{1.18}$$

In particular, the cardinality of the cycle class of the permutation Π defined by (1.4) equals $C\,(1, 0, 0, 1, 0) = 30$.

1.3 Fixed Points and the Inclusion-Exclusion Principle

Permutations of $N-$sets that have no fixed points are called *derangements*. A frequent problem is to count the number of derangements as a function of the number of elements of the set, often with additional constraints. The problem of counting derangements was solved by N. Bernoulli with the use of the inclusion–exclusion principle.

Let $|\mathcal{X}|$ denotes the cardinality of a set $\mathcal{X}$ and $\mathcal{X}_1$, $\mathcal{X}_2$ are its subsets. It is clear that

$$|\mathcal{X}_1 \cup \mathcal{X}_2| = |\mathcal{X}_1| + |\mathcal{X}_2| - |\mathcal{X}_1 \cap \mathcal{X}_2|. \tag{1.19}$$

In accordance with the *inclusion–exclusion principle*, the above equality can be extended to the case of any finite collection of subsets

$$\mathcal{X}_1 \cup \mathcal{X}_2 \cup \ldots \cup \mathcal{X}_n = \mathcal{X}.$$

The method of mathematical induction is used to justify the inclusion–exclusion principle. Let us assume that the principle holds true for $n - 1$ subsets. Then,

$$
\begin{aligned}
|\mathcal{X}_1 \cup \ldots \cup \mathcal{X}_{n-1}| + |\mathcal{X}_n| = {} & \sum_{i=1}^{n} |\mathcal{X}_i| \\
& - \sum_{1 \le i_1 < i_2 \le n-1} |\mathcal{X}_{i_1} \cap \mathcal{X}_{i_2}| \\
& + \ldots \\
& + (-1)^{k-1} \sum_{1 \le i_1 < \ldots < i_k \le n} |\mathcal{X}_{i_1} \cap \mathcal{X}_{i_2} \cap \ldots \cap \mathcal{X}_{i_k}| \\
& + \ldots \\
& + (-1)^{n-2} |\mathcal{X}_{i_1} \cap \mathcal{X}_{i_2} \cap \ldots \cap \mathcal{X}_{n-1}|. \tag{1.20}
\end{aligned}
$$

Furthermore,

$$
\begin{aligned}
|\mathcal{X}_1 \cup \ldots \cup \mathcal{X}_{n-1} \cap \mathcal{X}_n| = {} & |(\mathcal{X}_1 \cap \mathcal{X}_n) \cup \ldots \cup (\mathcal{X}_{n-1} \cap \mathcal{X}_n)| \\
= {} & \sum_{i=1}^{n-1} |\mathcal{X}_i \cap \mathcal{X}_n| \\
& - \sum_{1 \le i_1 < i_2 \le n-1} |\mathcal{X}_{i_1} \cap \mathcal{X}_{i_2} \cap \mathcal{X}_n| \\
& + \ldots \\
& + (-1)^{k-1} \sum_{1 \le i_1 < \ldots < i_k \le n-1} |\mathcal{X}_{i_1} \cap \ldots \cap \mathcal{X}_{i_k} \cap \mathcal{X}_n| \\
& + \ldots \\
& + (-1)^{n-2} |\mathcal{X}_1 \cap \ldots \cap \mathcal{X}_n|. \tag{1.21}
\end{aligned}
$$

Subtracting the latter equation from (1.20) and taking into account that

$$\sum_{1 \le i_1 < ... < i_2 \le n-1} |\mathcal{X}_{i_1} \cap ... \cap \mathcal{X}_{i_k}|$$

$$+ \sum_{1 \le i_1 < ... < i_{k-1} \le n-1} |\mathcal{X}_{i_1} \cap ... \cap \mathcal{X}_{i_{k-1}} \cap \mathcal{X}_n|$$

$$= \sum_{1 \le i_1 < ... < i_k \le n} |\mathcal{X}_{i_1} \cap ... \cap \mathcal{X}_{i_k}| \qquad (1.22)$$

we conclude that for $\mathcal{X} = \mathcal{X}_1 \cup ... \cup \mathcal{X}_n$,

$$|\mathcal{X}_1 \cup \mathcal{X}_2 ... \cup \mathcal{X}_n| = \sum_{j=1}^{n} (-1)^{k-1} \sum_{\{i_1, i_2, ..., i_k\}} |\mathcal{X}_{i_1} \cap \mathcal{X}_{i_2} \cap ... \cap \mathcal{X}_{i_k}| \qquad (1.23)$$

where $\{i_1, i_2, ..., i_k\}$ runs through all k–element subsets of the set $\mathcal{X}$. The relation (1.23) is called the *sieve formula*.

In some problems we need to compute how many elements have or do not have any of the given properties. In the classical work of Whitney (1932), it was demonstrated that the number of such elements can be calculated by using the inclusion–exclusion principle (1.23). Given a N–set,

$$\mathcal{X} = \{x_1, x_2, ..., x_N\},$$

we consider all transformations $U : \mathcal{X} \to \mathcal{X}$ such that $x_1 = U(x_1)$, $x_2 = U(x_2)$, etc., have x_1, x_2, etc., respectively as the fixed points. Clearly,

$$|U(x_{i_1}) \cap ... \cap U(x_{i_k})| = N^{N-k}, \quad 1 \le k \le N. \qquad (1.24)$$

Using the sieve formula (1.23) we can calculate the cardinality of the set of all transformations of the N–set which have at least one fixed point,

$$|U(x_1) \cup ... \cup U(x_N)|$$

$$= \sum_{k=1}^{N} (-1)^{k-1} \binom{N}{k} N^{N-k}$$

$$= N^N - (N-1)^N. \qquad (1.25)$$

Thus, the number of transformations of the N–set which have no fixed points at all equals

$$\mathfrak{H} = N^N - |U(x_1) \cup \ldots \cup U(x_N)|$$

$$= \sum_{k=0}^{N} (-1)^k \binom{N}{k} N^{N-k}$$

$$= (N-1)^N. \tag{1.26}$$

1.4 Finite Markov Chains

Random (or stochastic) processes deal with many possible scenarios of how the process might evolve in time. The notion of a stochastic process captures the indeterminacy of future evolution of a system by means of probability distributions describing some paths as more probable and others as less.

Given a probability space with state space $\mathfrak{X}$, a stochastic process $\mathcal{P} = \{X_t \in \mathfrak{X} : t \in \mathfrak{T}\}$ amounts to a sequence of random variables indexed by a set $\mathfrak{T}$. The random processes in which all information about the future states $X_{\tau > t}$ is contained in the present state X_t are called *Markov chains*. The concept of chains first appeared in Markov's 1906 paper (Markov 1906), in which he defined the simple chain as "an infinite sequence $x_1, x_2, \ldots, x_k, x_{k+1}, \ldots$, of variables connected in such a way that x_{k+1} for any k is independent of $x_1, x_2, \ldots, x_{k-1}$, in case x_k is known" (as cited by Basharin et al. 2004). Markov called the chain *homogeneous* if the conditional distributions of x_{k+1} given x_k were independent of k. The very term "Markov chain" was coined by S.N. Bernstein in 1926 (Bernstein 1926).

A finite Markov chain is a random stochastic process

$$\mathcal{P} = \{X_t \in \mathfrak{X} : t \in \mathbb{Z}_+\}$$

that takes on a finite number of possible values, i.e.. its state space $\mathfrak{X}$ is finite. Let us assume that the set of possible values $\mathfrak{X} = \{1, 2, \ldots N\}$. The Markov chain $\mathcal{P}$ is said to be in state s at time t if $X_t = s, s \in \mathfrak{X}$. The stochastic process evolves with time by changing its state from the current state s to some another state (or remain in the same state) according to some probability distribution. The changes of current state in the Markov chain are called *transitions*, and the probabilities associated with various state-changes are called *transition probabilities*. If $X_t = s$, we assume that the process moves from state s to state k with a fixed transition probability P_{sk},

$$\Pr(X_{t+1} = k | X_t = s, X_{t-1} = s_{t-1}, \ldots, X_1 = s_1, X_0 = s_0)$$

$$= \Pr(X_{t+1} = k | X_t = s) \tag{1.27}$$

$$= P_{sk},$$

for all states $s_0, s_1, \ldots, s_{t-1}, s \in \mathcal{X}$ and for all $t \geq 0$. Note that the transition probability P_{sk} does not involve $s_0, s_1, \ldots, s_{t-1}$ and is independent of t, and the stochastic process constitutes a homogeneous Markov chain.

The matrix with elements P_{sk} is stochastic, since all

$$P_{sk} \geq 0, \quad \forall s, k \in \mathcal{X}, \quad \text{and} \quad \sum_{s \in \mathcal{X}} P_{sk} = 1. \tag{1.28}$$

The Markov chain described by (1.27) is called a discrete time *random walk* over the finite set $\mathcal{X}$. Let $\mathbf{P}$ be the transition matrix of a Markov chain (1.27) with N states. State s is said to be *accessible* from state k if $(\mathbf{P}^n)_{sk} > 0$ for some $n \geq 1$. Markov chains establish an equivalence relation between the states, $i \sim j$ if an only if $(\mathbf{P}^n)_{ij}$ for some $n \geq 0$ and $(\mathbf{P}^m)_{ij}$ for some $m \geq 0$, and have all their states in one equivalence class. The Markov chain is said to be *irreducible* if its transition matrix (1.27) is irreducible that is equivalent to saying that with positive probability the process moves from any state to any other state in finitely many steps. If P_{ij}^n denotes the $(i, j) - element$ of the power matrix $\mathbf{P}^n$, we have

$$(\mathbf{P}^n)_{i,j} = \sum_t P_{it}^{(n-r)} P_{tj}^{(r)}, \quad r = 1, 2, \ldots, n - 1. \tag{1.29}$$

These above relation is known as the *Kolmogorov-Chapman* equation.

1.5   Birkhoff–von Neumann Theorem

A non-negative matrix $\mathbf{P}$ is said to be a *doubly stochastic* matrix if both $\mathbf{P}$ and $\mathbf{P}^\top$ are stochastic matrices. If a doubly stochastic matrix $\mathbf{P}$ defines a homogeneous Markov chain, the matrix $\mathbf{P}^\top$ describes the backward time homogeneous Markov chain.

The Birkhoff–von Neumann theorem relates doubly stochastic matrices to permutation matrices.

Theorem 1.1 (Birkhoff-von Neumann). *Let $\mathbf{P}$ be a doubly stochastic matrix. Then $\mathbf{P}$ is a convex combination of finitely many permutation matrices.*

Proof. Let us suppose that the doubly stochastic matrix $\mathbf{P}$ itself is not a permutation matrix, then there should exist a permutation Π over the finite set of indexes $\{1, 2, \ldots, N\}$ such that the product

$$1 > P_{1,\Pi(1)} P_{2,\Pi(2)} \ldots P_{N,\Pi(N)} > 0.$$

Let also denote

$$\lambda_1 = \min\left\{ P_{1,\Pi(1)}, P_{2,\Pi(2)} \ldots P_{N,\Pi(N)} \right\}$$

and let $\mathbf{\Pi}_1$ be the permutation matrix with $1's$ in the $(i, \Pi(i)) -$position for $i = 1, \ldots, N$. One can check that

$$\mathbf{P}_1 = \frac{\mathbf{P} - \lambda_1 \mathbf{\Pi}_1}{1 - \lambda_1} \tag{1.30}$$

is also a doubly stochastic matrix which has at least one more zero entry than the matrix $\mathbf{P}$ had. Moreover,

$$\mathbf{P} = \lambda_1 \mathbf{\Pi}_1 + (1 - \lambda_1)\mathbf{P}_1. \tag{1.31}$$

If $\mathbf{P}_1$ is not a permutation matrix itself, then we can repeat the above arguments and find another number λ_2, $0 < \lambda_2 < 1$, such that there exists a permutation matrix $\mathbf{\Pi}_2$, and

$$\mathbf{P}_2 = \frac{\mathbf{P}_1 - \lambda_2 \mathbf{\Pi}_2}{1 - \lambda_2} \tag{1.32}$$

is a doubly stochastic matrix again, with at least one more zero entry than $\mathbf{P}_1$. Then

$$\mathbf{P} = \lambda_1 \mathbf{\Pi}_1 + (1 - \lambda_1)\{\lambda_2 \mathbf{\Pi}_2 + (1 - \lambda_2)\mathbf{\Pi}_2\}. \tag{1.33}$$

Clearly this procedure terminates after a finite number of steps.

The set of all doubly stochastic matrices of order N forms a convex polytope in $\mathbb{R}^{N^2}$, known as the *Birkhoff polytope* of dimension $(N-1)^2$. Let denote the convex hull of permutation matrices $\mathbf{P}_\sigma$, $\sigma \in \mathcal{S}_N$, by

$$\mathcal{B} = \sum_{\sigma \in \mathcal{S}_n} c_\sigma \mathbf{\Pi}_\sigma, \quad \sum_{\sigma \in \mathcal{S}_n} c_\sigma = 1, \quad c_\sigma > 0. \tag{1.34}$$

Accordingly to the Birkhoff-von Neumann theorem, $\mathcal{B}$ is a doubly stochastic matrix, which, together with the transposed matrix $\mathcal{B}^\top$, respectively define the time-forward and time-backward Markov chains on a finite $N-$set.

1.6   Generating Functions

Counting problems in combinatorics often lead to a recursive answer representing a counting sequence. It is customary in combinatorial enumeration to represent such the sequences by means of a formal power series

$$F_a(x) = \sum_{k=0}^{\infty} a_n x^n, \tag{1.35}$$

with coefficients

$$a_n = \frac{1}{n!} \frac{d^n}{dx^n} F_a(x)\Big|_{x=0} \tag{1.36}$$

encoding information on the sequence. The formal power series (1.35) is not necessarily equal to the Taylor series of some function.

The most fundamental of all is the constant sequence

$$a = \{1, 1, 1, 1, ...,\}$$

whose generating function is

$$\sum_{n=0}^{\infty} x^n = \frac{1}{1-x}. \tag{1.37}$$

Computing the square of the generating function (1.37),

$$\frac{1}{(1-x)^2} = \sum_{n=0}^{\infty} (n+1)x^n, \tag{1.38}$$

we obtain the increasing sequence of natural numbers $\{1, 2, 3, \ldots\}$ as the coefficients of (1.38). Recursive formulas that are obtained in many combinatorial enumeration problems can often be transformed into some resolvable equations for the formal power series (1.35) usually by multiplying both sides of the recursions by some powers of the argument x and consecutive summing over all non-negative n.

The essential convenience of the generating functions formalism becomes evident when we have to convolve sequences. Let

$$\{a_n\}_{n\geq 0} \quad \text{and} \quad \{b_n\}_{n\geq 0}$$

are two sequences counting the numbers of ways to build the two different structures over an $n-$element set, and let

$$F_a(x) = \sum_{k=0}^{\infty} a_n x^n \quad \text{and} \quad F_b(x) = \sum_{k=0}^{\infty} b_n x^n$$

are their generating functions. What is the sequence which corresponds to the product of the generating functions, $F_a(x) \cdot F_b(x)$?

When we multiply the infinite sums $F_a(x)$ and $F_b(x)$, the typical product is of the form

$$a_i x^i \cdot b_j x^j$$

which contributes to the term proportional x^n if and only if $j = n - i$. Therefore, the generating function

$$F_c(x) = F_a(x) \cdot F_b(x)$$

$$= \sum_{k=0}^{\infty} c_n x^n \tag{1.39}$$

called the *product formula* for generating functions is characterized by the sequence of coefficients

$$c_n = \sum_{i=0}^{\infty} a_i b_{n-i}$$

accounting the number of ways to build the first structure on the i−partition of n elements, while the second structure is built on the $(n - i)$−partition of the n−element set.

Many objects of classical combinatorics present themselves naturally as labeled structures. Labeled constructions translate over *exponential generating functions,*

$$\mathcal{F}(x) = \sum_{k=0}^{\infty} \varphi_k \frac{x^k}{k!}, \tag{1.40}$$

with coefficients

$$\varphi_k = \left. \frac{d^k}{dx^k} \mathcal{F}(x) \right|_{x=0}. \tag{1.41}$$

In the previous section, we considered permutations over the finite sets of N elements that have the counting sequence $\{1!, 2!, \ldots, N!\}$. The appropriate exponential generating function for such a sequence is

$$\begin{aligned}
\sum_{k=0}^{\infty} k! \frac{x^k}{k!} &= \sum_{k=0}^{\infty} x^k \\
&= \frac{1}{1 - x}.
\end{aligned} \tag{1.42}$$

Note that the exponential generating function (1.42) is formally identical to the power generating function for the constant sequence $\{1, 1, \ldots, 1\}$ given by (1.37).

The product of two exponential generating functions has a natural combinatorial meaning. Namely, given the two sequences,

$$\{a_n\}_{n \geq 0} \quad \text{and} \quad \{b_n\}_{n \geq 0},$$

and their exponential generating functions,

$$\mathcal{F}_a(x) = \sum_{k=0}^{\infty} a_n x^n / n! \quad \text{and} \quad \mathcal{F}_b(x) = \sum_{k=0}^{\infty} b_n x^n / n!,$$

it is easy to check that the coefficients of the product

$$\begin{aligned}
\mathcal{F}_c(x) &= \mathcal{F}_a(x)\mathcal{F}_b(x) \\
&= \sum_{k=0}^{\infty} c_n \frac{x^n}{n!}
\end{aligned}$$

are given by

$$c_n = \sum_{k=0}^{n} \binom{n}{k} a_k b_{n-k} \tag{1.43}$$

where

$$\binom{n}{k} = \frac{n!}{(n-k)!k!}$$

is the binomial coefficient.

1.7   Partitions

An idealized thought experiment in which some objects are distributed over some containers (or urns) is called an *urn problem*. In the classical urn problems, we are interested in counting the number of admissible distributions of the M labeled/unlabeled balls over the N labeled/unlabeled urns. Consequently, we have to investigate four different cases:

1.7.1   Compositions

A *composition* of a positive integer M is a way of writing M as a sum of N strictly positive integers. Each composition corresponds to an allocation of the M *unlabeled*, identical balls over the N *labeled*, different urns and is uniquely determined by a solution of the equation

$$\alpha_1 + \alpha_2 + \ldots + \alpha_N = M \tag{1.44}$$

over the set of non-negative integer numbers. The sums (1.44) that differ in the order of their summands deemed to be different compositions.

Let us denote the number of possible compositions specified by the numbers M and N as $C_{M,N}$. Formally, we can write

$$C_{M,N} = \sum_{\alpha_1 + \alpha_2 + \ldots + \alpha_N = M} 1 \tag{1.45}$$

where the summation is over all admissible solutions of the (1.44). We denote the generating function of the numbers $C_{M,N}$ as

$$f_M^{(C)}(x) = \sum_{N=0}^{\infty} C_{M,N} x^N. \tag{1.46}$$

Accordingly to (1.45), the generating function (1.46) can be expressed as a product of M identical parentheses,

$$
\begin{aligned}
f_M^{(C)}(x) &= (1 + x + x^2 + \ldots) \ldots (1 + x + x^2 + \ldots) \\
&= \frac{1}{(1-x)^M},
\end{aligned}
\tag{1.47}
$$

so that

$$
C_{M,N} = \binom{N + M - 1}{N}, \quad N = 0, 1, \ldots, \quad M = 1, 2, \ldots .
\tag{1.48}
$$

1.7.2 Multi-set Permutations

Different allocations of the M *labeled* balls over the N *labeled* urns are called *multi-set permutations*. The number of ways to order linearly α_k objects of type k, for all $k = 1, \ldots, M$, equals

$$
D_{M,N}(\alpha_1, \alpha_2, \ldots, \alpha_M) = \frac{N!}{\alpha_1! \alpha_2! \ldots \alpha_M!}
\tag{1.49}
$$

where all numbers α_i sum to

$$
\alpha_1 + \alpha_2 + \ldots + \alpha_N = M.
\tag{1.50}
$$

The summation of $D_{M,N}$ over all possible combinations of non-negative terms α_i satisfying (1.50) gives the total number of integer N−vectors over M−sets,

$$
\sum_{\substack{\alpha_i \geq 0, \\ \alpha_1 + \ldots + \alpha_N = M}} \frac{N!}{\alpha_1! \alpha_2! \ldots \alpha_M!} = M^N.
\tag{1.51}
$$

The exponential generating function for the numbers $D_{M,N}$ is given by

$$
f_M^{(MP)}(x) = \sum_{N=0}^{\infty} D_{M,N} \frac{x^N}{N!},
\tag{1.52}
$$

and can be expressed by a product of M identical parentheses:

$$
\begin{aligned}
f_M^{(MP)}(x) &= \left(1 + \frac{x}{1!} + \frac{x^2}{2!} + \ldots\right) \ldots \left(1 + \frac{x}{1!} + \frac{x^2}{2!} + \ldots\right) \\
&= \exp Mx.
\end{aligned}
\tag{1.53}
$$

1.7.3 Weak Partitions

Let us discuss the M *labeled* balls distributed over the N *unlabeled* urns called a *weak partition*. In fact, we have already considered such an urn problem in Sect. 1.1 while counting the number of different permutation cycle classes. The number of different cycle classes in the symmetric group $\mathcal{S}_N$ equals the number of possible partitions of a positive integer N into non-negative integers called the Nth-*Bell number*. The generating function for the Bell numbers is

$$\sum_{N=0}^{\infty} B_N \frac{x^N}{N!} = e^{e^x - 1}. \tag{1.54}$$

By differentiating the both sides of (1.54) and equating the coefficients before $x^M / M!$, we obtain the recurrence formula for the Bell numbers,

$$B_{N+1} = \sum_{s=0}^{N} \binom{N}{s} B_{N-s}, \quad B_0 = 1, \tag{1.55}$$

the same as for the Nth-moment of a Poisson probability distribution, with expected value 1. Then, tailoring the series in the r.h.s. of (1.55) and calculating the coefficients for $x^M / M!$, we arrive at the *Dobinski formula* for the Bell numbers,

$$B_N = \frac{1}{e} \sum_{k=0}^{\infty} \frac{k^N}{k!}. \tag{1.56}$$

The number of partitions of a set into blocks grows very fast with the order of the set. In particular, $B_{10} = 115,975$ and $B_{20} = 51,724,158,235,372$.

Let us consider an important particular case of weak partitions. We denote a sequence $A = \{\alpha_1, \alpha_2, \ldots\} \in \mathbb{N}$ and call a partition of the $N-$set into the blocks of fixed sizes α_1, α_2, *etc.* prescribed by the sequence A as the $A-$*partition* of the $N-$set. Let us denote the number of such the $A-$partitions of the $N-$set by T_N^A and consider the exponential generating function of the sequence A,

$$\mathcal{A}(x) = \sum_{\alpha \in A} \frac{x^\alpha}{\alpha!}. \tag{1.57}$$

Then, the exponential generating function for the numbers T_N^A is given by

$$\sum_{N=0}^{\infty} T_N^A \frac{x^N}{N!} = e^{\mathcal{A}(x)}. \tag{1.58}$$

By differentiating the both sides of the above equation and calculating the coefficients before the typical term $x^N/N!$, we obtain the recurrence relation for T_N^A analogous to that of (1.55),

$$T_{N+1}^A = \sum_{\alpha \in A} \binom{N}{\alpha - 1} T_{N-\alpha+1}^A. \tag{1.59}$$

1.7.4 Integer Partitions

Finally, we consider the case of an *integer partition* when both the N balls and the M urns are indistinguishable. Each of these allocations corresponds to a solution of the noted equation

$$\alpha_1 + \alpha_2 + \ldots + \alpha_N = M, \tag{1.60}$$

over natural numbers. Two sums (1.60) that differ only in the order of their summands are considered as belonging to the same partition.

The graphical representations of integer partitions are conveniently given by the *Ferrers diagrams*. In Fig. 1.2, the five circles are lined up in five columns, each having the size of a part of the partition. It is obvious that the three first (left) Ferrers diagrams presented in Fig. 1.2 are nothing else but the mirror reflections of those three diagrams from the right side.

Such the partitions are said to be *conjugate* of one another. The symmetric partition represented by the Ferrers diagram in the middle is said to be *self-conjugate*. The number of self-conjugate partitions is the same as the number of partitions with distinct odd parts. The Ferrers diagrams (Fig. 1.2) can be used to prove various partition identities.

Let us denote the number of integer partitions of N by $\mathrm{Ip}(N)$. It is clear that the generating function for the numbers $\mathrm{Ip}(N)$ should have the typical monomial $x^{k\alpha_k}$, with various $1 \leq k \leq N$, and thus can be presented by the following product of parentheses,

$$(1+x+x^2+x^3+\ldots)(1+x^2+x^4+\ldots)\ldots(1+x^k+x^{2k}+x^{3k}+\ldots)\ldots \tag{1.61}$$

Converting the product (1.61) into a formal power series, we obtain

$$\sum_{N=0}^{\infty} \mathrm{Ip}(N)x^N = \prod_{k=1}^{\infty} \frac{1}{1-x^k}, \tag{1.62}$$

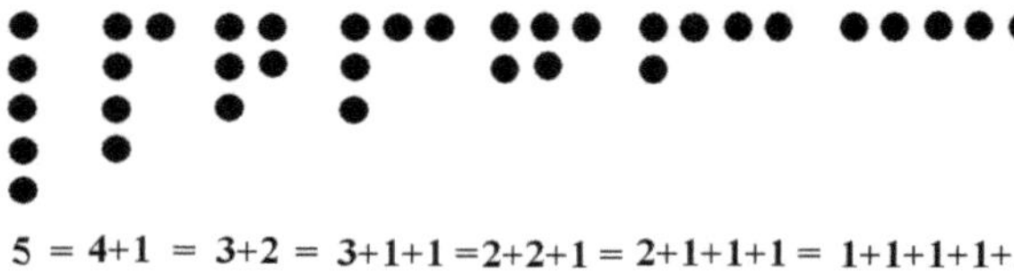

Fig. 1.2 The Ferrers diagrams for the seven partitions of the number 5

$$5 \ = \ 4{+}1 \ = \ 3{+}2 \ = \ 3{+}1{+}1 \ = 2{+}2{+}1 \ = \ 2{+}1{+}1{+}1 \ = \ 1{+}1{+}1{+}1{+}1$$

that is nothing else but the reciprocal of the Euler function, in the r.h.s. of (1.62). An asymptotic expression for Ip(N) is given by the *Hardy formula*,

$$p(N) \simeq_{N \to \infty} \frac{\sqrt{3}}{12} \frac{1}{N} \exp\left(\pi \sqrt{\frac{2N}{3}}\right). \tag{1.63}$$

1.8   Concluding Remarks and Further Reading

Combinatorics is the most classical area of mathematics, as counting the number of certain discrete (and usually finite) objects. Although the basic combinatorial concepts have appeared throughout the ancient world, they became known in Europe only in the thirteenth century through the works of Leonardo Fibonacci and Jordanus de Nemore (Biggs et al. 1996). Leonard Euler had developed a school of authentic combinatorial mathematics at the beginning of the eighteenth century. Graph theory was a permanent revival source of interest in combinatorics.

There are a number of textbooks covering combinatorics together with other topics of discrete mathematics. A comprehensive overview of the area is given in Graham et al. (1995). For a thorough introduction to the topics, we would recommend the works (Bollobas 1979; Skiena 1990; Conway and Guy 1996; Sachkov 1996; Sedgewick 1977; Trotter 2001; Bollobas and Thomason 2004; Diestel 2005; Harris et al. 2005; Brightwell et al. 2007), and the book (Bona 2004) essentially, as more appropriate for undergraduates.

Markovian systems (Nummelin 2004) appear extensively in mathematics, physics, and applied science. In economics, the random walk hypothesis is used to model share prices and other factors (Keane 1983). In population genetics, random walk describes the statistical properties of genetic drift (Cavalli-Sforza 2000). Random walk can be used to sample from a state space which is unknown or very large, for example to pick a random page of the Internet or, for research of working conditions of a random illegal worker (Hughes 1996). Random walks are often used in order to reach the 'obscure' parts of large sets and estimate the probable access times to them (Lovász 1993). Sampling by random walk was motivated by important algorithmic applications to computer science (see Deyer et al. 1986; Diaconis 1988; Jerrum and Sinclair 1989). There are a number of other processes that can be efficiently described by various types of diffusions of a large number of random walkers moving on a network at discrete time steps (Bilke and Peterson 2001). The Birkhoff - von Neumann theorem (Hall 1998; Schrijver 2002) establishes a profound relation between finite homogeneous Markov chains and permutations of objects of a finite set.

Chapter 2
Worth Another Binary Relation: Graphs

In Chap. 1, we have accounted for the different ways we can enumerate elements in a finite set. In particular, we have mentioned that a permutation $\Pi : \mathcal{G} \to \mathcal{G}$ of the finite set $\mathcal{G}$, defines an equivalence relation $\sim$ that partitions $\mathcal{G}$ into a set of equivalence classes $\mathcal{G}/\sim$. In the present chapter, we discuss worth another binary relation,

$$v \smile u, \quad v, u \in \mathcal{G} \tag{2.1}$$

called *adjacency*, and its graph.

2.1 Binary Relations and Their Graphs

A *binary relation* defined on a finite set $\mathcal{G}$ is a collection of ordered pairs $G \subseteq V \times U$ of elements from the arbitrary subsets $V, U \subseteq \mathcal{G}$. The sets $V \subseteq \mathcal{G}$ and $U \subseteq \mathcal{G}$ are called the *domain* and the *codomain* of the relation (2.1), while the collection of ordered pairs G is called its *graph*. In particular, if $V = U = \mathcal{G}$, we simply say that the binary relation (2.1) is defined over $\mathcal{G}$, and its graph is $G = (V, E)$ where V is the set of identical elements called *vertices* (or *nodes*), and $E \subseteq V \times V$ is a collection of pairs of elements from V called *edges*. Graphs are traditionally represented by diagrams in the following way: *vertices are shown by points and edges are the lines connecting vertices if they are related by (2.1).*

Given the two graphs, G and $\bar{G}$, defined on the same set of vertices V, the graph $\bar{G}$ is the *complement* of the graph G if its edge set consists of the edges not present in G.

Given a graph G, we may replace each edge of G by a vertex of some new graph $\mathcal{L}_G$ in such a way that two vertices of $\mathcal{L}_G$ are adjacent if and only if their corresponding edges share a common vertex ("are adjacent") in G. The graph $\mathcal{L}_G$ is called the *line* graph of G. It is obvious that properties of a graph G that depend only on adjacency between edges may be translated into equivalent properties in $\mathcal{L}_G$ that depend on adjacency between vertices. It is worth a mention that taking the line

P. Blanchard and D. Volchenkov, *Random Walks and Diffusions on Graphs and Databases*, Springer Series in Synergetics 10, DOI 10.1007/978-3-642-19592-1_2,
© Springer-Verlag Berlin Heidelberg 2011

Fig. 2.1 (**a**) A looped graph.
(**b**) A cyclic triple

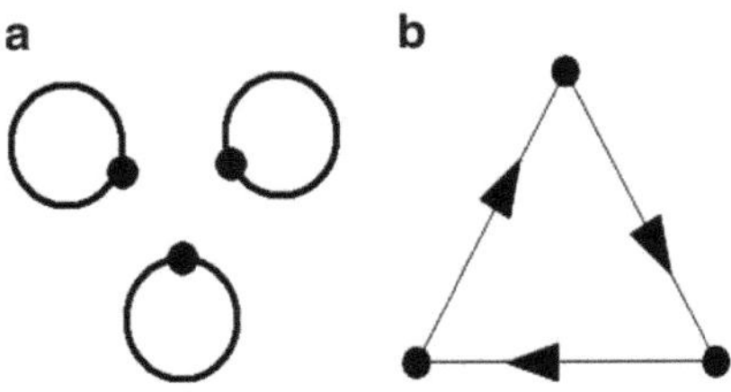

graph twice does not return the original graph G, unless it is a cycle graph, with the cycle length $k \geq 3$.

The graph G is

- *connected directed* if the relation (2.1) is *trichotomous* (for all $v \in \mathfrak{G}$ and $u \in \mathfrak{G}$, exactly one of $v \smile u$, $u \smile v$ or $u = v$ holds).
- *undirected* if the relation (2.1) is *symmetric*: for all $v \in \mathfrak{G}$ and $u \in \mathfrak{G}$, it holds that if $v \smile u$ then $u \smile v$.
- *looped* (its edges connect vertices to themselves, see Fig. 2.1a) if the relation (2.1) is *reflexive* (for all $v \in \mathfrak{G}$, it holds that $v \smile v$) and *coreflexive* (for all $v \in \mathfrak{G}$ and $u \in \mathfrak{G}$, it holds that if $v \smile u$ then $v = u$).
- *non looped* if the relation (2.1) is *irreflexive* (for all $v \in \mathfrak{G}$, it holds that $v \not\smile v$).
- *oriented*, if it has no symmetric pair of directed edges, that is, if the relation (2.1) is *antisymmetric* (for all $v \in \mathfrak{G}$ and $u \in \mathfrak{G}$, it holds that if $v \smile u$ and $u \smile v$ then $v = u$) and *asymmetric* (for all $v \in \mathfrak{G}$ and $u \in \mathfrak{G}$, it holds that if $v \smile u$, then $u \not\smile v$).
- *complete directed*, if the relation (2.1) is *total* (or *linear*) (for all $v \in \mathfrak{G}$ and $u \in \mathfrak{G}$, it holds that either $v \smile u$, or $u \smile v$, or both).
- consisting of a number of *cyclic triples* (or *transitive triples*) if the relation (2.1) is *transitive* (for all $v \in \mathfrak{G}$, $u \in \mathfrak{G}$, and $w \in \mathfrak{G}$, it holds that if $v \smile u$ and $u \smile w$ then $v \smile w$, see Fig. 2.1b).
- *complete* (with self-loops), if (2.1) constitutes an *equivalence* relation, i.e., it is reflexive, symmetric and transitive.
- a *partial order* if the relation (2.1) is reflexive, antisymmetric and transitive.
- a *chain* if it is a total partial order.

2.2 Representation of Graphs by Matrices

Graphs are conveniently represented by matrices. The major advantage of using matrices is that calculations of various graph characteristics can be performed by means of the well known operations with matrices.

For any finite set V of $|V| = N$ vertices, we introduce the canonical orthonormal basis $\{\mathbf{e}_j\}_{j=1}^{N}$ by assigning a unit vector

$$\mathbf{e}_i = (0, 0, \ldots, 1_i, \ldots 0),$$

with 1 at the i-th position, for every vertex $i \in V$. The set of orthonormal vectors constitutes a basis of the space of real functions on V, which we denote as $\mathfrak{F}(V)$. The inner product of two functions f and g from $f, g \in \mathfrak{F}(V)$ is then defined as

$$(f, g) = \sum_{i \in V} f(i)g(i). \tag{2.2}$$

We introduce the linear *adjacency operator* $\mathcal{A}$ on $\mathfrak{F}(V)$ by

$$(\mathcal{A}f)(i) = \sum_{j \sim i} f(j), \quad f \in \mathfrak{F}(V) \tag{2.3}$$

where $j \sim i$, iff $(i, j) \in E$. Therefore, the adjacency operator is unique for each graph $G(V, E)$, with fixed enumeration of its vertices. The $N \times N-$matrix $\mathbf{A}$ representing the adjacency operator $\mathcal{A}$ with respect to the canonical basis is called the *adjacency matrix* of the graph G. The off-diagonal entry A_{ij}, $i \neq j$, equals the number of edges linking the vertex $i \in V$ to the vertex $j \in V$. The diagonal entry A_{ii} equals the number of loops at the vertex $i \in V$. The adjacency matrix is unique for each graph (up to permuting rows and columns). In the special case of a finite *simple graph*, the adjacency matrix is a $(0, 1)$-matrix with zeros on its diagonal. The adjacency matrix of a complete graph is all 1's except for 0's on the diagonal. For example, the *Petersen graph* is represented by its adjacency matrix (see Fig. 2.2).

A *weighted graph* is a graph in which each edge $(i, j) \in E$ has an assigned weight w_{ij}, a real or complex number. The matrix $\mathbf{W}$ of elements w_{ij} describing the weights of all edges in the graph G is called its *affinity matrix*. Clearly, the adjacency matrix is a particular case of the affinity matrix, with all nontrivial weights $w_{ij} = 1$.

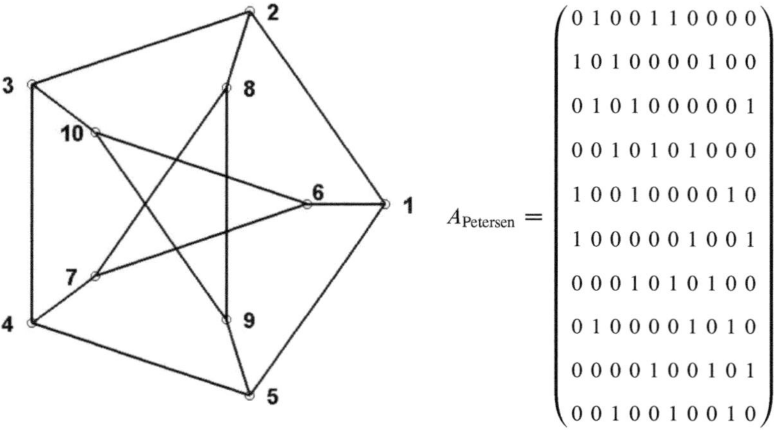

$$A_{\text{Petersen}} = \begin{pmatrix}
0 & 1 & 0 & 0 & 1 & 1 & 0 & 0 & 0 & 0 \\
1 & 0 & 1 & 0 & 0 & 0 & 0 & 1 & 0 & 0 \\
0 & 1 & 0 & 1 & 0 & 0 & 0 & 0 & 0 & 1 \\
0 & 0 & 1 & 0 & 1 & 0 & 1 & 0 & 0 & 0 \\
1 & 0 & 0 & 1 & 0 & 0 & 0 & 0 & 1 & 0 \\
1 & 0 & 0 & 0 & 0 & 0 & 1 & 0 & 0 & 1 \\
0 & 0 & 0 & 1 & 0 & 1 & 0 & 1 & 0 & 0 \\
0 & 1 & 0 & 0 & 0 & 0 & 1 & 0 & 1 & 0 \\
0 & 0 & 0 & 0 & 1 & 0 & 0 & 1 & 0 & 1 \\
0 & 0 & 1 & 0 & 0 & 1 & 0 & 0 & 1 & 0
\end{pmatrix}$$

Fig. 2.2 The Petersen graph and its adjacency matrix

The adjacency matrix allows us to formalize the intuitive idea of connectivity of a vertex in a graph. In various applications, higher connectivity vertices play the role of hubs being traversed by more paths between various origin/destination pairs than those with less connectivity. The *degree* of a vertex $i \in V$ is the number of other vertices adjacent to i in the graph G,

$$\deg(i) = \mathrm{card}\{j \in V : i \smile j\}$$
$$= \sum_{j \in V} A_{ij}. \tag{2.4}$$

The notion of a vertex degree can be readily generalized for the weighted graph as

$$\deg(i) = \sum_{j \in V} w_{ij}. \tag{2.5}$$

The vector of vertex degrees in the unweighted graph G can be calculated with the help of the vector $\mathbf{j} = (1, 1, \ldots, 1)^{\top}$ as

$$\mathbf{A}_G \mathbf{j} = (\deg(1), \deg(2), \ldots \deg(N)) . \tag{2.6}$$

The graph is *regular* if each vertex has the same degree. Since each edge is accounted twice while calculating the sum of degrees over all vertices in the graph, it is clear that

$$\sum_{i=1}^{N} \deg(i) = 2|E| \tag{2.7}$$

where $|E|$ is the cardinality of the set of edges.

A graph G can be alternatively represented by the *incidence* matrix $\mathbf{B}_G$ that shows the relationship between vertices and edges. The matrix $\mathbf{B}_G$ has one row for each vertex of the graph G and one column for each edge. The entry in the row i and the column j of the incidence matrix $\mathbf{B}_G$ is 1 if the edge j is incident to the vertex i in the graph G and is 0 if it is not. Therefore, the inner product of two columns of the incidence matrix of the graph G is nonzero if and only if the corresponding edges have a common vertex. For example, the incidence matrix of the Petersen graph is given in Fig. 2.3 (right).

The incidence matrix $\mathbf{B}_G$ of a graph G is the rectangular $N \times M$- matrix, and the product $\mathbf{B}_G^{\top} \mathbf{B}_G$ is a positive symmetric matrix that can be related to the adjacency matrix of the line graph $\mathcal{L}_G$ by

$$\mathbf{A}_{\mathcal{L}_G} = \mathbf{B}_G^{\top} \mathbf{B}_G - 2 \cdot \mathbf{1}. \tag{2.8}$$

The adjacency matrix of the line graph $\mathbf{A}_{\mathcal{L}_G}$ is the square $M \times M$ −matrix describing the line graph that consists of M vertices and

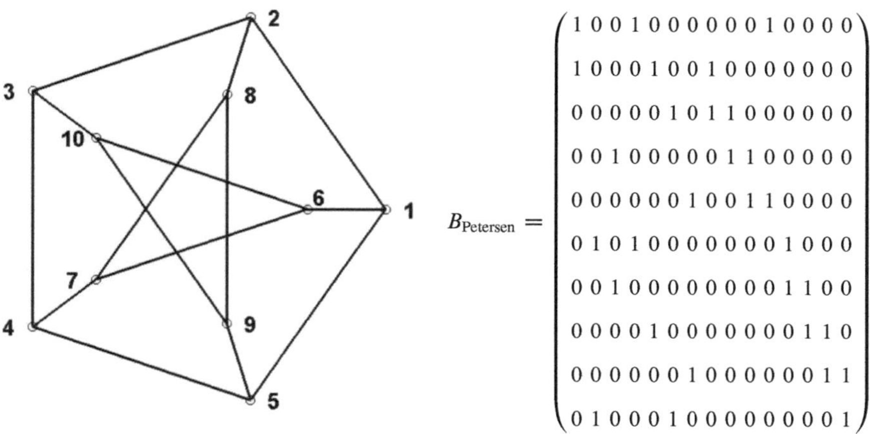

$$B_{\text{Petersen}} = \begin{pmatrix} 1\,0\,0\,1\,0\,0\,0\,0\,0\,0\,1\,0\,0\,0\,0 \\ 1\,0\,0\,0\,1\,0\,0\,1\,0\,0\,0\,0\,0\,0\,0 \\ 0\,0\,0\,0\,0\,1\,0\,1\,1\,0\,0\,0\,0\,0\,0 \\ 0\,0\,1\,0\,0\,0\,0\,0\,1\,1\,0\,0\,0\,0\,0 \\ 0\,0\,0\,0\,0\,0\,1\,0\,0\,1\,1\,0\,0\,0\,0 \\ 0\,1\,0\,1\,0\,0\,0\,0\,0\,0\,0\,1\,0\,0\,0 \\ 0\,0\,1\,0\,0\,0\,0\,0\,0\,0\,0\,1\,1\,0\,0 \\ 0\,0\,0\,0\,1\,0\,0\,0\,0\,0\,0\,0\,1\,1\,0 \\ 0\,0\,0\,0\,0\,0\,1\,0\,0\,0\,0\,0\,0\,1\,1 \\ 0\,1\,0\,0\,0\,1\,0\,0\,0\,0\,0\,0\,0\,0\,1 \end{pmatrix}$$

Fig. 2.3 The Petersen graph and its incidence matrix

$$M' = \frac{1}{2} \sum_{i \in V} \deg(i)^2 - M \tag{2.9}$$

edges.

2.3   Algebraic Properties of Adjacency Operators

A function $f \in \mathfrak{F}(V)$ is an eigenvector for $\mathbf{A}_G$ if there is a constant θ such that, for each vertex $i \in V$,

$$\theta f(i) = \sum_{j \sim i} f(j), \tag{2.10}$$

that means

$$\mathbf{A}_G f = \theta f.$$

The eigenvalue θ of the eigenfunction f is the root of the characteristic polynomial

$$Q_A = \det(\theta \cdot \mathbf{1} - \mathbf{A}_G). \tag{2.11}$$

It follows from (2.10) that

$$|\theta|\,|f(i)| = \left| \sum_{j \sim i} f(j) \right|$$
$$\leq \sum_{j \sim i} |f(j)|, \tag{2.12}$$

and therefore,

$$|\theta| \le \sum_{j \sim i} \frac{|f(j)|}{|f(i)|}, \tag{2.13}$$

where the equality holds if and only if f is a constant function.

Let us note that the determinant of the adjacency matrix $\mathbf{A}_G$ can be expanded into the sum of contributions from all possible permutations $\Pi \in \mathcal{S}_N$ involving $n = 2, \ldots, N$ nodes of the graph G,

$$\det(\mathbf{A}_G) = \sum_{\Pi \in \mathcal{S}_n} \text{sign}(\Pi) \cdot \prod_{i=1}^{n} A_{i,\Pi(i)} \tag{2.14}$$

where $\text{sign}(\Pi)$ is the sign of the permutation Π. It is obvious that a permutation Π contributes into the sum (2.14) if and only if $(i, \Pi(i)) \in E$. Since any permutation can be decomposed into a product of disjoint cycles (see Sect. 1.2), the permutation Π such that $(i, \Pi(i)) \in E$ induces a *cycle cover* γ of the graph G, the partition of the vertex set into disjoint cycles.

Let us denote the number of connected components in γ as $\text{comp}(\gamma)$, the number of cycles (of the length greater than 2) in that as $\text{cyc}(\gamma)$, and the number of cycles in the cycle cover (including the cycles of length one) as $\text{cyc}(\Pi)$. As the number of odd cycles in Π is congruent (modulo 2) to n, it follows that $n + \text{cyc}(\Pi)$ is congruent (modulo 2) to the number of even cycles in Π. Therefore,

$$\text{sign}(\Pi) = (-1)^{n + \text{cyc}(\Pi)}. \tag{2.15}$$

We also note that since direct and inverse cycles equally contribute into (2.14), there are $2^{\text{cyc}(\gamma)}$ permutations of the same sign in (2.14). Finally, we conclude that for a connected undirected graph G,

$$\det(\mathbf{A}_G) = \sum_{\gamma \subset G} (-1)^{n + \text{cyc}(\Pi)} 2^{\text{cyc}(\gamma)} \tag{2.16}$$

where the summation is over all subgraphs $\gamma \subset G$ of $n = 1, \ldots N$ nodes.

2.4 Perron–Frobenius Theory for Adjacency Matrices

For more information on the spectral properties of adjacency matrices, let us mention that a matrix $\mathbf{A}$ is called *irreducible* if it is not similar to a block upper triangular matrix via a permutation, i.e., there is no any permutation matrix Π such that the matrix $\Pi^{-1} \mathbf{A} \Pi$ is of the block upper triangular form. It is worth a mention that if an undirected graph underlying the adjacency matrix $\mathbf{A}$ is connected, then the matrix is always irreducible. Spectral properties of irreducible matrices with non-negative entries are described by the famous *Perron -Frobenius theorem* which asserts that any irreducible real square matrix with non-negative entries has a unique largest real eigenvalue and that the corresponding eigenvector has strictly positive components.

A matrix $\mathbf{A}$ with entries a_{ij} is said to be nonnegative if $a_{ij} \geq 0$, and $\mathbf{A}$ is said to be positive if $a_{ij} > 0$.

Theorem 2.1 (Perron-Frobenius theorem). *Let* $\mathbf{A}$ *be a non-negative irreducible matrix. Then*

1. *The largest eigenvalue* $v > 0$
2. *There exists a positive vector* $\mathbf{u} \in \mathbb{R}^n$ *such that* $\mathbf{A}\mathbf{u} = v\mathbf{u}$,
3. *There exists a positive vector* $\mathbf{v} \in \mathbb{R}^n$ *such that* $\mathbf{A}^\top \mathbf{v} = v\mathbf{v}$.
4. *The algebraic multiplicity of* v *as an eigenvalue of* $\mathbf{A}$ *is equal to one.*

The complete proof of the Perron-Frobenius theorem can be found in many text books such as Bapat and Raghavan (1997) and Dym (2007). Following Ninio (1976), we give below a simple proof of the theorem for positive symmetric matrices $\mathbf{A} = \mathbf{A}^\top$ describing undirected graphs.

Proof. Since the eigenvalues of $\mathbf{A}$ are real and their sum equals Tr $\mathbf{A} > 0$, it follows that the largest eigenvalue $v > 0$. Let $\mathbf{u}_i$ be any real normalized eigenvector belonging to v,

$$v u_i = \sum_j a_{ij} u_j, \quad i = 1, 2, \ldots, n, \tag{2.17}$$

and set $x_j = |u_j|$. Then

$$0 < v = \sum_{ij} a_{ij} \mathbf{u}_i \mathbf{u}_j$$

$$= \left| \sum_{ij} a_{ij} \mathbf{u}_i \mathbf{u}_j \right| \tag{2.18}$$

$$\leq \sum_{ij} a_{ij} x_i x_j.$$

By the variational theorem, the right-hand side is less than or equal to v, with equality if and only if x_j is an eigenvector belonging to the largest eigenvalue v. We therefore have

$$v x_i = \sum_j a_{ij} x_j, \quad i = 1, 2, \ldots, n. \tag{2.19}$$

Now if $x_i = 0$ for some i, then on account of $a_{ij} > 0$ for all j, it follows every $x_j = 0$, which cannot be. Thus, every $x_j > 0$. Finally, if v is a multiple eigenvalue, we can find (since $\mathbf{A}$ is real symmetric) two orthonormal eigenvectors $\mathbf{u}_j, \mathbf{v}_j$ belonging to v. Suppose that $u_i < 0$ for some i. Adding (2.1) and (2.2), we obtain

$$0 = v \left(u_i + |u_i| \right)$$
$$= \sum_j a_{ij} \left(u_j + |u_j| \right),$$

and as above, it follows that

$$u_j + |u_j| = 0,$$

for every j. In other words, we have either

$$u_j = |u_j| > 0$$

for every j, or

$$u_j = -|u_j| < 0$$

for every j. The same applies to $\mathbf{v}_j$. Hence,

$$\sum_j v_j u_j = \pm \sum_j |v_j u_j| \neq 0$$

so that $\mathbf{u}$ and $\mathbf{v}$ cannot be orthogonal, and therefore v is non-degenerate. Let now assume that $\mathbf{w}_j$ be a normalized eigenvector belonging to $\mu < v$,

$$\sum_j a_{ij} w_j = \mu w_i.$$

The variational property and the non-degeneracy of v then yields

$$v > \sum_{ij} a_{ij} |w_i| \cdot |w_j| \geq \left| \sum_{ij} a_{ij} w^* w_j \right| = |\mu|. \tag{2.20}$$

$\square$

2.5 Spectral Decomposition of Adjacency Operators

If the graph G is undirected, the corresponding adjacency operator is self-adjoint with respect to the scalar product (2.2), and therefore the adjacency matrix is symmetric; its eigenvalues are real and eigenvectors form an orthogonal basis for $\mathfrak{F}(V)$. For a regular graph G (where each vertex has the same number of neighbors), it is easy to check that the vector $\mathbf{j}$ consisting of all 1's is an eigenvector of the adjacency matrix A, with the eigenvalue $\theta = \deg_G$, the common degree of vertices in that.

A simple complete graph on N nodes also has the eigenvector $\mathbf{j}$ belonging to the eigenvalue $\theta = (N - 1)$. Another eigenvalue of the complete graph characterizes those eigenfunctions $f(i) \in \mathfrak{F}(V)$ satisfying

$$\sum_{i \in V} f(i) = 0. \tag{2.21}$$

From (2.21), it is obvious that such the eigenfunctions satisfy the relation

$$\sum_{j \sim i} f(j) = -f(i), \tag{2.22}$$

and therefore $\theta' = -1$ is the correspondent eigenvalue, with the multiplicity $(N-1)$.

The Petersen graph (see Fig. 2.2) is regular, $\deg_{\text{Pet}} = 3$; its maximal eigenvalue $\theta_{\max} = 3$ corresponds to the normalized eigenvector $\frac{1}{\sqrt{10}}\mathbf{j}$ and express the *mean value property*,

$$f(i) = \frac{1}{3} \sum_{j \sim i} f(j). \tag{2.23}$$

The next eigenvalue, $\theta' = 1$, with multiplicity $m_{\theta'} = 5$, describes the five configurations $f \in \mathfrak{F}(V)$, for which

$$f(i) = \sum_{j \sim i} f(j). \tag{2.24}$$

Let $\mathbf{U}_{\theta'}$ be a $N \times m_{\theta'}$ matrix whose columns form an orthogonal basis for the eigenspace belonging to θ',

$$\mathbf{U}_{\theta'}^{\mathsf{T}} = \begin{pmatrix}
-\frac{\sqrt{6}}{6} & -\frac{\sqrt{6}}{6} & 0 & 0 & -\frac{\sqrt{6}}{6} & \frac{\sqrt{6}}{6} & \frac{\sqrt{6}}{6} & 0 & 0 & \frac{\sqrt{6}}{6} \\[4pt]
-\frac{\sqrt{5}\sqrt{2}}{10} & -\frac{\sqrt{5}\sqrt{2}}{10} & 0 & \frac{\sqrt{5}\sqrt{2}}{5} & \frac{\sqrt{5}\sqrt{2}}{10} & -\frac{\sqrt{5}\sqrt{2}}{10} & \frac{\sqrt{5}\sqrt{2}}{10} & 0 & 0 & -\frac{\sqrt{5}\sqrt{2}}{10} \\[4pt]
-\frac{\sqrt{12}\sqrt{5}}{20} & \frac{\sqrt{12}\sqrt{5}}{30} & \frac{\sqrt{12}\sqrt{5}}{12} & \frac{\sqrt{12}\sqrt{5}}{60} & -\frac{\sqrt{12}\sqrt{5}}{30} & -\frac{\sqrt{12}\sqrt{5}}{20} & -\frac{\sqrt{12}\sqrt{5}}{30} & 0 & 0 & \frac{\sqrt{12}\sqrt{5}}{30} \\[4pt]
-\frac{\sqrt{9}\sqrt{4}}{36} & -\frac{\sqrt{9}\sqrt{4}}{18} & -\frac{\sqrt{9}\sqrt{4}}{36} & -\frac{\sqrt{9}\sqrt{4}}{36} & \frac{\sqrt{9}\sqrt{4}}{18} & -\frac{\sqrt{9}\sqrt{4}}{36} & -\frac{\sqrt{9}\sqrt{4}}{18} & 0 & \frac{\sqrt{9}\sqrt{4}}{9} & \frac{\sqrt{9}\sqrt{4}}{18} \\[4pt]
-\frac{\sqrt{2}}{6} & \frac{\sqrt{2}}{6} & -\frac{\sqrt{2}}{6} & -\frac{\sqrt{2}}{6} & -\frac{\sqrt{2}}{6} & -\frac{\sqrt{2}}{6} & \frac{\sqrt{2}}{6} & \frac{\sqrt{2}}{2} & \frac{\sqrt{2}}{6} & -\frac{\sqrt{2}}{6}
\end{pmatrix}. \tag{2.25}$$

Each column of $\mathbf{U}_{\theta'}$ is an eigenvector of $\mathbf{A}_{\text{Petersen}}$, and therefore

$$\theta' \mathbf{U}_{\theta'} = \mathbf{A}_{\text{Petersen}} \mathbf{U}_{\theta'}.$$

If we denote the ith–row of the matrix $\mathbf{U}_{\theta'}$ by $u_{\theta'}(i)$, then the above equation can be rewritten in the following form,

$$\theta u_{\theta'}(i) = \sum_{j \sim i} u_{\theta'}(j), \tag{2.26}$$

which allow us to interpret the function $u_{\theta'}(i) : V \to \mathbb{R}^{m_{\theta'}}$ as a *low-dimensional representation* of the vertex $i \in V$ belonging to the eigenvalue θ'.

Columns of the matrix $\mathbf{U}_{\theta'}$ are the orthonormal vectors, so that

$$\mathbf{U}_{\theta'}^{\mathsf{T}} \mathbf{U}_{\theta'} = \mathbf{1} \in \mathbb{R}^{m_{\theta'}}, \tag{2.27}$$

while another product,

$$\boldsymbol{\mathcal{P}}_{\theta'} = \mathbf{U}_{\theta'}\mathbf{U}_{\theta'}^{\top}, \tag{2.28}$$

is the symmetric matrix, $\boldsymbol{\mathcal{P}}_{\theta'} = \boldsymbol{\mathcal{P}}_{\theta'}^{\top}$, an orthogonal projection onto the column space of $\mathbf{U}_{\theta'}$, independent upon the orthonormal basis of vectors $\mathbf{U}_{\theta'}$, as being an *invariant* of the eigenspace belonging to the eigenvalue θ'.

Analogously, the remaining eigenvalue of the adjacency matrix of the Petersen graph, $\theta'' = -2$, with multiplicity $m_{\theta''} = 4$, describes the configurations $f \in \mathfrak{F}(V)$, for which

$$f(i) = -\frac{1}{2}\sum_{j \sim i} f(j). \tag{2.29}$$

The eigenspace belonging to this eigenvalue has the orthogonal basis

$$\mathbf{U}_{\theta''}^{\top} = \begin{pmatrix} -\frac{\sqrt{6}}{6} & \frac{\sqrt{6}}{6} & -\frac{\sqrt{6}}{6} & 0 & \frac{\sqrt{6}}{6} & 0 & 0 & 0 & -\frac{\sqrt{6}}{6} & \frac{\sqrt{6}}{6} \\[2mm] \frac{\sqrt{6}}{6} & 0 & \frac{-\sqrt{6}}{6} & \frac{\sqrt{6}}{6} & \frac{-\sqrt{6}}{6} & \frac{-\sqrt{6}}{6} & 0 & 0 & 0 & \frac{\sqrt{6}}{6} \\[2mm] \frac{\sqrt{9}\sqrt{2}}{18} & \frac{-\sqrt{9}\sqrt{2}}{9} & \frac{\sqrt{9}\sqrt{2}}{18} & \frac{-\sqrt{9}\sqrt{2}}{18} & \frac{\sqrt{9}\sqrt{2}}{18} & \frac{-\sqrt{9}\sqrt{2}}{18} & 0 & \frac{\sqrt{9}\sqrt{2}}{9} & \frac{-\sqrt{9}\sqrt{2}}{9} & \frac{\sqrt{9}\sqrt{2}}{18} \\[2mm] \frac{\sqrt{5}\sqrt{2}}{30} & \frac{\sqrt{5}\sqrt{2}}{30} & \frac{\sqrt{5}\sqrt{2}}{30} & \frac{-2\sqrt{5}\sqrt{2}}{15} & \frac{\sqrt{5}\sqrt{2}}{30} & \frac{-2\sqrt{5}\sqrt{2}}{15} & \frac{\sqrt{5}\sqrt{2}}{5} & \frac{-2\sqrt{5}\sqrt{2}}{15} & \frac{\sqrt{5}\sqrt{2}}{30} & \frac{\sqrt{5}\sqrt{2}}{30} \end{pmatrix}. \tag{2.30}$$

Let us denote the orthogonal projection onto the column space of $\mathbf{U}_{\theta''}$ as

$$\boldsymbol{\mathcal{P}}_{\theta''} = \mathbf{U}_{\theta''}\mathbf{U}_{\theta''}^{\top}. \tag{2.31}$$

It is obvious that

$$\boldsymbol{\mathcal{P}}_{\theta}\boldsymbol{\mathcal{P}}_{\tau} = \boldsymbol{\mathcal{P}}_{\theta}\delta_{\theta\tau}, \tag{2.32}$$

and $\boldsymbol{\mathcal{P}}_{\theta'}\boldsymbol{\mathcal{P}}_{\theta''} = 0$, in particular. Generalizing this example, we conclude that

$$\mathbf{A}_{G}\boldsymbol{\mathcal{P}}_{\theta} = \theta\boldsymbol{\mathcal{P}}_{\theta}, \tag{2.33}$$

and therefore, there is a *spectral decomposition* for the adjacency matrix $\mathbf{A}_{G}$,

$$\mathbf{A}_{G} = \sum_{\theta} \theta\boldsymbol{\mathcal{P}}_{\theta} \tag{2.34}$$

where the summation is over all eigenvalues of $\mathbf{A}_{G}$.

2.6 Adjacency and Walks on a Graph

A *walk* W_{ℓ} of length $\ell \geq 1$ in a graph G is an ordered sequence of vertices of G,

$$W_{\ell} = \{v_0, v_1, \ldots v_{\ell}\},$$

such that $v_{k-1} \smile v_k$, $k = 1, \ldots, \ell$. If the first and the last vertices of the walk coincide, then W_ℓ is a cycle.

The nonnegative integer powers of a matrix $\mathbf{A}_G$ of order N are defined by

$$\mathbf{A}_G^0 = \mathbf{1}, \quad \mathbf{A}_G^1 = \mathbf{A}_G,$$

and

$$\mathbf{A}_G^k = \mathbf{A}_G \cdot \mathbf{A}_G^{k-1},$$

for $k > 1$. Provided $\mathbf{A}_G$ is the adjacency matrix of the graph G, the elements of its positive integer power, A_{Gij}^k, equal the numbers of walks of length k connecting the vertices $i \in V$ and $j \in V$ in the graph G. This is obviously true for $k = 1$ since the graph G has precisely one walk connecting i and j if $A_{Gij} = 1$, but the vertices are not connected if $A_{Gij} = 0$. For $k > 1$, we can justify the above statement by the inductive assumption. Namely, let us assume that A_{Gij}^k equals the number of all walks of length k connecting the two vertices i and j in G. For the elements of the forthcoming matrix, we have

$$\begin{aligned}
A_G^{k+1}{}_{ij} &= A_{i1} \cdot A_{G1j}^k + \ldots + A_{iN} \cdot A_{GNj}^k \\
&= \sum_{l \sim i} A_{glj}^k
\end{aligned} \tag{2.35}$$

where the latter sum is nothing else but the *total number of all walks* of length k between the vertex j and all vertices $l \in V$ directly connected to i in the graph G. Hence, $A_G^{k+1}{}_{ij}$ equals the number of all walks of length $k+1$ connecting the vertices i and j, completing the induction.

The number of closed walks of length k in G equals the sum of diagonal elements in the matrix $\mathbf{A}_G^k$,

$$\mathrm{Tr}\, \mathbf{A}^k = \sum_\theta \theta^k, \tag{2.36}$$

where the last sum is over all eigenvalues θ, with the account of their multiplicity. Hence we get the following simple results:

$$\mathrm{Tr}\, \mathbf{A}_G = 0 \quad \text{iff} \quad G \quad \text{is} \quad \text{simple}; \tag{2.37}$$

$$\mathrm{Tr}\, \mathbf{A}_G^2 = 2E, \tag{2.38}$$

where E is the number of edges,

$$\mathrm{Tr}\, \mathbf{A}_G^k = k!\,\mathrm{cyc}_k(G), \tag{2.39}$$

where $\mathrm{cyc}_k(G)$ is the number of cycles of length k in the graph G.

Let $\mathbf{U}$ is the orthogonal matrix of eigenvectors of the adjacency matrix $\mathbf{A_G}$. Then

$$A_{Gij}^k = \sum_{l=1}^n \theta_l^k u_{il} u_{jl}.$$

The number of all walks of length k in G equals

$$N_k(G) = \sum_{i,j \in V} A_{Gij}^k$$

$$= \sum_{l=1}^{N} \left(\sum_{i=1}^{N} u_{il} \right)^2 \theta_l^k$$

$$= \sum_{l=1}^{N} \xi_l \theta_l^k$$

where $\xi_l \equiv \left(\sum_{i=1}^{N} u_{il} \right)^2$. The generating function for the numbers $N_k(G)$ is

$$H(t) = \sum_{k=0}^{\infty} N_k t^k$$

$$= \sum_{l=1}^{N} \frac{\xi_l}{1 - t\theta_l}.$$

2.7 Principal Invariants of the Graph Adjacency Matrix

An *isomorphism* between the two undirected non-weighted graphs G_1 and G_2 is an edge-preserving bijection f between their vertex sets. Namely, any two vertexes $v \smile u$ adjacent in G_1, are mapped by f into the two vertexes $f(v) \smile f(u)$, adjacent in G_2. Isomorphic graphs are said to have the same structure, as sharing all *graph invariants* which depend on neither a labeling of the graph vertexes, nor a drawing. The important structural characteristics of a graph, such as the order of the graph N, the number of 1-loops in the graph N_o, the size of the graph E, the number of triangles $N_\triangle$, and, in general, the number of the k–cycles $\mathrm{cyc}_k(G)$, for all $k = 1, \ldots, N$, are the graph invariants, as they are preserved under the action of graph isomorphisms, but changed if the graph transformation is not an isomorphism. It is then obvious that the values of polynomials in the above structural characteristics are also preserved under the action of graph isomorphisms, though in the common case they could be the same even for two non-isomorphic graphs.

In the present section, we show that for each undirected graph there are some polynomials in the structural characteristics which remains invariant under the graph isomorphisms. They are related to the *principal invariants* of the graph adjacency matrix $\mathbf{A}_G$, that are the coefficients $I_k(\mathbf{A}_G)$, $k = 1, \ldots, N$, of its characteristic polynomial,

$$\det (\mathbf{A}_G - \theta \cdot \mathbf{1}) = \sum_{k=0}^{N} I_k(\mathbf{A}_G)(-\theta)^{N-k} \tag{2.40}$$

$$= 0,$$

where θ is an eigenvalue of $\mathbf{A}_G$. The principal invariants $I_k(\mathbf{A}_G)$, can be expressed, in terms of the moments $\mathrm{Tr}\,\mathbf{A}_G^k$ (see Gantmacher 1959, Chap. 4), with the use of Newton's identities resulting in the k-th symmetric polynomials,

$$I_k\left(\mathbf{A}_G\right) = \frac{(-1)^{k-1}}{k}\sum_{l=0}^{k-1}(-1)^l I_l(\mathbf{A}_G)\mathrm{Tr}\,\mathbf{A}_G^l, \tag{2.41}$$

where we assume that $I_0 = 1$. In particular, accordingly to (2.38, 2.39)

$$\begin{aligned} I_1\left(\mathbf{A}_G\right) &= \mathrm{Tr}\,\mathbf{A}_G \\ &= N_{\circ}, \end{aligned} \tag{2.42}$$

$$\begin{aligned} I_2\left(\mathbf{A}_G\right) &= \frac{1}{2}\left((\mathrm{Tr}\,\mathbf{A}_G)^2 - \mathrm{Tr}\,\mathbf{A}_G^2\right) \\ &= \frac{N_{\circ}^2}{2} - E, \end{aligned} \tag{2.43}$$

$$\begin{aligned} I_3\left(\mathbf{A}_G\right) &= \frac{1}{3}\left((\mathrm{Tr}\,\mathbf{A}_G)^3 - 3\mathrm{Tr}\,\mathbf{A}_G^2\,\mathrm{Tr}\mathbf{A}_G + 2\mathrm{Tr}\,\mathbf{A}_G^3\right) \\ &= \frac{N_{\circ}^3}{3} - 2E\cdot N_{\circ} + 4N_{\triangle}, \end{aligned} \tag{2.44}$$

etc. It can be shown that the expressions (2.41) correspond to the non-negative integer partitions of the number k,

$$k = 1\cdot m_1 + 2\cdot m_2 + \ldots + k\cdot m_k, \tag{2.45}$$

in which m_i is the number of subsets containing precisely i elements in the corresponding partition, as each such a partition contributes into (2.41) by the product of moments,

$$\begin{aligned} T_{m_1,\ldots,m_k} &\equiv \mathrm{Tr}\left(\mathbf{A}_G^{m_1}\right)\ldots\mathrm{Tr}\left(\mathbf{A}_G^{m_k}\right) \\ &= m_1!\ldots m_k! N_{\circ}\cdot E\cdot N_{\triangle}\cdot N_{\square}\ldots\mathrm{cyc}_k(G) \end{aligned} \tag{2.46}$$

where $\mathrm{cyc}_k(G)$ is the number of the k-cycles in the graph G. We have used (2.39) to derive the last equality in (2.46). Since the partition labels a conjugate class in the symmetric group of permutations of k elements $\mathcal{S}_k$, we conclude from (1.18) that the number of elements in the conjugate class is equal to

$$C_{m_1,\ldots,m_k} = \frac{k!}{m_1!\ldots m_k!1^{m_1}\ldots k^{m_k}}, \tag{2.47}$$

and taking into account the parities of partitions, we derive the combinatorial expression for the principal invariants of the graph,

$$I_k\left(\mathbf{A}_G\right) = \frac{1}{k!} \sum_{\{\sum_i im_i=k\}} (-1)^{\sum_i (i-1)m_i} C_{m_1,\ldots,m_k} T_{m_1,\ldots,m_k}, \tag{2.48}$$

in which the summation is defined over all nonnegative–integer partitions (2.45).

Alternatively, in order to obtain the expressions (2.48) we can use the generating function approach proposed by Zhang et al. 2008. Let us define the two generating functions $\mathfrak{F}(z)$ and $\mathfrak{G}(z)$ for the infinite sequences $\{I_k\}_{k=1}^\infty$ and $\left\{\operatorname{Tr}\mathbf{A}_G^k\right\}_{k=1}^\infty$ respectively,

$$\mathfrak{F}(z) = \sum_{k=0}^\infty z^k I_k, \quad I_0 = 1, \tag{2.49}$$

and

$$\mathfrak{G}(z) = \sum_{k=0}^\infty z^k \operatorname{Tr}\mathbf{A}_G^k. \tag{2.50}$$

Analyzing the recursive relations (4.15) between the principal invariants I_k, we can conclude that the generating functions (2.49) and (2.50) satisfy the differential equation

$$\frac{d}{dz}\mathfrak{F}(z) = -\mathfrak{F}(z)\mathfrak{G}(z) \tag{2.51}$$

supplied by the initial condition $\mathfrak{F}(0) = 1$. The solution of (2.51) is

$$\begin{aligned}
\mathfrak{F}(z) &= \exp\left(-\int_0^z \mathfrak{G}(z)dz\right) \\
&= \exp\left(-\sum_{k=1}^\infty \frac{z^k}{k}\operatorname{Tr}\mathbf{A}_G^k\right) \\
&= \prod_{k=1}^\infty \exp\left(-\frac{z^k}{k}\operatorname{Tr}\mathbf{A}_G^k\right) \\
&= \sum_{k=0}^\infty z^k \cdot \sum_{\{\sum_{l=1}^k lm_l=k\}} \prod_{l=1}^k \frac{(-1)^{m_l}}{m_l!}\left(\frac{\operatorname{Tr}\mathbf{A}_G^l}{l}\right)^{m_l}.
\end{aligned} \tag{2.52}$$

Thus, we obtain, for the principal invariants of the graph adjacency matrix, the expression equivalent to (2.48):

$$I_k\left(\mathbf{A}_G\right) = \sum_{\{\sum_{l=1}^k l\cdot m_l=k\}} \prod_{l=1}^k \frac{(-1)^{m_l}}{m_l!}\left(\frac{\operatorname{Tr}\mathbf{A}_G^l}{l}\right)^{m_l}, \tag{2.53}$$

where the summation is defined over all partitions (2.45).

Plugging the expressions (2.46) and (2.47) back into (2.48), we obtain the general expression for the principal invariants of the adjacency matrix $\mathbf{A}_G$ in terms of the numbers of l–cycles in the graph G,

$$I_k(\mathbf{A}_G) = \sum_{\{\sum_i i m_i = k\}} (-1)^{\sum_i (i-1) m_i} \prod_{l=1,\dots,k\,;\,m_l \neq 0} \frac{\mathrm{cyc}_l(G)}{l^{m_l}}. \tag{2.54}$$

2.8   Euler Characteristic and Genus of a Graph

Any graph can be drawn as a set of points in $\mathbb{R}^3$ and of continuous arcs connecting some pairs of them. Aiming at a convenient visualization of certain graph's properties, we can draw the graph in many different ways supposing that good graph drawing algorithms allow for as few edge crossings as possible. Those graphs which can be drawn on a plane without edge crossings are called *planar*, as they can be embedded in the plane. As the arcs of a planar graph can be drawn without edge crossings, they divide that plane into some number of regions called *faces*.

The relations between the order (the number of vertices) N, the size (the number of edges) E, and the number of faces F in a planar polygon

$$N - E + F = 1, \tag{2.55}$$

and its direct generalization to a convex polyhedron, a geometric solid in three dimensions with flat faces and straight edges,

$$N - E + F = 2 \tag{2.56}$$

have been known since Descartes (1639). Leonard Euler was published the formula (2.56) in 1751, while proving that there are exactly five Platonic solids. The remarkable fact is that the result of the sign alternating sums in (2.55, 2.56) called the *Euler characteristic* is independent of both the particular figure and the way it is bent, as being sensitive merely to its topological structure: any change to the graph that creates an additional face would keep the value $N - E + F = 2$ an invariant.

For a general connected graph, the Euler characteristic χ can be defined axiomatically as its unique additive characteristic over its subgraphs,

$$\chi(X \cup Y) = \chi(X) + \chi(Y) - \chi(X \cap Y), \tag{2.57}$$

normalized in such a way that $\chi(\emptyset) = 0$ and $\chi(\text{Polygon}) = 1$, for any polygon. It can be considered as a version of the inclusion-exclusion principle and meets the sieve formula (1.23). In particular, the Euler characteristic can be defined for a finite connected graph G by the alternating sum,

$$\chi(G) = \sum_{k=1}^{k_{\max}} (-1)^{k-1} q_k, \tag{2.58}$$

in which $k_{\max}$ is the maximal degree of vertices in the graph G, $q_1 = E$ is the number of edges in G, q_2 is the number of couples of incident edges (sharing a common vertex), q_3 is the number of triples of incident edges, etc., until $q_{k_{\max}}$ is the number of $k_{\max}$–tuples of incident edges. It can be demonstrated readily that the definition (2.58) coincides with

$$\chi(G) = N - E \tag{2.59}$$

where N is the number of vertices and E is the number of edges in G. To prove the formula (2.59), let us classify vertices of the graph G accordingly to their degrees,

$$c_k = \{v \in V : \deg(v) = k\}, \quad k = 1, \ldots, k_{\max}, \tag{2.60}$$

where $k_{\max}$ is the maximal degree of nodes in the graph G and note that

$$\sum_{k=1}^{k_{\max}} k \cdot |c_k| = 2E. \tag{2.61}$$

Let us calculate the number q_2 of couples of edges sharing a common vertex in G. Clearly, each vertex of degree 2 corresponds to a pair of edges contributing to q_2. Moreover, each vertex of degree 3 corresponds to the $\binom{3}{2}$ pairs of edges accounted in q_2. Analogously, each vertex of degree 4 corresponds to the $\binom{4}{2}$ pairs of edges accounted in q_2, etc. Consequently, we obtain

$$q_2 = \sum_{m=2}^{k_{\max}} \binom{m}{2} |c_m|. \tag{2.62}$$

Similarly, we conclude that

$$q_3 = \sum_{m=3}^{k_{\max}} \binom{m}{3} |c_m|, \ldots, \quad q_{k_{\max}} = \binom{k_{\max}}{k_{\max}} |c_{k_{\max}}|. \tag{2.63}$$

The above equations establish a duality between the cardinalities of degree classes of vertices in a finite undirected graph and their analogs for edges by means of the linear transformation involving the matrix of binomial coefficients,

$$
\begin{pmatrix} q_2 \\ q_3 \\ \vdots \\ q_{k_{max}-1} \\ q_{k_{max}} \end{pmatrix} = \begin{pmatrix} 1 & \binom{3}{2} & \cdots\cdots & \binom{k_{max}}{2} \\ 0 & 1 & \cdots\cdots & \binom{k_{max}}{3} \\ \vdots & \vdots & \ddots \;\; \vdots & \vdots \\ 0 & 0 & \cdots\; 1 & \binom{k_{max}}{k_{max}-1} \\ 0 & 0 & \cdots\; 0 & 1 \end{pmatrix} \begin{pmatrix} |c_2| \\ |c_3| \\ \vdots \\ |c_{k_{max}-1}| \\ |c_{k_{max}}| \end{pmatrix}.
\tag{2.64}
$$

Now, if we substitute the above relations back into (2.58) and take into account that for any $n \in \mathbb{N}$,

$$
\sum_{l=1}^{n}(-1)^{l-1}\binom{n}{l} = 1,
\tag{2.65}
$$

we obtain the formula (2.59).

A natural generalization of planar graphs are graphs which can be drawn on a surface of a given *genus* that is the number of non-intersecting cycles on the graph. Genera are used in topological theory for classifying surfaces, as two surfaces can be deformed one into the other if and only if they have the same genus; surfaces of higher genus have correspondingly more holes. Spheres have genus zero, as having no holes. Surfaces of genus one are tori. Surfaces of genus two and higher are associated with the hyperbolic plane. The genus $g(G)$ of a graph G can be defined in terms of the Euler characteristic (2.58) via the relationship

$$
\chi(G) = 2 - g(G).
\tag{2.66}
$$

It is easy to check that planar graphs have genus one, and convex polyhedra have genus zero.

2.9   Euler Characteristics and Genus of Complex Networks

In many real-world networks represented by large highly inhomogeneous graphs the distribution describing the fractions $P(k)$ of nodes having precisely k connections to other nodes exhibit a heavy tail (Newman 2003a). The well-known example is a *scale-free network*, in which the degree distribution (asymptotically) follows a power law,

$$
P(k) \propto k^{-\gamma},
$$

for some $\gamma > 1$. Scale-free graphs noteworthy ubiquitous to many empirically observed networks (Albert and Barabási 2002). It is important to note that the Euler characteristic of a graph defined by (2.58, 2.59) is simply related to the *mean degree* of nodes calculated with respect to the degree statistics. Provided the degree distribution in the graph is $P(k)$, we note that the number of vertices having precisely k neighbors equals

$$|c_k| = N \cdot P(k),$$

so that the relation (2.61) reads as

$$N \cdot \sum_{k=1}^{k_{\max}} k \cdot P(k) = N \langle k \rangle \tag{2.67}$$
$$= 2E.$$

where $k_{\max}$ is the maximal node degree in the graph G and $\langle \ldots \rangle$ denotes the mean degree in the graph, with respect to the given degree distribution $P(k)$. Then, it follows from the definition of the Euler characteristic (2.59) that for such a graph

$$\frac{\chi(G)}{N} = 1 - \frac{\langle k \rangle}{2}. \tag{2.68}$$

If the mean degree of a node in the network is $\langle k \rangle > 2$, it follows from (2.68) that $\chi(G) < 0$. In particular, the genus of the graph underlying a complex network in such a case equals to

$$g(G) = 2 - \frac{N}{2}(2 - \langle k \rangle) > 2 \tag{2.69}$$

indicating that the graph can rather be embedded into a surface associated with a hyperbolic geometry. In Krioukov et al. (2009), it has been found that the Internet represented on the level of autonomous systems exhibits a remarkable congruency with the Poincaré disc model of hyperbolic geometry.

2.10 Coloring a Graph

Graph coloring is an assignment of colors to vertices of a graph subject to such a constraint that no two adjacent vertices share the same color. The most famous result in the graph coloring theory know as the *four color map theorem* states that given any separation of a plane into contiguous regions, called a map, the regions can be colored using at most four colors so that no two adjacent regions have the same color. The number of graph colorings $\mathcal{P}_G(z)$ as a function of the number of colors z (known as the *chromatic polynomial* of a graph) was originally defined by G.D. Birkhoff for planar graphs, in an attempt to prove the four color map theorem. The Birkhoff's chromatic polynomial has been generalized to the case of general graphs by Whitney (1932).

Following his work, we consider coloring of N vertices in a graph G in such a way that any two vertices which are joined by an arc are of different colors. It is clear that there are $n = z^N$ possible colorings, formed by giving each vertex in succession any one of z colors at our disposal. Accordingly to the inclusion-exclusion principle (1.23), the cardinality of the set of admissible colorings is

$$
\begin{aligned}
\mathcal{P}_G(z) = z^N & \\
& - \left[n(e_{ij}) + \ldots + n(e_{kl}) \right] \\
& + \left[n(e_{ij} e_{kl} + \ldots) \right] \\
& - \ldots \\
& + (-1)^E n(e_{ij} \ldots e_{kl}),
\end{aligned}
\tag{2.70}
$$

where e_{ij} denotes those colorings with the property that i and j are of the same color (associated with the arc e_{ij} in G), and $n(\ldots)$ denotes the cardinality of those colorings. The first term in (2.70) corresponds to the subgraph containing no arcs, the second term stands for the subgraph of disjoint arcs, eventually the last term corresponds to the whole graph G.

A typical term $n(e_{ij} \ldots e_{kl})$ in (2.70) is the number of ways of coloring G in z or fewer colors in such a way that i and j are of the same color, k and l are of the same color, etc. Since any two vertices that are joined by an arc in the corresponding subgraph must be of the same color, all the vertices in a single connected piece of the subgraph are of the same color. Consequently, if there are p connected pieces in that, the absolute value of the correspondent typical term is therefore z^p. Moreover, if the subgraph contains b arcs in p connected pieces, its contribution into (2.70) is

$$
(-1)^b n(e_{ij} \ldots e_{kl}) = (-1)^b z^p.
\tag{2.71}
$$

Let (p, b) denote the number of subgraphs of b arcs in p connected pieces, summing over all values of p and b, we obtain the chromatic polynomial in z,

$$
\mathcal{P}_G(z) = \sum_{p,b} (-1)^b (p, b) z^p.
\tag{2.72}
$$

If we assume that the graph G is connected and for any $G' \subseteq G$ let

$$
r(G') = |G'| - p(G'),
$$

where $p(G')$ is the number of connected components in G', we further transform (2.72) to

$$
\mathcal{P}_G(z) = \sum_{G' \subseteq G} (-1)^{|G'| - r(G')} (-z)^{r(G) - r(G')}
\tag{2.73}
$$

that is usually written in the form of the two variable polynomial,

$$
\mathcal{R}_G(u, v) = \sum_{G' \subseteq G} u^{|G'| - r(G')} v^{r(G) - r(G')},
\tag{2.74}
$$

called the *Whitney rank generating function.*

A root of a chromatic polynomial (a *chromatic root*) is a value z where $\mathcal{P}_G(z) = 0$. It is obvious that $z = 0$ is always a chromatic root for any graph, as no graph can be 0-colored. Furthermore, $z = 1$ is a chromatic root for every graph with at least an edge, as only edgeless graphs can be 1-colored. An edge e can be colored only with two colors, therefore

$$\mathcal{P}_e(z) = z(z - 1).$$

The triangle K_3 can be colored with three colors, so that

$$\mathcal{P}_{K_3}(z) = z(z - 1)(z - 2),$$

etc. In general, for a connected graph G on N vertices, the chromatic polynomial $\mathcal{P}_G(z)$ is a polynomial of degree N. Although, for some basic graph classes, recurrent formulas for the chromatic polynomials are known, computational problems associated with the finding the chromatic polynomial for a given graph often require nondeterministic polynomial time to be solved. The *chromatic number*, the smallest number of colors needed to color the vertices of the graph so that no two adjacent vertices share the same color, is obviously the smallest positive integer that is not a chromatic root. Under a simple change of variables, the Whitney rank generating function (2.74) is transformed into the *Tutte polynomial* in two dual variables, $x = u - 1$ and $y = 1 - v$,

$$
\begin{aligned}
\mathcal{R}_G(u - 1, v - 1) &\equiv \mathcal{T}_G(x, y) \\
&= \sum_{G' \subseteq G} (x - 1)^{p(G')-p(G)} (y - 1)^{p(G')+N'-N}
\end{aligned}
\tag{2.75}
$$

where $p(G')$ denotes the number of connected components in the graph G', N' is the order of the graph G', and N is the order of G. The Tutte polynomial of a graph is its invariant that factors into graph's connected components: given $G = G' \bigcup G''$,

$$\mathcal{T}_G(x, y) = \mathcal{T}_{G'}(x, y) \times \mathcal{T}_{G''}(x, y).$$

2.11 Shortest Paths in a Graph

A connected graph is called a *tree* if it contains no cycles. A *spanning tree* of a connected, undirected graph G is a connected tree composed of all the vertices of G. Much of the research in the various applications such as communications, road network design, and engineering has involved problems in which the network to be designed is a tree connecting a collection of sites and satisfying the different optimization criteria. For instance, one can mention the *travelling salesman problem*, in which the cheapest round-trip route is searched such that the salesman visits each city exactly once and then returns to the starting city (Dantzig et al. 1954). The algorithms for constructing the minimum spanning trees and searching the shortest path originated in 1926 for the purpose of efficient electrical coverage of Bohemia (Nesetril et al. 2000). It is clear that a single connected graph G can have many different spanning trees. The number $t(G)$ of spanning trees of the graph G is its important invariant, which can be calculated using *Kirchhoff's matrix-tree theorem.*

Kirchhoff's theorem relies on the notion of the *canonical Laplace matrix* of a graph G (de Verdiére 1998),

$$\mathbf{L_c} = \mathbf{D} - \mathbf{A}_G, \tag{2.76}$$

where $\mathbf{D}$ is the diagonal graph's degree matrix and its adjacency matrix $\mathbf{A}_G$. The matrix (2.76) has the property that the sum of its entries across any row and any column is 0, so that the vector of all ones $\mathbf{j}$ spans the null-space of $\mathbf{L_c}$,

$$\mathbf{L_c j} = 0. \tag{2.77}$$

The Laplacian matrix (2.76) can be factored into the product of the incidence matrix and its transpose,

$$\mathbf{L_c} = \mathbf{B}_G \mathbf{B}_G^\top. \tag{2.78}$$

Let $\mathbf{B}'_G$ be the incidence matrix with its first row deleted, so that

$$\mathbf{B}'_G \mathbf{B}'^\top_G = \text{Minor}$$

where Minor is a minor of the Laplace operator (2.76) (let us note that all minors of the Laplace operator are equal, as its null-space is one dimensional). Then, using the Cauchy-Binet formula (see, for example, Roman 2005; Shores 2006) we can write

$$\begin{aligned}
\det(\text{Minor}) &= \sum_{\mathfrak{S}} \det(\mathbf{B}'_{\mathfrak{S}}) \det(\mathbf{B}'^\top_{\mathfrak{S}}) \\
&= \sum_{\mathfrak{S}} \det(\mathbf{B}'_{\mathfrak{S}})^2
\end{aligned} \tag{2.79}$$

where $\mathbf{B}'_{\mathfrak{S}}$ denotes the $(N-1) \times (N-1)$ matrix whose columns are those of $\mathbf{B}'_G$ with index in the $(N-1)$-subset $\mathfrak{S}$. Since any of such subsets specifies $(N-1)$ edges of the original graph, and any set of edges forming a cycle gives zero contribution into the determinant (2.79), those edges induce a spanning tree, with the determinant $\det(\mathbf{B}'_{\mathfrak{S}}) = \pm 1$. Summing over all possible subsets $\mathfrak{S}$ in (2.79), we conclude that the total number of spanning trees $T(G)$ of the connected graph G is

$$\begin{aligned}
T(G) &= \det(\text{Minor}) \\
&= \frac{1}{N} \prod_{\lambda_k \neq 0} \lambda_k
\end{aligned} \tag{2.80}$$

where the product is over all the non-zero eigenvalues of the Laplacian matrix (2.76). The result (2.80) is known as Kirchhoff's matrix tree theorem.

In graph theory, the *shortest path problem* consists of finding the quickest way to get from one location to another on a graph. The number of edges in a shortest path connecting two vertices, $i \in V$ and $j \in V$, in the graph is the *shortest path distance* between them, d_{ij}. It is obvious that $d_{ij} = 1$ if $i \sim j$. We may find the shortest path from the node to any other node by performing a breadth first search of eligible arcs on a graph spanning tree. It is obvious that the shortest path can be not unique, as there might be many spanning trees for the graph. However, we can choose one shortest path for each node, so that the resulting graph of eligible arcs for a breadth first search of all shortest paths from the node forms a tree.

There are a number of other graph properties defined in terms of distance. The diameter of a graph is the greatest distance between any two vertices,

$$\mathfrak{D}_G = \max_{i,j \in V} d_{ij}.$$

The radius of a graph is given by

$$\mathfrak{R}_G = \min_{i \in V} \max_{j \in V} d_{ij}.$$

The distance would determine the relative importance of a vertex within the graph. The *mean shortest path distance* from vertex i to any other vertex in the graph is

$$\ell_i = \frac{1}{N-1} \sum_{j \in V} d_{ij}. \tag{2.81}$$

In graph theory, the relation (2.81) expresses the *closeness* being a *centrality measure* of the vertex within a graph. *Betweenness* is another centrality measure of a vertex within a graph. It captures how often in average a vertex may be used in journeys from all vertices to all others in the graph. Vertices that occur on many shortest paths between others have higher betweenness than those that do not. Betweenness is estimated as the ratio

$$\text{Betweenness}(i) = \frac{\{\#\text{shortest paths through } i\}}{\{\#\text{all shortest paths}\}}. \tag{2.82}$$

Betweenness is, in some sense, a measure of the influence a node has over the spread of information through the graph. The betweenness centrality is essential in the analysis of many real world networks and social networks, in particular, but costly to compute. The Dijkstra algorithm and the Floyd-Warshall algorithm (Cormen et al. 2001), may be used in order to calculate betweenness.

2.12 Concluding Remarks and Further Reading

When a finite ordered set is endowed with an additional internal structure descried
by a binary relation of adjacency, the collection of order pairs from this set is a graph.
Each graph can be uniquely represented by its adjacency operator characterized by
the adjacency matrix, with respect to the canonical basis of vectors in Hilebrt space.
Spectral properties of the adjacency operator are related to walks and cycles of the
correspondent graph.

There are many handbooks of graph theory, perhaps the most popular topic
in discrete mathematics. Suggested readings are Harary (1969), Bollobas (1979),
Chartrand (1985), Gould (1988), Biggs et al. (1996), Tutte (2001), Bona (2004),
Diestel (2005), Harris et al. (2005) and Gross (2008). The textbooks (Bona 2004;
Harris et al. 2005) are essentially appropriate for undergraduates. The classical
surveys on the relationship between structural and spectral properties of graphs are
Chung (1997) and Cvetkovic et al. (1997, 1980). An introduction to algebraic graph
theory concerned with the interplay between algebra and graph theory can be found
in Biggs (1993), Chan and Godsil (1997) and Godsil and Royle (2001).

Many interesting invariants of graphs can be computed from the Tutte polynomials (Tutte 1954). For a wealth of information on the applications of Tutte
polynomial, see Brylawski and Oxley (1992) and Bollobas (1979). Problems involving some form of geometric minimum or maximum spanning tree are discussed in
geometric network design theory (Eppstein 1999; Wu and Chao 2004).

Chapter 3
Permutations Sieved Through Adjacency: Graph Automorphisms

A group action is called *transitive* if it possesses only a single group orbit, so that for every pair of elements a and b, there is a group element γ, such that $a = \gamma b$.

Consider a transitive permutation group $\mathfrak{P}(V) \subseteq \mathcal{S}_N$ defined on a finite set V, $|V| = N$. We can define an induced action of $\mathfrak{P}(V)$ on the set of all 2-subsets $V \times V$ by

$$\Pi(v, u) = (\Pi v, \Pi u), \quad \Pi \in \mathfrak{P}(V), \quad v, u \in V. \tag{3.1}$$

Given a binary relation "$\smile$" (2.1) defined on the set V, we denote its graph by G. Among all possible permutation groups $\mathfrak{P}(V) \subseteq \mathcal{S}_N$, there is one compatible with the binary relation "$\smile$" called the *automorphism group* $\mathrm{Aut}(G)$ of the graph G.

3.1 Graph Automorphisms

For any $\Pi \in \mathrm{Aut}(G)$,

$$\Pi u \smile \Pi v, \quad \text{iff } u \smile v, \tag{3.2}$$

and therefore the automorphism group maps vertices to vertices preserving their adjacency, so that edges are mapped to edges. Clearly, the degree of a node (2.4) is invariant under the action of graph automorphisms,

$$\deg(\Pi v) = \deg(v), \quad \Pi \in \mathrm{Aut}(G). \tag{3.3}$$

The automorphism group of a graph characterizes its symmetries and arises in the enumeration of graphs, specifically in the relations between counting labeled and unlabeled graphs. A finite N-set can be labeled in $N!$ different ways. Since the action of $\mathrm{Aut}(G)$ preserves the graph structure, the number of different labellings of an unlabeled G is $N!/|\mathrm{Aut}(G)|$.

Although it is known that for every group $\mathcal{G}$ there exists a graph G whose automorphism group $\mathrm{Aut}(G)$ is isomorphic to $\mathcal{G}$ (Skiena 1990), it is usually a difficult task to decide whether a graph has nontrivial automorphisms. A graph possessing a

P. Blanchard and D. Volchenkov, *Random Walks and Diffusions on Graphs and Databases*, Springer Series in Synergetics 10, DOI 10.1007/978-3-642-19592-1_3,
© Springer-Verlag Berlin Heidelberg 2011

single identity automorphism is called an *asymmetric* graph. Accordingly to the famous result of Erdös and Rényi (1963), almost all graphs have no non-trivial automorphisms, as the proportion of graphs on N vertices which have a non-trivial automorphism tends to zero as $N \to \infty$. It is obvious that $\mathrm{Aut}(G) \subset \mathcal{S}_N$, excepting for the complete graph K_N where $\mathrm{Aut}(G) \simeq \mathcal{S}_N$, since every permutation of K_N is its automorphism. The Petersen graph (Fig. 2.2) is the complement of the line graph of the complete graph K_5, and thus its automorphism group is the symmetric group $\mathcal{S}_5$.

Given $f \in \mathcal{F}(V)$, let us consider the composition of f and $\Pi \in \mathrm{Aut}(G)$,

$$(f \circ \Pi)(v) = f(\Pi v), \quad v \in V \tag{3.4}$$

First of all, we note that the map

$$f \mapsto f \circ \Pi \tag{3.5}$$

is an invertible linear mapping of $\mathcal{F}(V)$ onto itself. It is easy to demonstrate that since

$$\sum_{u \sim u}(f \circ \Pi)(u) = \sum_{u \sim v} f(\Pi u)$$
$$= \sum_{l \sim \Pi v} f(l)$$
$$= \theta(f \circ \Pi)(v),$$

any eigenvector of the adjacency matrix $\mathbf{A}_G$ of the graph G belonging to the eigenvalue θ is mapped by $\Pi \in \mathrm{Aut}(G)$ into other eigenvector $f \circ \Pi$ of $\mathbf{A}_G$ belonging to the same eigenvalue θ.

Therefore, multiple eigenvalues of the adjacency matrix $\mathbf{A}_G$ indicate the existence of non identity automorphisms for the graph G. Moreover, the linear automorphisms associated to some eigenvalue θ with multiplicity $m_\theta > 1$ have to be represented by some $m_\theta \times m_\theta$ real orthogonal matrices transforming the different vectors of the eigenspace $\mathbf{U}_\theta$ into each other (Chan and Godsil 1997). Indeed, given $f \in \mathcal{F}(V)$, it follows from (3.4) that

$$\sum_{u \sim v} f(\mathbf{\Pi} u) = \sum_{l \sim \Pi v} f(l), \quad \mathbf{\Pi} \in \mathrm{Aut}(G) \tag{3.6}$$

and therefore

$$(\mathbf{A}_G \mathbf{\Pi}) f(i) = \mathbf{A}_G f(\Pi(j)). \tag{3.7}$$

The last equality means that the adjacency matrix $\mathbf{A}_G$ of the graph G *commutes* with any permutation matrix $\mathbf{\Pi} \in \mathrm{Aut}(G)$,

$$\mathbf{A}_G \mathbf{\Pi} - \mathbf{\Pi} \mathbf{A}_G \equiv [\mathbf{A}_G, \mathbf{\Pi}] = 0, \quad \mathbf{\Pi} \in \mathrm{Aut}(G). \tag{3.8}$$

Using the spectral decomposition (2.34) for the adjacency matrix $\mathbf{A}_G$, we may conclude from (3.8) that any graph automorphism $\mathbf{\Pi} \in \mathrm{Aut}(G)$ also commutes with any orthogonal projection onto the column space $\mathbf{U}_\theta$ belonging to any eigenvalue θ,

$$[\mathcal{P}_\theta, \mathbf{\Pi}] = 0. \tag{3.9}$$

The latter equation allows to calculate the representation of the automorphism group $\mathrm{Aut}(G)$ in the class of $m_\theta \times m_\theta-$real orthogonal matrices. Since

$$\begin{aligned} \mathbf{\Pi}\mathbf{U}_\theta \cdot \mathbf{1} &= \mathbf{\Pi}\mathbf{U}_\theta \mathbf{U}_\theta^\top \mathbf{U}_\theta \\ &= \mathbf{U}_\theta \mathbf{U}_\theta^\top \mathbf{\Pi}\mathbf{U}_\theta, \end{aligned} \tag{3.10}$$

we obtain for any low-dimensional representation $u_\theta(v)$ of a vertex $v \in V$ that

$$u_\theta\left(\mathbf{\Pi}\, v\right) = u_\theta(v)\, \mathbf{U}_\theta^\top \mathbf{\Pi}\mathbf{U}_\theta. \tag{3.11}$$

The matrix $\mathbf{U}_\theta^\top \mathbf{\Pi}\mathbf{U}_\theta$ in the r.h.s. of (3.11) is orthogonal,

$$\left(\mathbf{U}^\top \mathbf{\Pi}\mathbf{U}\right)^\top = \mathbf{U}^\top \mathbf{\Pi}^\top \mathbf{U}_\theta,$$

and $\mathbf{\Pi}^\top = \mathbf{\Pi}^{-1}$, so that (3.11) gives the $m_\theta-$dimensional representation of $\mathrm{Aut}(G)$.

For example, let us return to the Petersen graph (Fig. 3.1) and consider one of its automorphisms consisting of the clockwise permutation of its vertices:

$$\tilde{\Pi} = (1 \rightarrow 5 \rightarrow 4 \rightarrow 3 \rightarrow 2)(6 \rightarrow 9 \rightarrow 7 \rightarrow 10 \rightarrow 8). \tag{3.12}$$

The permutation (3.12) is described by the permutation matrix.

Now we calculate the representation of the permutation (Fig. 3.1) associated to the eigenvalue $\theta' = 1$ with multiplicity $m_{\theta'} = 5$:

$$\mathbf{U}_{\theta'}^\top \tilde{\mathbf{\Pi}}\mathbf{U}_{\theta'} = \begin{pmatrix} \frac{1}{2} & -\frac{\sqrt{6}\sqrt{5}\sqrt{2}}{20} & \frac{\sqrt{6}\sqrt{12}\sqrt{5}}{60} & \frac{\sqrt{6}\sqrt{9}\sqrt{4}}{36} & \frac{\sqrt{6}\sqrt{2}}{6} \\ \frac{\sqrt{6}\sqrt{5}\sqrt{2}}{20} & \frac{1}{10} & \frac{3\sqrt{2}\sqrt{12}}{20} & -\frac{\sqrt{5}\sqrt{2}\sqrt{9}\sqrt{4}}{60} & -\frac{\sqrt{5}}{5} \\ -\frac{\sqrt{6}\sqrt{12}\sqrt{5}}{60} & -\frac{\sqrt{2}\sqrt{12}}{10} & \frac{3}{20} & -\frac{11\sqrt{12}\sqrt{5}\sqrt{9}\sqrt{4}}{720} & \frac{\sqrt{12}\sqrt{5}\sqrt{2}}{30} \\ \frac{\sqrt{6}\sqrt{9}\sqrt{4}}{36} & \frac{\sqrt{5}\sqrt{2}\sqrt{9}\sqrt{4}}{30} & -\frac{\sqrt{12}\sqrt{5}\sqrt{9}\sqrt{4}}{240} & -\frac{5}{12} & \frac{\sqrt{9}\sqrt{4}\sqrt{2}}{18} \\ \frac{\sqrt{6}\sqrt{2}}{6} & -\frac{\sqrt{5}}{5} & -\frac{\sqrt{12}\sqrt{5}\sqrt{2}}{20} & -\frac{\sqrt{9}\sqrt{4}\sqrt{2}}{36} & -\frac{1}{3} \end{pmatrix}. \tag{3.13}$$

3.2   Nontrivial Graph Automorphisms and the Structure of Eigenvectors of the Adjacency Matrix

The group of graph automorphisms consists of all permutation matrices $\mathbf{\Pi}$ which commute with the adjacency matrix $\mathbf{A}_G$. Following Chan and Godsil (1997) and Godsil and Royle (2001), let us consider a graph G, with a non-trivial

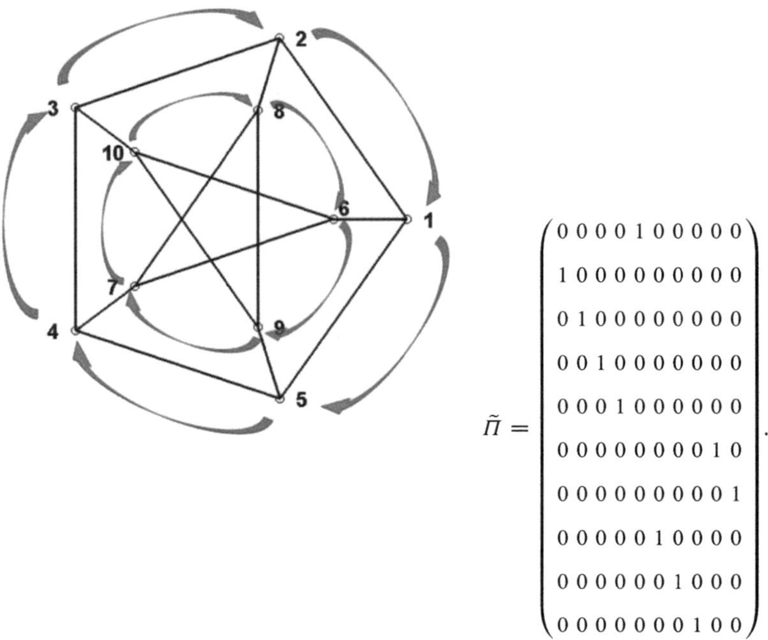

$$\tilde{\Pi} = \begin{pmatrix} 0\ 0\ 0\ 0\ 1\ 0\ 0\ 0\ 0\ 0 \\ 1\ 0\ 0\ 0\ 0\ 0\ 0\ 0\ 0\ 0 \\ 0\ 1\ 0\ 0\ 0\ 0\ 0\ 0\ 0\ 0 \\ 0\ 0\ 1\ 0\ 0\ 0\ 0\ 0\ 0\ 0 \\ 0\ 0\ 0\ 1\ 0\ 0\ 0\ 0\ 0\ 0 \\ 0\ 0\ 0\ 0\ 0\ 0\ 0\ 0\ 1\ 0 \\ 0\ 0\ 0\ 0\ 0\ 0\ 0\ 0\ 0\ 1 \\ 0\ 0\ 0\ 0\ 0\ 1\ 0\ 0\ 0\ 0 \\ 0\ 0\ 0\ 0\ 0\ 0\ 1\ 0\ 0\ 0 \\ 0\ 0\ 0\ 0\ 0\ 0\ 0\ 1\ 0\ 0 \end{pmatrix}.$$

Fig. 3.1 An automorphism of the Petersen graph and its permutation matrix

automorphism group $\mathrm{Aut}(G)$, let $\mathbf{u}$ be an eigenvector of the adjacency matrix $\mathbf{A}_G$, with corresponding eigenvalue ν,

$$\mathbf{A}_G \mathbf{u} = \nu\, \mathbf{u}.$$

Then it is obvious that

$$\mathbf{A}_G\, \Pi\, \mathbf{u} = \Pi \mathbf{A}_G\, \mathbf{u} \tag{3.14}$$

if $\Pi \in \mathrm{Aut}(G)$, and moreover

$$\Pi \mathbf{A}_G\, \mathbf{u} = \Pi \nu\, \mathbf{u} = \nu\, \Pi \mathbf{u}, \tag{3.15}$$

so that all the vectors $\Pi \mathbf{u}$ are also the eigenvectors of the matrix $\mathbf{A}_G$ belonging to the same eigenvalue ν. Let us assume that ν is a simple eigenvalue of $\mathbf{A}_G$ and $\mathbf{u} = (u_1, \ldots, u_N)^\top$ is its real eigenvector. It is clear that any permutation $\Pi \in \mathrm{Aut}(G)$ maps the eigenvector $\mathbf{u}$ into a linearly dependent vector

$$\Pi \mathbf{u} = \kappa \mathbf{u}, \tag{3.16}$$

with some $\kappa \in \mathbb{R}$. Given $C = (1, 2, \ldots, t)$ a cycle of $\Pi \in \mathrm{Aut}(G)$, the latter equation implies that

$$C \mathbf{u}' = \kappa \mathbf{u}', \tag{3.17}$$

for a non-zero partial vector $\mathbf{u}' = (u_1, \ldots, u_t)^\top$ of $\mathbf{u}$, and whence

$$C'\mathbf{u}' = \mathbf{u}' = \kappa^t \mathbf{u}' \tag{3.18}$$

it is clear that

$$\kappa^t = 1, \quad \kappa = \begin{cases} +1, & t \text{ is odd,} \\ \pm 1, & t \text{ is even.} \end{cases} \tag{3.19}$$

In particular, if t is odd, then t elements, in the eigenvector $\mathbf{u}$ belonging to a simple eigenvalue, are all equal,

$$u_1 = u_2 = \ldots u_t;$$

if t is even, then either those t elements are all equal,

$$u_1 = u_2 = \ldots u_t,$$

or they are of alternating sign

$$u_1 = -u_2 = \ldots = u_{t-1} = -u_t.$$

Consequently, if all eigenvalues of $\mathbf{A}_G$ are simple, then

$$\mathbf{\Pi}\mathbf{u} = \pm\mathbf{u},$$

and consequently

$$\mathbf{\Pi}^2\mathbf{u} = \mathbf{u},$$

so that any permutation $\mathbf{\Pi} \in \mathrm{Aut}(G)$ is an involution.

3.3   Automorphism Invariant Linear Functions of a Graph

The degree of a node is not the unique automorphism invariant function which can be defined on a graph G. In the present section, following Smola and Kondor (2003) and Blanchard and Volchenkov (2008b, 2009a), we consider the general linear transformation of the adjacency matrix $\mathbf{A}_G$,

$$\mathcal{F}\left(A_{ij}\right) = \sum_{s,l=1}^{N} \Phi_{ijsl}\, A_{sl}, \quad \Phi_{ijsl} \in \mathbb{R}, \tag{3.20}$$

invariant under any permutation $\mathbf{\Pi} \in \mathrm{Aut}(G)$,

$$\mathbf{\Pi}^\top \mathcal{F}(\mathbf{A}_G)\, \mathbf{\Pi} = \mathcal{F}\left(\mathbf{\Pi}^\top \mathbf{A}_G\, \mathbf{\Pi}\right), \tag{3.21}$$

and preserving connectivity for any vertex $i \in V$

$$\sum_{j \in V} A_{ij} = \deg(i) = \sum_{j \in V} \mathcal{F}(\mathbf{A}_G)_{ij} \,. \tag{3.22}$$

The first relation (3.21) means that the transformation (3.20) preserves the conjugate classes of index partition structures, so that the tensor Φ_{ijsl} is permutation invariant,

$$\Phi_{\Pi(i)\,\Pi(j)\,\Pi(s)\,\Pi(l)} = \Phi_{ijsl}, \tag{3.23}$$

provided $\Pi \in \mathrm{Aut}(G)$. Any tensor Φ_{ijsl} enjoying (3.23) has to be expressed as a linear combination of the tensors preserving the conjugate classes of index partition structures and admissible by symmetry:

$$\mathcal{F}(A_{ij}) = c_1 + \delta_{ij}\,(c_2 + c_3\,\deg(j)) + \beta\,A_{ij}, \tag{3.24}$$

in which $c_{1,2,3}$ and $\beta > 0$ being arbitrary constants. It is important to mention that not all functions (3.24) preserve connectivity of vertices. The relation (3.22) requires

$$c_1 = c_2 = 0,$$

since the contributions of $c_1 N$ and c_2 are indeed incompatible with that. Moreover, the remaining constants should satisfy the relation

$$c_3 + \beta = 1.$$

As a result, we obtain the general form for an automorphism invariant linear function of a graph,

$$\begin{aligned}
\mathcal{F}(\mathbf{A}_G)_{ij} &= (1 - \beta)\,\delta_{ij}\,\deg(j) + \beta\,A_{ij} \\
&= ((1 - \beta)\mathbf{D} + \beta\mathbf{A})_{ij}
\end{aligned} \tag{3.25}$$

where D is the diagonal graph's degree matrix. It is obvious from (3.25) that the automorphism invariant linear function of a graph commutes with its adjacency matrix,

$$[\mathcal{F}(\mathbf{A}_G), \mathbf{A}_G] = 0. \tag{3.26}$$

3.3.1 Automorphism Invariant Stochastic Processes

The automorphism invariant linear functions that satisfy the *probability conservation relation*,

$$1 = \frac{1}{\deg(i)} \sum_{j \in V} \mathcal{F}(A_{ij}), \quad \forall i \in V, \tag{3.27}$$

can be naturally interpreted as a Markov stochastic process (Markov 1906) determined by the following matrix of transition probabilities,

$$T_{ij}^{(\beta)} = \frac{\mathcal{F}(A_{ij})}{\deg(i)}$$

$$= \Pr\left[v_{t+1} = j \,|\, v_t = i\right] > 0 \Leftrightarrow i \sim j,$$

$$= (1 - \beta)\,\delta_{ij} + \beta\,\frac{A_{ij}}{\deg(i)} \tag{3.28}$$

$$= \left((1 - \beta)\mathbf{1} + \beta\mathbf{D}^{-1}\mathbf{A}_G\right)_{ij}.$$

The operator (3.28), for $0 < \beta \le 1$, defines a generalized *lazy random walk*, in which a random walker stays in the initial vertex with probability $1 - \beta$, while it moves to another node randomly chosen among the nearest neighbors with probability $\beta/\deg(i)$. In particular, for $\beta = 1$, the operator $\mathbf{T}_{ij}^{(\beta)}$ describes the fair random walks extensively studied in classical surveys (see Lovász 1993; Lovász and Winkler 1995 and references therein). Random walks provide the theory of graphs with rich probabilistic interpretations.

It is worth a mention that the matrix $\mathbf{T}^{(\beta)}$ belongs to a convex set of doubly stochastic matrices, since for any value $0 < \beta \le 1$, it has all the entries nonnegative and for any j,

$$\sum_{j \in V} T_{ij}^{(\beta)} = 1 = \sum_{i \in V} T_{ij}^{(\beta)}.$$

Therefore, accordingly to the Birkhoff-von Neumann theorem, it can be represented by a convex combination of at most $N^2 - 2N + 2$ permutation matrices Π_k,

$$\mathbf{T}^{(\beta)} = \sum_{k=1}^{m} \alpha_k \mathbf{\Pi}_k, \qquad \sum_{k=1}^{m} \alpha_k = 1,$$
$$0 \le \alpha_1, \alpha_2, \ldots, \alpha_m \le 1, \ 1 \le m \le N^2 - 2N + 2. \tag{3.29}$$

However, in spite of

$$\left[\mathbf{T}^{(\beta)}, \mathbf{A}_G\right] = 0, \tag{3.30}$$

the permutation matrices $\mathbf{\Pi}_k$ contributing into its Birkhoff-von Neumann decomposition (3.29) may not belong to the group of graph automorphisms $\mathrm{Aut}(G)$. Moreover, the transition matrix $\mathbf{T}^{(\beta)}$ can certainly be defined for a graph which has no non-trivial automorphisms.

3.3.2   Automorphism Invariant Harmonic Functions

Should we require that the linear function of a graph $\mathcal{F}(\mathbf{A}_G)$ to be *harmonic*,

$$\sum_{j \in V} f(A_{ij}) = 0, \quad \forall i \in V, \tag{3.31}$$

the linear function (3.25) defines a diffusion process on the graph G described by the *generalized Laplace operator* Smola and Kondor 2003,

$$\mathcal{L}_{ij} = -\frac{\alpha_2}{N} + \delta_{ij}\left(\alpha_2 + \alpha_3 k_i\right) - \alpha_3 A_{ij}, \tag{3.32}$$

characterized by the conservation of mass. The choice of the constants α_2 and α_3 in (3.32) depends upon the details of the model. The constant α_2 describes a *zero-level transport mode* and is usually taken as $\alpha_2 = 0$, in absence of additional sources of mass. The Laplace operator (3.32) where $\alpha_2 = 0$ and $\alpha_3 = 1$ is called the *canonical Laplace operator* (de Verdiére 1998),

$$\mathbf{L_c} = \mathbf{D} - \mathbf{A}_G, \tag{3.33}$$

where $\mathbf{D}$ is the diagonal graph's degree matrix.

The canonical Laplace operator (3.33) arises in theory of electric circuits (Doyle and Snell 1984; Tetali 1991; Wu 2004) consisting of N nodes, in which each pair of nodes, i and j, are characterized by the resistance $r_{ij} > 0$ (or conductance, $c_{ij} \equiv r_{ij}^{-1}$), so that $r_{ij} = \infty$ and $c_{ij} = 0$ if there is no resistor connecting i and j.

We denote the electric potential at the ith node by U_i and the net current flowing into the network at the ith node by I_i. Since there exists no sinks or sources of currents in the network, the conservation relation (3.31) holds,

$$\sum_{i=1}^{N} I_i = 0. \tag{3.34}$$

Then the *Kirchhoff law* states that

$$\mathbf{Lu} = \mathbf{i} \tag{3.35}$$

where $\mathbf{u} = (u_1, u_2, \ldots, u_N)$ is the vector of electric potentials, $\mathbf{i} = (i_1, i_2, \ldots, i_N)$ is the vector of net currents, and

$$\mathbf{L} = \begin{pmatrix} c_1 & -c_{12} & \ldots & -c_{1N} \\ -c_{21} & c_2 & \ldots & -c_{2N} \\ \vdots & \vdots & \ddots & \vdots \\ c_{N1} & -c_{N2} & \ldots & c_N \end{pmatrix} \tag{3.36}$$

is the Laplace matrix (also known as the *Kirchhoff matrix*), in which the diagonal elements

$$c_i = \sum_{j=1, i \neq j}^{N} c_{ij}.$$

If nonzero resistances are uniformly equal to 1, the matrix (3.36) coincides with that of (3.33), otherwise it describes an undirected graph, in which some weight c_{ij} is specified for each edge. Theory of electric circuits gives us worth another look on graph theory consistently extending it to the case of complex edges connecting the vertices in a graph. Indeed, we know that connections in electric circuits can be characterized by the complex impedance $z = r + i\chi$ where r is the resistive part and χ is the reactive part. The probabilistic interpretations of some phenomena related to ac circuits (such as LC resonances) are difficult.

Finally, let us note that the conservation relation (3.31) is defined as

$$\frac{1}{\deg(i)} \sum_{j \in V} \mathcal{F}(A_{ij}) = 0, \quad \forall i \in V, \tag{3.37}$$

Then, the linear function (3.25) determines yet another *combinatorial* Laplace operator,

$$\begin{aligned} \mathbf{L}_T &= \mathbf{D}^{-1}\mathbf{L}_c \\ &= \mathbf{1} - \mathbf{D}^{-1}\mathbf{A}_G, \end{aligned} \tag{3.38}$$

which is simply related to the transition matrix of random walks $\mathbf{T}^{(\beta)}$, for $\beta = 1$, by

$$\mathbf{L}_T = \mathbf{1} - \mathbf{T}, \tag{3.39}$$

For $\beta < 1$, we obtain a family of Laplace operators describing *lazy diffusions* on the graph G,

$$\mathbf{L}_\beta = \beta\mathbf{L}_T.$$

The structural properties of graphs can be described by algebraic properties of automorphism invariant linear functions defined on them.

3.4   Relations Between Eigenvalues of Automorphism Invariant Linear Functions

Spectra of the various automorphism invariant functions of a graph are simply related to each other. The eigenvalues of random walks

$$\mathbf{T} = \mathbf{D}^{-1}\mathbf{A}_G \tag{3.40}$$

are the roots of the characteristic polynomial

$$Q_T \equiv \det\left(\mu \cdot \mathbf{1} - \mathbf{D}^{-1}\mathbf{A}_G\right). \tag{3.41}$$

The maximal eigenvalue of the transition matrix (3.40) is $\mu_1 = 1$, and the vector of all ones $\mathbf{j}$ is the correspondent eigenvector,

$$\mathbf{T}\mathbf{j} = \mathbf{j}. \tag{3.42}$$

The characteristic polynomial (3.41) is invariant under any orthogonal transformation of the matrix $\mathbf{T}$. Let us consider such a transformation,

$$\begin{aligned}
\hat{T}_{ij} &= \left[\mathbf{D}^{1/2}\left(\mathbf{D}^{-1}\mathbf{A}_G\right)\mathbf{D}^{-1/2}\right]_{ij} \\
&= \frac{A_{ij}}{\sqrt{\deg(i)\,\deg(j)}}
\end{aligned} \tag{3.43}$$

with respect to the diagonal graph's degree matrix $\mathbf{D}$. The matrix $\hat{\mathbf{T}}$ is symmetric and its characteristic polynomial $Q_{\hat{T}}$ is identical to that of (3.41), so that all the eigenvalues $\{\mu_1, \mu_2, \ldots, \mu_N\}$ and the eigenvectors corresponding to them are real. Moreover, it follows from the Perron-Frobenius theorem that the maximal eigenvalue $\mu_1 = 1$ is simple and dominates all others,

$$1 = \mu_1 > \mu_2 \geq \ldots \mu_N \geq -1.$$

It can be shown that the $\mu_N = -1$ only for bipartite graphs.

The characteristic polynomial of the automorphism invariant harmonic functions are

$$\begin{aligned}
Q_{L_c} &= \det\left(\Lambda \cdot \mathbf{1} - \mathbf{L}_c\right) \\
&= \det\left(\Lambda \cdot \mathbf{1} - \mathbf{D} + \mathbf{A}_G\right),
\end{aligned} \tag{3.44}$$

for the canonical Laplace operator (3.33), and

$$\begin{aligned}
Q_{L_{\mathrm{T}}} &= \det\left(\lambda \cdot \mathbf{1} - \mathbf{L}_{\mathrm{T}}\right) \\
&= \det\left(\lambda \cdot \mathbf{1} - \mathbf{1} + \mathbf{D}^{-1}\mathbf{A}_G\right),
\end{aligned} \tag{3.45}$$

for the combinatorial Laplace operator (3.38). It follows immediately from

$$Q_{L_c} = \det(\mathbf{D}) \cdot \det\left(\mathbf{D}^{-1}\Lambda \cdot \mathbf{1} - \mathbf{1} + \mathbf{D}^{-1}\mathbf{A}_G\right) \tag{3.46}$$

that the roots of the characteristic polynomials Q_{L_c} and $Q_{L_{\mathrm{T}}}$ are simply related by

$$\lambda_k = \frac{\Lambda_k}{\deg(k)}. \tag{3.47}$$

The combinatorial Laplace operator can also be made symmetric by the orthogonal transformation (3.43):

$$\begin{aligned}
\hat{L}_{ij} &= \delta_{ij} - \frac{A_{ij}}{\sqrt{\deg(i)\,\deg(j)}} \\
&= \left(\mathbf{1} - \hat{\mathbf{T}}\right)_{ij}.
\end{aligned} \tag{3.48}$$

The characteristic polynomial for the *normalized* Laplace operator (3.48) is obviously identical to Q_{L_T}; all its eigenvalues are real.

There is also an obvious relation between the eigenvalues of random walks and of Laplace operator, as $\mathbf{L}_T = \mathbf{1} - \mathbf{T}$,

$$\lambda_k = 1 - \mu_k, \quad k = 1, \ldots, N. \tag{3.49}$$

The maximal eigenvalue $\mu_1 = 1$ of the random walk transition matrix is transformed by (3.49) into the minimal eigenvalue of the Laplace operator, $\lambda_1 = 0$, whence all other eigenvalues of the Laplace operator are positive.

We conclude the section with an interesting relation between the eigenvalues of the canonical Laplace operators defined on a graph and its complement (see Cvetkovic et al. 1980). Given a connected graph $G(V, E)$ and its complement $\bar{G}(V, K_N \setminus E)$, we can define the canonical Laplace operators on both of them by

$$\mathbf{L}_c(G) = \mathbf{D} - \mathbf{A}_G \tag{3.50}$$

and

$$\mathbf{L}_c(\bar{G}) = (N \cdot \mathbf{1} - \mathbf{D}) - \mathbf{A}_{\bar{G}}. \tag{3.51}$$

respectively. Clearly,

$$\begin{aligned} \mathbf{L}_c(G) + \mathbf{L}_c(\bar{G}) &= \mathbf{L}_c(K_N) \\ &= (N \cdot \mathbf{1} - \mathbf{J}). \end{aligned} \tag{3.52}$$

where $\mathbf{L}_c(K_N)$ is the canonical Laplace operator defined on the complete graph K_N and $\mathbf{J}$ is the matrix of all ones. Since the vector $\mathbf{j} = (1, 1, \ldots, 1)$ is an eigenvector of the operator $\mathbf{L}_c(K_N)$ belonging to the minimal eigenvalue $\Lambda = 0$, the same vector $\mathbf{j}$ is also an eigenvector for the both operators,

$$\mathbf{L}_c(G)\mathbf{j} = 0 = \mathbf{L}_c(\bar{G})\mathbf{j}. \tag{3.53}$$

Now, let $\mathbf{z}$ be such that

$$\mathbf{L}_c(G)\mathbf{z} = \Lambda\mathbf{z}, \tag{3.54}$$

but $\mathbf{z} \neq \mathbf{j}$, then $\mathbf{z}$ should be orthogonal to $\mathbf{j}$

$$(\mathbf{z}, \mathbf{j}) = 0$$

and therefore

$$\mathbf{J}\mathbf{z} = 0.$$

Thus,

$$\begin{aligned} N \cdot \mathbf{z} &= (N \cdot \mathbf{1} - \mathbf{J})\mathbf{z} \\ &= \left(\mathbf{L}_c(G) + \mathbf{L}_c(\bar{G})\right)\mathbf{z} \\ &= \Lambda\mathbf{z} + \mathbf{L}_c(\bar{G})\mathbf{z}, \end{aligned} \tag{3.55}$$

from where it follows that

$$(N - \Lambda)\mathbf{z} = \mathbf{L}_c(\bar{G})\mathbf{z}.$$

Comparing the latter equality with (3.54), we conclude that

$$\Lambda'_k = N - \Lambda_k, \tag{3.56}$$

where Λ'_k are the nonzero eigenvalues of the operator $\mathbf{L}_c(\bar{G})$ and Λ_k are the nonzero eigenvalues of the operator $\mathbf{L}_c(G)$.

3.5 Summary

All permutations preserving the relations of adjacency in a connected graph constitute the group of its automorphisms. The group of the graph automorphisms consists of all permutation matrices which commute with the adjacency matrix of the graph. All the eigenvalues of a graph possessing a single identity automorphism (asymmetric graph) are simple, while multiple eigenvalues indicate the existence of non-identity graph automorphisms. A random walk on a graph is the only stochastic process invariant with respect to graph automorphisms. The harmonic functions invariant with respect to graph automorphisms describe diffusions on a graph.

Chapter 4
Exploring Undirected Graphs by Random Walks

A finite connected undirected graph $G(V, E)$ can be seen as a *discrete time dynamical system* possessing a finite number of states (nodes) (Prisner 1995). The behavior of such a dynamical system can be studied by means of a transfer operator which describes the time evolution of distributions in phase space. The transfer operator can be represented by a stochastic matrix determining a discrete time random walk on the graph in which a walker picks at each node between the various available edges with equal probability. An obvious benefit of the approach based on random walks to graph theory is that the relations between individual nodes and subgraphs acquire a precise quantitative probabilistic description that enables us to attack applied problems which could not even be started otherwise.

Random walks defined on an undirected graph by a transition probability matrix can be mathematically decomposed into shape and orientation information, determined by the eigenvalues and eigenvectors, respectively. Here, we refer the *graph shape* to those degrees of freedom by the changes in the eigenvectors, while keeping the eigenvalues fixed. The complementary degrees of freedom associated with changes in the eigenvectors, while keeping the eigenvalues fixed, might be referred to as *orientation of a graph*. Such the definitions of orientation and shape are compatible with the standard visualization of a tensor as an ellipsoid, in which the orientation is determined by aligning the axes with the eigenvectors. The eigenvectors of the combinatorial Laplace matrices defined on graph have received much attention in Biyikoglu et al. (2004, 2007), in concern with the nodal domain theorems which give bounds on the number of connected subgraphs on which the components of an eigenvector do not change sign. The Fiedler eigenvector corresponding to the second smallest eigenvalue of the Laplace operator matrix of a graph is used in spectral bisection of the graph (Chung 1997).

In the present chapter, we discuss the properties of random walks defined on undirected connected graphs and give an account to each group of nodes with respect to the entire graph structure by means of random currents traversing the graph.

P. Blanchard and D. Volchenkov, *Random Walks and Diffusions on Graphs and Databases*, Springer Series in Synergetics 10, DOI 10.1007/978-3-642-19592-1_4,
© Springer-Verlag Berlin Heidelberg 2011

4.1 Graphs as Discrete Time Dynamical Systems

Given a finite connected undirected graph $G(V, E)$, let us consider a transformation

$$\mathcal{S} : V \to V$$

mapping any subset of nodes $U \subset V$ into the set of their direct neighbors,

$$\mathcal{S}(U) = \{ w \in V \mid v \in U, v \sim w \}. \tag{4.1}$$

We denote the result of $t \geq 1$ consequent applications of $\mathcal{S}$ to $U \subset V$ as $\mathcal{S}_t(U)$. The iteration of the map $\mathcal{S}$ leads to a study of possible paths in the graph G beginning at $v \in V$. However, we rather discuss the time evolution of smooth functions under iteration, than the individual trajectories $\mathcal{S}_t(v)$. Given a discrete *density function*

$$f(v) \geq 0, \quad v \in V,$$

defined on a undirected connected graph $G(V, E)$ such that

$$\sum_{v \in V} f(v) = 1,$$

the dynamics of the map (4.1) is described by the norm-preserving transformation

$$\sum_{v \in U} f(v)\, \mathbf{T}^t = \sum_{\mathcal{S}_t^{-1}(U)} f(v), \tag{4.2}$$

where $\mathbf{T}^t$ is the *Ruelle-Perron-Frobenius transfer operator* corresponding to the transformation $\mathcal{S}_t$. The uniqueness of the Ruelle-Perron-Frobenius operator for a given transformation $\mathcal{S}_t$ is a consequence of the Radon-Nikodym theorem extending the concept of probability densities to probability measures defined over arbitrary sets (Shilov and Gurevich 1978). It was shown by Mackey (1991) that the relation (4.2) is satisfied by a homogeneous Markov chain $\{v_t\}_{t \in \mathbb{N}}$ determining a random walk of the nearest neighbor type defined on the connected undirected graph $G(V, E)$ by the transition matrix

$$\begin{aligned}
T_{ij} &= \Pr[v_{t+1} = j \mid v_t = i] > 0 \Leftrightarrow i \sim j, \\
&= \mathbf{D}^{-1}\mathbf{A}, \quad \mathbf{D} = \mathrm{diag}(\deg(1), \ldots \deg(N)) \\
&= 1/\deg(i), \quad \text{iff} \quad (i, j) \in E,
\end{aligned} \tag{4.3}$$

where $\mathbf{A}$ is the adjacency matrix of the graph, so that the probability of transition from i to j in $t > 0$ steps equals

$$p_{ij}^{(t)} = \left(\mathbf{T}^t\right)_{ij}. \tag{4.4}$$

The discrete time random walks on graphs have been studied in details in Lovász (1993), Lovász and Winkler (1995), Saloff-Coste (1997) and by many other authors.

A Markov chain $\{v_t\}_{t\in\mathbb{N}}$ is called *ergodic* if it is possible to go from every state to every state (not necessarily in one move). The Markov chain defined on a connected undirected graph by (4.3) is always ergodic.

4.2   Generating Functions of the Transition Probabilities

The *generating function* (or the *Green function*) of the transition probabilities (4.4) is a power series representation

$$\begin{aligned}
\mathfrak{G}_{ij}(z) &= \sum_{t\geq 0} p_{ij}^{(t)} z^t \\
&= (1 - z\mathbf{T})^{-1},
\end{aligned} \tag{4.5}$$

where $p_{ij}^{(t)}$ is the probability mass function of transition from i to j in $t > 0$ steps. The power series with non-negative coefficients (4.5) converges absolutely inside the unit circle $|z| < 1$, and therefore $z = 1$ is the spectral radius of $\mathbf{T}$. The probabilities $p_{ij}^{(t)}$ are recovered by taking derivatives of $\mathfrak{G}(z)$,

$$p_{ij}^{(t)} = \frac{1}{t!} \left.\frac{d\,\mathfrak{G}_{ij}(z)}{dz^t}\right|_{z=0}. \tag{4.6}$$

The *first-hitting probabilities* characterizing the statistics of the first passage (with no recurrences allowed) from i to j are

$$q_{ij}^{(t)} = \Pr\left[v_t = j, v_l \neq j, l \neq 1, \ldots, t-1 \,|\, v_0 = i\right], \quad q_{ij}^{(0)} = 0. \tag{4.7}$$

They are related to the transition probabilities $p_{ij}^{(t)}$ by

$$p_{ij}^{(t)} = \sum_{s=0}^{t} q_{ij}^{(s)} p_{jj}^{(t-s)} \tag{4.8}$$

and calculated by means of the generating function

$$\mathfrak{F}_{ij}(z) = \sum_{t\geq 0} q_{ij}^{(t)} z^t, \quad i, j \in V, \quad z \in \mathbb{C}, \tag{4.9}$$

with the generating property

$$q_{ij}^{(t)} = \frac{1}{t!} \left. \frac{d\,\mathfrak{F}_{ij}\,(z)}{dz^t} \right|_{z=0}. \tag{4.10}$$

From (4.7), it follows that the generating functions (4.5) and (4.9) are related to each other by the simple equation

$$\mathfrak{G}_{ij}\,(z) = \mathfrak{F}_{ij}\,(z)\mathfrak{G}_{jj}\,(z) \tag{4.11}$$

and therefore $\mathfrak{F}_{ij}\,(z)$ is nothing else but the Green function $\mathfrak{G}_{ij}\,(z)$ normalized in such a way that its diagonal entries become one (Lovász and Winkler 1995).

4.3 Cayley-Hamilton's Theorem for Random Walks

The Cayley-Hamilton theorem in linear algebra asserts that any $N \times N$ matrix is a solution of its associated characteristic polynomial. Given the transition matrix $\mathbf{T}$ of a random walk defined on a graph $G(V, E)$, its characteristic equation is

$$\det\,(\mathbf{T} - \mu \cdot \mathbf{1}) = 0$$
$$= \sum_{k=0}^{N} I_k \mu^{N-k} \tag{4.12}$$

where the roots μ are the eigenvalues of $\mathbf{T}$, and $\{I_k\}_{k=1}^{N}$ are its principal invariants (2.41), with $I_0 = 1$. Then, the transition matrix $\mathbf{T}$ itself satisfies the characteristic equation,

$$0 = \sum_{k=0}^{N} I_k \mathbf{T}^{N-k}. \tag{4.13}$$

The proofs of the Cayley-Hamilton theorem follows from the definition of eigenvalue of a matrix and can be found in any standard textbook in linear algebra (for example, see Kolman and Hill 2007; Golub and Van Loan 1996; Greub 1981). Thus, the higher powers $t \geq N$ of $\mathbf{T}$ can be expressed by a matrix polynomial of the lower powers. As the powers of $\mathbf{T}$ determines the probabilities of transitions (4.4), we obtain the following expression for the probability of transition from i to j in $t = N + 1$ steps as the sign alternating sum of the conditional probabilities

$$p_{ij}^{(N+1)} = \sum_{k=1}^{N} (-I_k)\, p_{ij}^{(N+1-k)}, \tag{4.14}$$

in which $p_{ij}^{(N+1-k)}$ takes the values of the probabilities to reach j starting from i faster than in $N+1$ steps as k runs from 1 to N, and $|I_k|$ are the k–steps *recurrence probabilities* of random walks in the graph G expressing the chance of the random walk returns to the initial node after k time steps. The principal invariants (2.44) of the transition matrix $\mathbf{T}$ are

$$
\begin{aligned}
I_k &= \frac{(-1)^{k-1}}{k} \sum_{l=0}^{k-1} (-1)^l\, I_l\, \mathrm{Tr}\,\mathbf{T}^l, \\
&= \sum_{\left\{ \sum_{l=1}^{k} l \cdot m_l = k \right\}} (-1)^{\sum_l (l-1)m_l} \left(\frac{\mathrm{Tr}\,\mathbf{T}^{m_1}}{m_1!\,1^{m_1}} \cdot \ldots \cdot \frac{\mathrm{Tr}\,\mathbf{T}^{m_k}}{m_k!\,k^{m_k}} \right),
\end{aligned}
\tag{4.15}
$$

in which the last summation is performed over all non-negative partitions $\sum_{l=1}^{k} l \cdot m_l = k$. In particular, $|I_1| = \mathrm{Tr}\,\mathbf{T}$ is the probability that a random walker stays at a node in one time step, and $|I_N| = |\det \mathbf{T}|$ expresses the probability that the random walks revisit an initial node in N steps. In regulatory networks, the probability of recurrence estimates the chance of establishing the circular sequences of interactions that have been shown to play the key dynamical roles (see Chap. 9 for details).

4.4   Stationary Distribution and Recurrence Time of Random Walks

For a random walk defined on a connected undirected graph, the Perron-Frobenius theorem (see Graham 1987; Minc 1988; Horn and Johnson 1990) asserts the unique strictly positive probability vector

$$
\boldsymbol{\pi} = (\pi_1, \ldots, \pi_N)
$$

(the left eigenvector of the transition matrix $\mathbf{T}$ belonging to the maximal eigenvalue $\mu = 1$) such that

$$
\boldsymbol{\pi}\,\mathbf{T} = 1 \cdot \boldsymbol{\pi}.
\tag{4.16}
$$

The vector $\boldsymbol{\pi}$ satisfies the condition of *detailed balance*,

$$
\pi_i\, T_{ij} = \pi_j\, T_{ji},
\tag{4.17}
$$

from which it follows that a random walk defined on an undirected graph is time reversible: it is also a random walk if considered backward, and it is not possible to determine, given the walker at a number of nodes in time after running the walk, which state came first and which state arrived later. It is then obvious that any row vector $\boldsymbol{\sigma}$ such that

$$
\boldsymbol{\sigma}\,\mathbf{T} = \boldsymbol{\sigma}
$$

is a multiple of $\boldsymbol{\pi}$, and any column vector $\boldsymbol{\omega}$ such that

$$
\mathbf{T}\boldsymbol{\omega} = \boldsymbol{\omega}
$$

Fig. 4.1 Random tours of a
knight on a chess-board
starting from a corner square

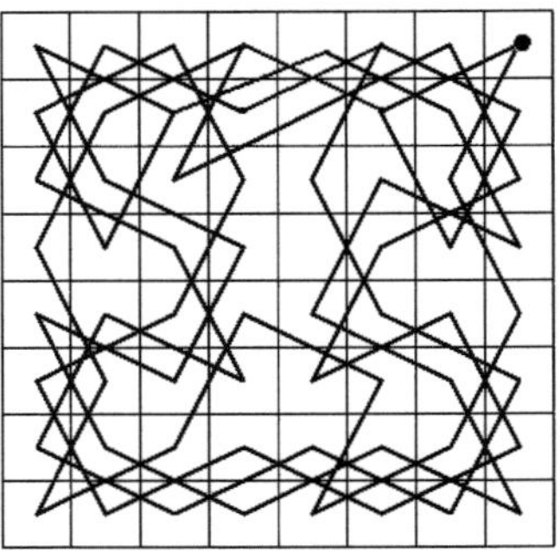

is a multiple of the vector of units $\mathbf{j} = (1, \ldots, 1)$ that is the right eigenvector of $\mathbf{T}$
belonging to the same eigenvalue $\mu = 1$.

For the nearest neighbor random walks defined on an undirected graph, the
stationary distribution of random walks on an undirected graph equals

$$\pi_i = \frac{\deg(i)}{2|E|}, \quad \sum_{i \in V} \pi_i = 1 \tag{4.18}$$

where $|E|$ is the total number of edges in the graph (Lovász 1993; Lovász and
Winkler 1995). Interestingly, the probability to observe a random walker at a node
does not depend upon neither the order of the entire graph, nor upon its structure but
only on the total number of its edges and the local property of the node – its degree.
In particular, the distribution (4.18) is uniform for a regular graph.

It is known that for a stationary, discrete-valued stochastic process the expected
recurrence time to return to a state is the reciprocal of the probability of this state
(Kac 1947). The expected *recurrence time* to a node which indicates how long a
random walker must wait to revisit the site is inverse proportional to π_i,

$$r_i = \frac{2|E|}{\deg(i)} = \frac{1}{\pi_i}. \tag{4.19}$$

We also conclude from (4.18) that for any edge $(i, j) \in E$ the expected number of
steps before a random walker passes through the same edge next time is equal to

$$\frac{1}{\pi_i T_{ij}} = 2|E|. \tag{4.20}$$

The expected number of steps before the random walker passes through the same
node is as twice as less. These considerations together give the solution to the
famous *knight tour problem* (Aldous and Fill 2002) aiming to count the mean
number of random moves a knight can do starting at a corner square of an empty
chess-board until it returns to the starting square (see Fig. 4.1). The knight is
performing random walks on a graph in which 64 squares are the nodes, and the
possible knight-moves are the edges. This graph is obviously connected and so, the

expected number of steps before the knight passes through the same corner square equals to the number of edges in such a graph, $M = 168$.

4.5   Entropy of Random Walks Defined on a Graph

While in $\mathbb{R}^3$ a walker has three basic directions to move at each point - these are the physical dimensions of our space. Simulating the diffusion equation

$$\dot{u} = \Delta u \tag{4.21}$$

for a scalar function u defined on a regular d-dimensional lattice $\mathcal{L}_a = a\mathbb{Z}^d$, with the lattice scale length a, one uses the discrete representation of the Laplace operator Δ,

$$u^{t+1}(x) = \frac{1}{k} \cdot \frac{1}{a^2} \left[\sum_{y \in U_x} u^t(y) - k \cdot u^t(x) \right], \tag{4.22}$$

where U_x is the neighborhood of the node x in the lattice $\mathcal{L}_a$. The degree of each site in the lattice uniformly equals to

$$k = \deg(i) = 2^d, \quad \forall i \in V, \tag{4.23}$$

and therefore the parameter d in (4.23) can be naturally interpreted as the physical dimension of space.

Being defined on an undirected graph $G(V, E)$, the discrete Laplace operator (4.22) has the pretty same form, excepting for the cardinality k which now variates upon the site,

$$k_i = \deg(i),$$

so that the parameter

$$\delta_i = \log_2 k_i = \log_2 \frac{2|E|}{r_i} \tag{4.24}$$

can be considered as the *local* analog of the physical dimension d at the node $i \in V$.

An interesting question arises in concern with (4.24), namely whether it is possible to define a universal, *global* dimension for the graph G that can be considered as generalizing the space dimension in lattices?

Below, we show that this can be done on a statistical ground, by estimating the spreading of a set of independent random walkers. In information theory (Cover and Thomas 1991), such a spreading is measured by means of the *entropy rate* that may be considered as the analog of the physical dimension of space in information theory.

It is clear that the number of possible paths of length n in an undirected graph $G(V, E)$,

$$|\mathcal{X}_n| = \sum_{i,j \in V} (\mathbf{A}^n)_{ij} , \tag{4.25}$$

where $\mathbf{A}^n$ is the nth power of the adjacency matrix of the graph G, grows up exponentially with the path length n. Therefore, the probability to observe a long enough typical random path

$$\{X_1 = i_1, \ldots X_n = i_n\} \in \mathcal{X}$$

decreases asymptotically exponentially with $n \gg 1$,

$$2^{-n\,(H(\mathcal{X})+\varepsilon)} \leq \Pr\left[\{X_1 = i_1, \ldots X_n = i_n\}\right] \leq 2^{-n\,(H(\mathcal{X})-\varepsilon)} \tag{4.26}$$

where the parameter $n\,H(\mathcal{X})$ measuring the uncertainty of paths in random walks (4.26) grows asymptotically linearly with n at a rate

$$H(\mathcal{X}) = \lim_{n \to \infty} \frac{\log_2 |\mathcal{X}_n|}{n}, \tag{4.27}$$

which is called the *entropy rate* of random walks.

In order to proceed further, we should answer the following question: which random path can asymptotically be considered as a typical one?

Given $\mathbf{x}_s$ and $\mathbf{y}_s$,

$$(\mathbf{y}_s, \mathbf{x}_{s'}) = \delta_{s,s'},$$

the left and right eigenvectors of the transition matrix (4.3),

$$\mathbf{x}_s \mathbf{T} = \mu_s \mathbf{x}_s, \quad \mathbf{T} \mathbf{y}_s = \mu_s \mathbf{y}_s, \tag{4.28}$$

the spectral decompositions of the transition matrix and its powers are

$$\mathbf{T} = \sum_{s=1}^{N} \mu_s \mathbf{x}_s \mathbf{y}_s, \quad \mathbf{T}^t = \sum_{s=1}^{N} \mu_s^t \mathbf{x}_s \mathbf{y}_s. \tag{4.29}$$

Let us consider a *continuous time Markov jump process*

$$\{w_t\}_{t \in \mathbb{R}_+} = \{v_{\mathrm{Po}(t)}\}$$

where $\mathrm{Po}(t)$ is the Poisson distribution instead of the discrete time Markov chain $\{v_t\}_{t \in \mathbb{N}}$. Supposing that the transition time τ is a discrete random variable distributed with respect to the Poisson distribution $\mathrm{Po}(\tau)$ with mean 1, we use the spectral decomposition (4.29) to write down the probability of transition (4.4) as

$$p_{ij}^t = \pi_j + \sum_{s=2}^{N} x_{si} y_{sj} \sum_{\tau=0}^{\infty} \mu_s^{\tau} \frac{t^\tau e^{-t}}{\tau!}$$

$$= \pi_j + \sum_{s=2}^{N} x_{si} y_{sj} e^{-t\lambda_l}, \qquad (4.30)$$

where $\lambda_l \equiv (1 - \mu_l)$ is the lth *spectral gap*. It is obvious that

$$\lim_{t \to \infty} p_{ij}^t = \pi_j, \qquad (4.31)$$

since $|\mu_s| < 1$, for $2 \le s \le N$. The characteristic *decay times* of the relaxation processes,

$$\tau_l = \frac{1}{\lambda_l}, \quad 2 \le l \le N, \qquad (4.32)$$

estimate how fast the stationary distribution π can be achieved. The rate of convergence (4.30) to the stationary distribution π is characterized by the *mixing rate*,

$$\eta = \lim_{t \to \infty} \sup \max_{i,j \in V} \left| p_{ij}^{(t)} - \pi_j \right|. \qquad (4.33)$$

The asymptotic rate of convergence (4.30) is determined by the largest spectral gap,

$$\lambda_2 = 1 - \mu_2.$$

The reciprocal *mixing time*,

$$\tau = -\frac{1}{\ln \eta}, \qquad (4.34)$$

estimates the expected number of steps required to achieve the stationary distribution for the given graph G.

Being interested in statistics of very long walks, we can consider those of them satisfying the stationary distribution of walkers π as being typical for the given walk. Should the one-step transition probability of the random walk between nodes is defined by (4.3) the entropy rate (4.26) is given by

$$H = -\sum_{i \in V} \pi_i \sum_{j \in V} T_{ij} \log_2 \left(T_{ij} \right)$$

$$= -\sum_{i \in V} \frac{k_i}{2M} \sum_{j \in V} \frac{A_{ij}}{k_i} \log_2 \left(\frac{A_{ij}}{k_i} \right), \qquad (4.35)$$

in which we assume that $0 \cdot \log(0) = 0$. From (4.35), it is clear that

$$H = \frac{1}{2M} \sum_{i \in V} k_i \log_2 k_i$$

$$= \sum_{i \in V} \frac{\delta_i}{r_i}$$

$$= \sum_{i \in V} \pi_i \, \delta_i \tag{4.36}$$

$$= \langle \delta_i \rangle$$

where $\langle \delta_i \rangle$ means the average of the local dimension (4.24) over the stationary distribution of random walks π. From (4.36), it follows that the entropy rate H is nothing else but the averaged "local physical dimension" δ_i of the graph $G(V, E)$ at the node $i \in V$.

In information theory (Cover and Thomas 1991), the entropy rate (4.36) is important as a measure of the average message size required to describe a stationary random walk π defined on the graph G. Provided we use the binary code, we need approximately nH bits in order to describe the typical long enough path of length n. The entropy rates have recently been used by Boccaletti et al. (2006) and Gomez-Gardenes and Latora (2008) as a measure characterizing topological properties of complex networks.

4.6 Hyperbolic Embeddings of Graphs by Transition Eigenvectors

The stationary distribution (4.18) of random walks defined on a connected undirected graph $G(V, E)$ determines a unique measure on V,

$$D = \sum_{j \in V} \deg(j)\delta_j, \tag{4.37}$$

with respect to which the transition operator (4.3) becomes self-adjoint and is represented by a symmetric transition matrix,

$$\widehat{T_{ij}} = \left(\mathbf{D}^{1/2}\,\mathbf{T}\,\mathbf{D}^{-1/2}\right)_{ij}$$

$$= \frac{A_{ij}}{\sqrt{\deg(i)\,\deg(j)}} \tag{4.38}$$

where $\mathbf{D}$ is the diagonal matrix of graph's degrees. The matrix (4.38) corresponds to the *normalized Laplace operator*,

$$\widehat{\mathbf{L}} = \mathbf{1} - \widehat{\mathbf{T}}. \tag{4.39}$$

The use of self-adjoint operators (4.38) and (4.39) becomes now standard in spectral graph theory, Chung (1997) and in studies devoted to random walks on graphs (Lovász 1993).

Diagonalizing the symmetric matrix (4.38), we obtain

$$\widehat{\mathbf{T}} = \boldsymbol{\Psi} \mathbf{M} \boldsymbol{\Psi}^{\top}, \tag{4.40}$$

where $\boldsymbol{\Psi}$ is an orthonormal matrix,

$$\boldsymbol{\Psi}^{\top} = \boldsymbol{\Psi}^{-1},$$

and $\mathbf{M}$ is a diagonal matrix with entries

$$1 = \mu_1 > \mu_2 \geq \ldots \geq \mu_N > -1 \tag{4.41}$$

(here, we do not consider bipartite graphs, for which $\mu_N = -1$). The rows $\boldsymbol{\psi}_k = \{\psi_{k,1}, \ldots, \psi_{k,N}\}$ of the orthonormal matrix

$$\boldsymbol{\Psi} = \{\boldsymbol{\psi}_1, \boldsymbol{\psi}_2, \ldots \boldsymbol{\psi}_N\}^{\top} \tag{4.42}$$

are the real eigenvectors of $\widehat{\mathbf{T}}$ that forms an orthonormal basis in Hilbert space $\mathcal{H}(V)$,

$$\boldsymbol{\psi}_k : V \to S_1^{N-1}, \quad k = 1, \ldots N, \tag{4.43}$$

where S_1^{N-1} is the $N - 1$-dimensional unit sphere. We consider the eigenvectors (4.42) ordered in accordance to the eigenvalues they belong to. For eigenvalues of algebraic multiplicity $\alpha > 1$, a number of linearly independent orthonormal ordered eigenvectors can be chosen to span the associated eigenspace.

The first eigenvector $\boldsymbol{\psi}_1$ belonging to the largest eigenvalue $\mu_1 = 1$ (which is simple) is the Perron-Frobenius eigenvector that determines the stationary distribution of random walks over the graph nodes,

$$\boldsymbol{\psi}_1 \widehat{\mathbf{T}} = \boldsymbol{\psi}_1, \quad \psi_{1,i}^2 = \pi_i, \quad i = 1, \ldots, N. \tag{4.44}$$

The squared Euclidean norm of the vector in the orthogonal complement of $\boldsymbol{\psi}_1$,

$$\sum_{s=2}^{N} \psi_{s,i}^2 = 1 - \pi_i > 0, \tag{4.45}$$

expresses the probability that a random walker is not in i. Since all elements of the first eigenvector are positive,

$$\boldsymbol{\psi}_1 = \{\psi_{1,1}, \ldots, \psi_{1,i}, \ldots \psi_{1,N}\}, \quad \psi_{1,i} > 0, \quad i = 1, \ldots, N,$$

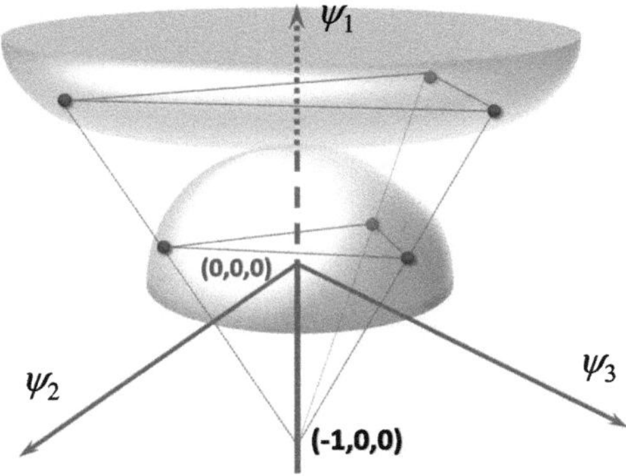

Fig. 4.2 A triangle embedded into the hyperbolic domains (4.46) and (4.48)

each node $i \in V$ of the graph $G(V, E)$ might be represented with respect to the orthonormal basis (4.43) by a point belonging to the hyperbolic domain in $\mathbb{R}^N$ – the surface of the (upper) hemisphere (Fig. 4.2),

$$\left\{ (\psi_{1,i}, \ldots, \psi_{N,i}) : \sum_{s=1}^{N} \psi_{s,i}^2 = 1 \text{ and } \psi_{1,i} > 0 \right\}. \tag{4.46}$$

Under the projective transformation

$$(y_{1,i}, \ldots, y_{N,i}) \mapsto \left(\frac{1}{\psi_{1,i}}, \frac{\psi_{2,i}}{\psi_{1,i}}, \ldots, \frac{\psi_{N,i}}{\psi_{1,i}} \right), \tag{4.47}$$

the hyperbolic domain (4.46) is isometrically equivalent (see Cannon et al. 1997) to the positive sheet of the $(N-1)$-dimensional hyperboloid,

$$\left\{ (y_{1,i}, \ldots, y_{N,i}) : y_{2,i}^2 + \ldots + y_{N,i}^2 - y_{1,i}^2 = -1 \text{ and } y_{1,i} > 0 \right\}. \tag{4.48}$$

The equation of the hyperboloid (4.48) follows directly from (4.45) if we divide it by $\pi_i = \psi_{1,i}^2 > 0$,

$$\begin{aligned}
\sum_{s=2}^{N} \left(\frac{\psi_{s,i}}{\psi_{1,i}} \right)^2 &= \psi_{1,i}^{-2} - 1 \\
&= \pi_i^{-1} - 1 \\
&= r_i - 1,
\end{aligned} \tag{4.49}$$

where r_i is the recurrence time of random walks to the node $i \in V$.

It is worth a mention that the vectors

$$\mathbf{y}_1 = \mathbf{j} \equiv (1, 1, \ldots, 1)$$

$$\mathbf{y}_s = \sum_{j \in V} \frac{\psi_{s,j}}{\psi_{1,j}} \tag{4.50}$$

where $s = 2, \ldots, N$ are the right eigenvectors of the transition matrix (4.3), while the vectors

$$\mathbf{x}_1 = \boldsymbol{\pi} \equiv (\psi_{1,1}^2, \psi_{1,2}^2, \ldots, \psi_{1,N}^2)$$

$$\mathbf{x}_s = \sum_{j \in V} \psi_{s,j} \, \psi_{1,j} \tag{4.51}$$

are its left eigenvectors.

Let us introduce the *Lorentzian inner product* (see Ratcliffe 1994) of two vectors, $\mathbf{v}$ and $\mathbf{w}$, in the hyperbolic domain (4.46) by

$$\mathbf{v} \circ \mathbf{w} = -\frac{1}{\psi_{1,v}} \frac{1}{\psi_{1,w}} + \sum_{s=2}^{N} \frac{\psi_{s,v}}{\psi_{1,v}} \frac{\psi_{s,w}}{\psi_{1,w}}. \tag{4.52}$$

The *Lorentzian norm* of a vector $\mathbf{v}$ is defined to be the complex number

$$\|\mathbf{v}\|_L = \sqrt{(\mathbf{v} \circ \mathbf{v})}. \tag{4.53}$$

For any graph node $v \in V$, its norm (4.53) is positive imaginary,

$$\begin{aligned}
\|\mathbf{v}\|_L &= \sqrt{-\frac{1}{\psi_{1,v}^2} + \sum_{s=2}^{N} \left(\frac{\psi_{s,v}}{\psi_{1,v}}\right)^2} \\
&= \sqrt{-r_v + r_v - 1} \\
&= \sqrt{-1} \equiv i.
\end{aligned} \tag{4.54}$$

The *Lorentzian distance* between the nodes $v, w \in V$ is defined by

$$\begin{aligned}
d_L(v, w) &= \|\mathbf{v} - \mathbf{w}\|_L \\
&= \sqrt{-\left(\frac{1}{\psi_{1,w}} - \frac{1}{\psi_{1,v}}\right)^2 + \sum_{s=2}^{N} \left(\frac{\psi_{s,v}}{\psi_{1,v}} - \frac{\psi_{s,w}}{\psi_{1,w}}\right)^2}.
\end{aligned} \tag{4.55}$$

The inner product (4.52) can be uniquely represented by

$$\mathbf{v} \circ \mathbf{w} = -\cosh \gamma(\mathbf{v}, \mathbf{w}), \tag{4.56}$$

with the positive *Lorenzian angle*,

$$\gamma(\mathbf{v}, \mathbf{w}) = \operatorname{arcosh}(-\mathbf{v} \circ \mathbf{w}). \tag{4.57}$$

Then, the *hyperbolic distance* between the vectors $\mathbf{v}$ and $\mathbf{w}$ can be defined as

$$d_H(\mathbf{v}, \mathbf{w}) = \gamma(\mathbf{v}, \mathbf{w})$$
$$= \ln\left(\mathbf{v} \circ \mathbf{w} + \sqrt{(\mathbf{v} \circ \mathbf{w})^2 - 1}\right). \tag{4.58}$$

The distance d_H is a metric on the hyperboloid (4.48), as being nonnegative, symmetric, nondegenerate, and satisfying the triangle inequality (Ratcliffe 1994).

The hyperbolic embedding of a graph might be useful for displaying and assessing highly inhomogeneous, hierarchical graph structures such as trees containing many generations of nodes. The hyperbolic space property allows to visualize those trees in an uncluttered manner, as in every generation a daughter node acquires almost the same amount of space as the parent one. The approach to use a hyperbolic tree for visualizing large hierarchies had been first proposed by Lamping et al. (1995). Recently, it has been shown by Krioukov et al. (2010) and Boguñá et al. (2010) that heterogeneous degree distributions and strong clustering in complex networks might also be apprised in terms of hyperbolic geometry.

We conclude the present section with a remark about the hyperbolic embedding of a star graph (see Fig. 4.3, left), which consists of a central node (hub) characterized by the uttermost connectivity and a number of terminal vertices (clients) linked to the hub. Due to the obvious structural symmetry of the star graph, all the client nodes are equidistant with respect to the hyperbolic distance (4.58), and the hyperbolic distances between the hub and any of the client nodes are also equal. Each star graph is therefore characterized by the two hyperbolic distances that grow with the graph order N, although those growth rates are different. On the right hand side of Fig. 4.3, we have shown the result of consequent calculations of the both hyperbolic distances for the star graphs in the order range $N = 3, \ldots, 200$. It is obvious that the hyperbolic distance between the client nodes grows faster with the order of the graph.

4.7 Exploring the Shape of a Graph by Random Currents

In contrast to the previous works of Biyikoglu et al. (2004, 2007) concerned the combinatorial Laplace operator of a graph, in the present section we consider the matrix of the transition eigenvectors $\boldsymbol{\Psi} \in O(N)$ of the symmetrized transition matrix $\widehat{\mathbf{T}}$. Below, we describe the rigorous and intuitive exposition of the graph shape by random currents over all subsets of graph nodes.

We discuss the determinants of the minors $\mathfrak{M}_{i_1,\ldots,i_k}^{s_1,\ldots,s_k}$ of the order k which cut down from the orthogonal matrix $\boldsymbol{\Psi}$ by removing the $s_1,\ldots,s_k$ rows and the $i_1,\ldots,i_k$ columns. The determinant is an antisymmetric multi-linear form, as interchanging any two rows (or columns) in that changes its sign. With respect to the property of the orthogonal transformation $\boldsymbol{\Psi}$, each undirected graph $G(V, E)$ refers to either proper ($\det(\boldsymbol{\Psi}) = +1$), or improper ($\det(\boldsymbol{\Psi}) = -1$) rotation (rotoreflection) that can be considered as a *signature* of the graph.

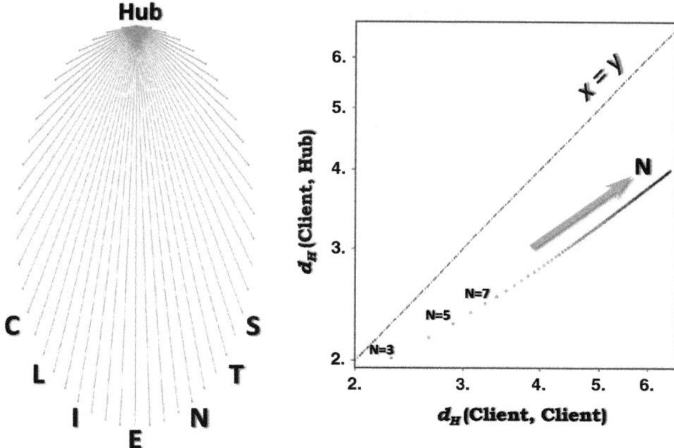

Fig. 4.3 A star graph consists of a hub and a number client nodes linked to that. Due to the obvious structural symmetry, a star graph is characterized by the two hyperbolic distances (between the client nodes and between the hub and any of the client nodes) that grow with the graph order N at different rates. The hyperbolic distance between the client nodes (of low hierarchy) grows faster with the order of the graph. The diagonal ($x = y$) is given for a reference

It is well known (see, for example Muir 1960, Chap. 14) that every element $\psi_{s,i}$ is numerically equal to its algebraic complement $\mathfrak{C}_{s,i}$ in the orthogonal matrix $\boldsymbol{\Psi}$,

$$\begin{aligned}
\det(\boldsymbol{\Psi}) \cdot \psi_{s,i} &= \mathfrak{C}_{s,i} \\
&= \det\left(\mathfrak{M}_i^s\right).
\end{aligned} \tag{4.59}$$

In particular, the elements of the Perron-Frobenius eigenvector $\psi_{1,i}$ might be calculated as

$$\begin{aligned}
\psi_{1,i} &= \det\left(\boldsymbol{\Psi}\right) \cdot \mathfrak{C}_{s,i} \\
&= \sqrt{\pi_i}.
\end{aligned} \tag{4.60}$$

Furthermore, it follows from (4.59) that the determinant of each minor $\det\left(\mathfrak{M}_{i_1,\ldots,i_k}^{s_1,\ldots,s_k}\right)$ of the kth order of $\boldsymbol{\Psi}$ is equal to the determinant of its algebraic complement $\det\left(\mathfrak{N}_{j_1,\ldots,j_{N-k}}^{z_1,\ldots,z_{N-k}}\right)$, multiplied by the $\det(\boldsymbol{\Psi})$,

$$\det\left(\mathfrak{M}_{i_1,\ldots,i_k}^{s_1,\ldots,s_k}\right) = \det\left(\mathfrak{N}_{j_1,\ldots,j_{N-k}}^{z_1,\ldots,z_{N-k}}\right) \cdot \det(\boldsymbol{\Psi}). \tag{4.61}$$

Multiplying the both sides of (4.61) by $\det\left(\mathfrak{M}_{i_1,\ldots,i_k}^{s_1,\ldots,s_k}\right)$, we obtain

$$\det\left(\mathfrak{M}_{i_1,\ldots,i_k}^{s_1,\ldots,s_k}\right)^2 = \det\left(\mathfrak{N}_{i_1,\ldots,i_k}^{s_1,\ldots,s_k}\right) \cdot \det\left(\mathfrak{M}_{i_1,\ldots,i_k}^{s_1,\ldots,s_k}\right) \cdot \det(\boldsymbol{\Psi}). \tag{4.62}$$

Summing the last equality over all possible $\binom{N}{k}$ ordered sets $\mathcal{L}_k$ of the k–indexes, we arrive at

$$\sum_{(i_1,\ldots,i_k)\in\mathcal{L}_k} \det\left(\mathfrak{M}_{i_1,\ldots,i_k}^{s_1,\ldots,s_k}\right)\cdot\det\left(\mathfrak{M}_{i_1,\ldots,i_k}^{s_1',\ldots,s_k'}\right) = \delta_{s_1,s_1'}\ldots\delta_{s_k,s_k'} \tag{4.63}$$

and

$$\sum_{(s_1,\ldots,s_k)\in\mathcal{L}_k} \det\left(\mathfrak{M}_{i_1,\ldots,i_k}^{s_1,\ldots,s_k}\right)\cdot\det\left(\mathfrak{M}_{i_1',\ldots,i_k'}^{s_1,\ldots,s_k}\right) = \delta_{i_1,i_1'}\ldots\delta_{i_k,i_k'}, \tag{4.64}$$

where δ_{ij} is the Kronecker delta symbol. We conclude that the determinants of the kth order minors $\mathfrak{M}_{i_1,\ldots,i_k}^{s_1,\ldots,s_k}$ define an orthonormal basis in the $\binom{N}{k}$ dimensional vector space $\bigwedge^k \mathbb{R}^N$ of contra–variant vectors of degree k.

It also follows from (4.63, 4.64) that the squares of these determinants define the probability distributions $\mathcal{P}(\mathcal{L}_k)$ over the $\binom{N}{k}$ ordered sets of k indexes. Namely,

$$\Pr\{s_1,\ldots,s_k\} = \frac{1}{k!}\sum_{\{i_1,\ldots,i_N\}} \det\left(\mathfrak{M}_{i_1,\ldots,i_k}^{s_1,\ldots,s_k}\right)^2, \tag{4.65}$$

$$\Pr\{i_1,\ldots,i_k\} = \frac{1}{k!}\sum_{\{s_1,\ldots,s_N\}} \det\left(\mathfrak{M}_{i_1,\ldots,i_k}^{s_1,\ldots,s_k}\right)^2, \tag{4.66}$$

satisfying the natural normalization condition,

$$\sum_{\{i_1,\ldots,i_N\}} \Pr\{i_1,\ldots,i_k\} = 1 = \sum_{\{s_1,\ldots,s_N\}} \Pr\{s_1,\ldots,s_k\}. \tag{4.67}$$

The simplest example of such a probability distribution is given by (4.60), as

$$\pi_i = \frac{\deg(i)}{2E} = \det\left(\mathfrak{M}_i^1\right)^2 \tag{4.68}$$

is the stationary distribution of random walks over the graph nodes. We conclude that the individual determinants of the k-order minors determine the normalized currents of random walks over the k-sets $\mathcal{L}_k$ in the graph $G(V,E)$.

For an alternative view of the above results, we could look instead at *exterior algebra* associated to random walks. For each $k = 0,\ldots,N$ we can construct a new vector space $\bigwedge^k \mathbb{R}^N$ over $\mathbb{R}$, which may be identified with a contra-variant vector of degree k. Thus,

$$\overset{0}{\bigwedge}\mathbb{R}^N = \mathbb{R}, \qquad \overset{1}{\bigwedge}\mathbb{R}^N = \mathbb{R}^N,$$

and $\bigwedge^k \mathbb{R}^N$ $(2 \le k \le N)$ consists of all sums

$$\sum_{i_1 < i_2 < \ldots < i_k} a_{i_1,i_2,\ldots i_k}\, \alpha_{i_1} \wedge \alpha_{i_2} \wedge \ldots \wedge \alpha_{i_k}, \quad a \in \mathbb{R}, \quad \alpha_i \in \mathbb{R}^N,$$

where the symbol $\wedge$ denotes the standard wedge product of vectors in $\mathbb{R}^N$. For an ordered set of indexes

$$\mathcal{I} = \{i_1, i_2, \ldots, i_n\}, \quad 1 \le i_1 < i_2 < \ldots < i_n \le N,$$

the forms

$$\boldsymbol{\psi}_{\mathcal{I}} = \boldsymbol{\psi}_{i_1} \wedge \ldots \wedge \boldsymbol{\psi}_{i_n} \tag{4.69}$$

over all possible $\binom{N}{k}$ ordered k-sets $\mathcal{I}$ define an orthonormal basis in $\bigwedge^k \mathbb{R}^N$.

Then, the inner product in Hilbert space $\mathcal{H}(V)$ induces an inner product in $\bigwedge^k \mathbb{R}^N$. Given two ordered set of indexes,

$$\mathcal{K} = \{k_1, \ldots, k_n\}, \quad k_1 < \ldots < k_n,$$

and

$$\mathcal{L} = \{l_1, \ldots, l_n\}, \quad l_1 < \ldots < l_n,$$

we define the inner product by

$$(\boldsymbol{\psi}_{\mathcal{K}}, \boldsymbol{\psi}_{\mathcal{L}}) = \pm \delta_{\mathcal{K},\mathcal{L}} \tag{4.70}$$

where $\delta_{\mathcal{K},\mathcal{L}} = \delta_{k_1,l_1} \ldots \delta_{k_n,l_n}$.

Let us denote the complete ordered set of indexes by

$$I = \{1, 2, \ldots, N\},$$

then the signature of the graph G is

$$\mathrm{Sgn}(G) = \boldsymbol{\psi}_I = \pm 1. \tag{4.71}$$

Since minors of the matrix $\boldsymbol{\Psi}$ are numerically equal to their complements, we can define the *Hodge star operator*

$$\star : \overset{n}{\bigwedge} \mathbb{R}^N \to \overset{N-n}{\bigwedge} \mathbb{R}^N$$

by its action on the basis forms,

$$\star\, \boldsymbol{\psi}_{\mathcal{I}} = (-1)^{n(N-n)} \left(\boldsymbol{\psi}_{I \backslash \mathcal{I}}, \boldsymbol{\psi}_{I \backslash \mathcal{I}} \right) \boldsymbol{\psi}_{I \backslash \mathcal{I}}, \tag{4.72}$$

in which $\mathcal{I}$ is an ordered subset of indexes, and $I \setminus \mathcal{I}$ is its complement in I. It is easy to deduce that

$$\star \left(\star \boldsymbol{\psi}_{\mathcal{I}} \right) = (-1)^{n(N-n)} \operatorname{Sgn}(G)\, \boldsymbol{\psi}_{\mathcal{I}}. \tag{4.73}$$

It follows from (4.72) that any eigenvector $\boldsymbol{\psi}_k$ can be deduced from its compliment in the ordered orthonormal complete set $\boldsymbol{\Psi}$, $\left\{ \boldsymbol{\psi}_1, \ldots \widehat{\boldsymbol{\psi}_k} \ldots \boldsymbol{\psi}_N \right\}$ where the "hat" denotes a missing vector in the set,

$$\boldsymbol{\psi}_k = (-1)^{k(N-k)}\, \boldsymbol{\psi}_1 \wedge \ldots \widehat{\boldsymbol{\psi}_k} \ldots \wedge \boldsymbol{\psi}_N. \tag{4.74}$$

In particular, the relation (4.68) follows from (4.72), as

$$\star\ \boldsymbol{\psi}_2 \wedge \boldsymbol{\psi}_3 \wedge \ldots \wedge \boldsymbol{\psi}_N = (-1)^{N-1} \boldsymbol{\psi}_1. \tag{4.75}$$

In a conclusion, we remark on the striking similarity between the above representation of the graph shape by the set of random currents in that (4.65, 4.66) and a *Slater determinant* used in quantum mechanics to describe the wavefunction of a multi-fermionic system that satisfies the Pauli exclusion principle (requiring antisymmetry upon exchange of fermions). The multi-particle wave function defined by the Slater determinant is antisymmetric and no longer distinguishes between the individual particles but returns the probability amplitude whose modulus squared represents a probability density over the entire multi-particle system.

4.8 Summary

We have considered a connected undirected graph as a discrete dynamical system in which any subset of nodes are mapped into the set of their direct neighbors, in each time step. The transfer operator in such a dynamical systems is given by powers of a random walk operator defined on the graph. Random walks on undirected graphs are time reversible and satisfies the condition of detailed balance. Entropy rate of random walks defines the "local physical dimension" at a node averaged over all nodes in the graph. Eigenvectors of the symmetrized random walk transition matrix embed graphs into high-dimensional hyperbolic space that might be useful for displaying and assessing highly inhomogeneous, hierarchical graph structures. In analogy with quantum mechanics, where the wavefunctions of multi-fermionic systems satisfying the Pauli exclusion principle are approximated by the Slater determinants, the normalized determinants over the elements of transition eigenvectors provide us with the probability amplitudes over all subsets of nodes and transition modes.

Chapter 5
Embedding of Graphs in Probabilistic Euclidean Space

In the present chapter, we discuss the Markov chains methods of analysis of non-random, connected graphs of a mesoscopic scale which might contain large but finite number of nodes. We use intensively the concept of generalized inverse (Meyer 1975; Campbell and Meyer 1979; Ben-Israel and Greville 2003) which plays an important role in studies of Markov chains, in electrical engineering, linear programming and in many other applications. We also discuss the relations between random walks and harmonic functions defined on finite connected graphs establishing the natural association between random walks and voltages, charges, effective resistances, and currents in an electrical network spanned by the graph G.

5.1 Methods of Generalized Inverses in the Study of Graphs

The concept of generalized inversion plays the important role in studies of Markov chains, in electrical engineering, linear programming and in many other applications (Campbell and Meyer 1979; Ben-Israel and Greville 2003). In particular, it has been shown that all the important characteristics of a finite Markov chain can be determined from the group inverse of the Laplace operator associated to that Meyer 1975; Campbell and Meyer 1979. Clearly, the generalized inverses can be efficiently used in graph theory.

A homogeneous ergodic Markov chain defined by the transition matrix (4.3) on a finite connected undirected graph $G(V, E)$ determines a diffusion process described by the Laplace operator,

$$\mathbf{L} = \mathbf{1} - \mathbf{T}. \tag{5.1}$$

which is irreducible (due to ergodicity of the Markov chain) and has the one-dimensional null space spanned by the vector of stationary distribution of random walks π. Let us note that $\mathrm{rank}(\mathbf{L}) = \mathrm{rank}(\mathbf{L}^2) = N - 1$.

On the one hand, as being a member of a multiplicative group under the ordinary matrix multiplication (Erdelyi 1967; Meyer 1975), the Laplace operator

P. Blanchard and D. Volchenkov, *Random Walks and Diffusions on Graphs and Databases*, Springer Series in Synergetics 10, DOI 10.1007/978-3-642-19592-1_5, © Springer-Verlag Berlin Heidelberg 2011

(5.1) possesses a *group inverse* (a special case of *Drazin inverse*, Drazin 1958; Ben-Israel and Greville 2003; Meyer 1975) with respect to this group, $L^\sharp$, which satisfies the conditions (Erdelyi 1967)

$$\mathbf{L}\mathbf{L}^\sharp\mathbf{L} = \mathbf{L}, \quad \mathbf{L}^\sharp\mathbf{L}\mathbf{L}^\sharp = \mathbf{L}^\sharp, \quad \text{and} \quad \left[\mathbf{L}, \mathbf{L}^\sharp\right] = 0 \tag{5.2}$$

where $[\mathbf{A}, \mathbf{B}] = \mathbf{A}\mathbf{B} - \mathbf{B}\mathbf{A}$ denotes the commutator of the two matrices. The last condition in (8.51) implies that $\mathbf{L}^\sharp$ describes a set of symmetries of the Laplace equation defined on a finite connected graph. The role of group inverses (8.51) in the analysis of Markov chains have been discussed in details in Meyer (1975), Campbell and Meyer (1979) and Meyer (1982).

The methods for computing the group generalized inverse for matrices of rank($\mathbf{L}$) $= N - 1$ have been developed in Robert (1968), Campbell et al. (1976) and by many other authors. Perhaps, the most elegant way is by considering the eigenprojection of the matrix $\mathbf{L}$ corresponding to the eigenvalue $\lambda_1 = 1 - \mu_1 = 0$ developed in Campbell et al. (1976), Hartwig (1976) and Agaev and Chebotarev (2002),

$$\mathbf{L}^\sharp = (\mathbf{L} + \mathbf{Z})^{-1} - \mathbf{Z}, \quad \mathbf{Z} = \prod_{\lambda_i \neq 0} \left(1 - \frac{1}{\lambda_i}\mathbf{L}\right), \quad \lambda_i = 1 - \mu_i \tag{5.3}$$

where the product in the idempotent matrix $\mathbf{Z}$ is taken over all nonzero eigenvalues of $\mathbf{L}$.

On the other hand, given a matrix $\mathbf{L}$ with rank $N - 1$, there is a unique *Moore-Penrose inverse* matrix (Penrose 1955; Ben-Israel and Greville 2003) $\mathbf{L}^\flat$ such that

$$\mathbf{L}\mathbf{L}^\flat\mathbf{L} = \mathbf{L}, \quad \mathbf{L}^\flat\mathbf{L}\mathbf{L}^\flat = \mathbf{L}^\flat, \quad \left(\mathbf{L}\mathbf{L}^\flat\right)^\top = \mathbf{L}\mathbf{L}^\flat, \quad \left(\mathbf{L}^\flat\mathbf{L}\right)^\top = \mathbf{L}^\flat\mathbf{L}, \tag{5.4}$$

where it is not mandatory that $\mathbf{L}\mathbf{L}^\flat = \mathbf{L}^\flat\mathbf{L}$. The matrix $\mathbf{L}$ has the *singular value decomposition* (Horn and Johnson 1990; Golub and Van Loan 1996)

$$\mathbf{L} = \mathbf{U}\Omega\mathbf{V}^\top \tag{5.5}$$

where $\Omega = \text{diag}\left(\sqrt{\omega_2}, \ldots, \sqrt{\omega_N}\right)$ is the diagonal matrix, in which $\omega_2, \ldots, \omega_N$ are the non-zero eigenvalues of the matrix $\mathbf{L}^\top\mathbf{L}$; $\mathbf{V}$ is the $N \times (N - 1)$ matrix with columns consisting of the corresponding $N - 1$ orthonormalized eigenvectors of $\mathbf{L}^\top\mathbf{L}$ (i.e., the left eigenvectors of $\mathbf{L}$); $\mathbf{U}$ is the $N \times (N - 1)$ matrix with columns being the $N - 1$ orthonormalized eigenvectors of $\mathbf{L}\mathbf{L}^\top$ (i.e., the right eigenvectors of $\mathbf{L}$). Then, the Moore-Penrose inverse of $\mathbf{L}$ can be computed as

$$\mathbf{L}^\flat = \mathbf{V}\Omega^{-1}\mathbf{U}^\top. \tag{5.6}$$

Following Ben-Israel and Charnes (1963), we may derive a representation for (5.6) based on the Lagrange-Sylvester interpolation polynomial,

$$\mathbf{L}^{\flat} = \sum_{\lambda_i \neq 0} \frac{\mathbf{L}^{\top}}{\lambda_i} \frac{\prod_{\theta \neq \lambda_i}(\mathbf{L}\mathbf{L}^{\top} - \theta \mathbf{1})}{\prod_{\theta \neq \lambda_i}(\lambda_i - \theta)}. \tag{5.7}$$

From (5.6), it is clear that

$$\mathbf{L}^{\flat}\mathbf{L} = \mathbf{V}\mathbf{V}^{\top}, \quad \mathbf{L}\mathbf{L}^{\flat} = \mathbf{U}\mathbf{U}^{\top}, \tag{5.8}$$

and therefore the Moore-Penrose inverse coincides with the group inverse, $\mathbf{L}^{\flat} = \mathbf{L}^{\sharp}$, if $\mathbf{V}\mathbf{V}^{\top} = \mathbf{U}\mathbf{U}^{\top}$, i.e., when the matrix $\mathbf{L}$ is symmetric.

Here, we often deal with orthonormal systems of vectors and therefore use the *Dirac's bra-ket notations* especially convenient for working with inner products and rank-one operators in Hilbert space. The inner product of two vectors is denoted by a *bracket*, $\langle a \mid b \rangle$, consisting of a left part (a row vector), $\langle a \mid$, called the *bra*, and a right part (a column vector),$\mid b \rangle$, called the *ket*. The normalized Laplace operator (4.39) has a unique generalized inverse; its spectral representation in the Dirac notation reads as following

$$\widehat{\mathbf{L}}^{\natural} \equiv \widehat{\mathbf{L}}^{\flat} = \widehat{\mathbf{L}}^{\sharp} = \sum_{k=2}^{N} \frac{|\boldsymbol{\psi}_k\rangle\langle\boldsymbol{\psi}_k|}{\lambda_k} \tag{5.9}$$

where the $\boldsymbol{\psi}_k$ are the eigenvectors of the normalized Laplace operator (4.39) belonging to the ordered eigenvalues $0 < \lambda_k < 2$.

5.2   Affine Probabilistic Geometry of Pseudo-inverses

Discovering of important nodes and quantifying differences between them in a graph is not easy, since the graph does not possess *a priori* the structure of Euclidean space. We use the algebraic properties of the self-adjoint operators in order to define an Euclidean metric on any finite connected undirected graph. Geometric objects, such as points, lines, or planes, can be given a representation as elements in projective space based on *homogeneous coordinates* (Möbius 1827). Given an orthonormal basis $\{\boldsymbol{\psi}_k : V \to S_1^{N-1}\}_{k=1}^{N}$ in $\mathbb{R}^N$, any vector in Euclidean space can be expanded into

$$\mathbf{v} = \sum_{k=1}^{N} \langle\mathbf{v}|\psi_k\rangle\,\langle\psi_k|, \tag{5.10}$$

Provided $\{\boldsymbol{\psi}_k\}_{k=1}^{N}$ are the eigenvectors of the symmetric matrix of the operator $\widehat{\mathbf{T}}$, we can define new basis vectors,

$$\boldsymbol{\Psi}' \equiv \left\{1, \frac{\psi_{2,2}}{\psi_{1,2}}, \ldots, \frac{\psi_{N,N}}{\psi_{1,N}}\right\}, \tag{5.11}$$

since we have always $\psi_{1,i} \equiv \sqrt{\pi_i} > 0$ for any $i \in V$. The basis vectors (5.11) span the projective space $P\mathbb{R}_\pi^{(N-1)}$, so that the vector $\mathbf{v}$ can be expanded into

$$\mathbf{v}\pi^{-1/2} = \sum_{k=2}^{N} \langle v | \psi_k' \rangle \langle \psi_k' |. \tag{5.12}$$

It is easy to see that the transformation (5.12) defines a stereographic projection on $P\mathbb{R}_\pi^{(N-1)}$ such that all vectors in $\mathbb{R}^N(V)$ collinear to the vector $| \psi_1 \rangle$ corresponding to the stationary distribution of random walks are projected onto a common image point. If the graph $G(V, E)$ has some isolated nodes $\iota \in V$, for which $\pi_\iota = 0$, they play the role of the plane at infinity with respect to (5.12), away from which we can use the basis $\mathbf{\Psi}'$ as an ordinary Cartesian system. The transition to the homogeneous coordinates (5.11) transforms vectors of $\mathbb{R}^N$ into vectors on the $(N-1)$-dimensional hyper-surface $\{\psi_{1,x} = \sqrt{\pi_x}\}$, the orthogonal complement to the vector of stationary distribution π.

The kernel of the generalized inverse operator (5.9) in the homogeneous coordinates (5.11) is given by

$$\widehat{\mathbf{L}}^\flat = \sum_{k=2}^{N} \frac{|\psi_k'\rangle\langle\psi_k'|}{\lambda_k}. \tag{5.13}$$

5.3   Reduction to Euclidean Metric Geometry

In order to obtain a Euclidean metric on the graph $G(V, E)$, one needs to introduce distances between points (nodes of the graph) and the angles between vectors pointing at them that can be done by determining the inner product between any two vectors $\xi, \zeta \in P\mathbb{R}_\pi^{(N-1)}$ by

$$(\xi, \zeta)_T = \left(\xi, \mathbf{L}^\flat \zeta\right). \tag{5.14}$$

The dot product (5.14) is a symmetric real valued scalar function that allows us to define the (squared) norm of a vector $\xi \in P\mathbb{R}_\pi^{(N-1)}$ by

$$\|\xi\|_T^2 = \left(\xi, \mathbf{L}^\flat \xi\right). \tag{5.15}$$

The angle $\theta \in [0, 180^\circ]$ between two vectors $\xi, \zeta \in P\mathbb{R}_\pi^{(N-1)}$ is then given by

$$\theta = \arccos\left(\frac{(\xi, \zeta)_T}{\|\xi\|_T \, \|\zeta\|_T}\right). \tag{5.16}$$

The Euclidean distance between two vectors $\xi, \zeta \in P\mathbb{R}_\pi^{(N-1)}$ is

$$\|\boldsymbol{\xi} - \boldsymbol{\zeta}\|_T^2 = \|\boldsymbol{\xi}\|_T^2 + \|\boldsymbol{\zeta}\|_T^2 - 2\,(\boldsymbol{\xi}, \boldsymbol{\zeta})_T$$
$$= \mathbb{P}_{\boldsymbol{\xi}}\,(\boldsymbol{\xi} - \boldsymbol{\zeta}) + \mathbb{P}_{\boldsymbol{\zeta}}\,(\boldsymbol{\xi} - \boldsymbol{\zeta}) \tag{5.17}$$

where

$$\mathbb{P}_{\boldsymbol{\xi}}\,(\boldsymbol{\xi} - \boldsymbol{\zeta}) \equiv \|\boldsymbol{\xi}\|_T^2 - (\boldsymbol{\xi}, \boldsymbol{\zeta})_T \quad \text{and} \quad \mathbb{P}_{\boldsymbol{\zeta}}\,(\boldsymbol{\xi} - \boldsymbol{\zeta}) \equiv \|\boldsymbol{\zeta}\|_T^2 - (\boldsymbol{\xi}, \boldsymbol{\zeta})_T$$

are the lengths of projections of the vector $(\boldsymbol{\xi} - \boldsymbol{\zeta}) \in P\,\mathbb{R}_\pi^{(N-1)}$ onto the unit vectors in the directions of $\boldsymbol{\xi}$ and $\boldsymbol{\zeta}$ respectively. It is clear that

$$\mathbb{P}_{\boldsymbol{\zeta}}\,(\boldsymbol{\xi} - \boldsymbol{\zeta}) = \mathbb{P}_{\boldsymbol{\xi}}\,(\boldsymbol{\xi} - \boldsymbol{\zeta}) = 0, \quad \text{if} \quad \boldsymbol{\xi} = \boldsymbol{\zeta}.$$

5.4   Probabilistic Interpretation of Euclidean Geometry

The Euclidean structure introduced in the previous section can be related to a length structure $V \times V \to \mathbb{R}_+$ defined on the class of all admissible paths $\mathcal{P}$ between pairs of nodes in G. It is clear that every path $\mathfrak{p}(i, j) \in \mathcal{P}$ is characterized by some probability to be followed by a random walker depending on the weights $w_{ij} > 0$ of all edges necessary to connect i to j. Therefore, the path length statistics is a natural candidate for the length structure on G.

Let us consider the vector $\mathbf{e}_i = \{0, \ldots 1_i, \ldots 0\}$ that represents the node $i \in V$ in the canonical basis as a density function. In accordance to (5.15), the vector $\mathbf{e}_i$ has the squared norm of $\mathbf{e}_i$ associated to random walks is

$$\|\mathbf{e}_i\|_T^2 = \frac{1}{\pi_i} \sum_{s=2}^{N} \frac{\psi_{s,i}^2}{\lambda_s}. \tag{5.18}$$

It is remarkable that in the theory of random walks (Lovász 1993) the r.h.s. of (5.18) is known as the spectral representation of the *first passage time* to the node $i \in V$, the expected number of steps required to reach the node $i \in V$ for the first time starting from a node randomly chosen among all nodes of the graph accordingly to the stationary distribution π. The first passage time, $\|\mathbf{e}_i\|_T^2$, can be directly used in order to characterize the level of accessibility of the node i.

The Euclidean distance between any two nodes of the graph G calculated in the $(N-1)$−dimensional Euclidean space associated to random walks,

$$K_{ij} = \|\mathbf{e}_i - \mathbf{e}_j\|_T^2 = \sum_{s=2}^{N} \frac{1}{\lambda_s} \left(\frac{\psi_{s,i}}{\sqrt{\pi_i}} - \frac{\psi_{s,j}}{\sqrt{\pi_j}} \right)^2, \tag{5.19}$$

also gets a clear probabilistic interpretation as the spectral representation of the *commute time*, the expected number of steps required for a random walker starting at $i \in V$ to visit $j \in V$ and then to return back to i (Lovász 1993).

The commute time can be represented as a sum, $K_{ij} = H_{ij} + H_{ji}$, in which

$$H_{ij} = \|\mathbf{e}_i\|_T^2 - (\mathbf{e}_i, \mathbf{e}_j)_T \tag{5.20}$$

is the *first-hitting time* which quantifies the expected number of steps a random walker starting from the node i needs to reach j for the first time (Lovász 1993).

The first-hitting time satisfies the equation

$$H_{ij} = 1 + \sum_{i \sim v} H_{vj} T_{vi} \tag{5.21}$$

reflecting the fact that the first step takes a random walker to a neighbor $v \in V$ of the starting node $i \in V$, and then it has to reach the node j from there. In principle, the latter equation can be directly used for computing of the first-hitting times, however, H_{ij} are not the unique solutions of (5.21); the correct definition requires an appropriate diagonal boundary condition, $H_{ii} = 0$, for all $i \in V$ (Lovász 1993). The spectral representation of H_{ij} given by

$$H_{ij} = \sum_{s=2}^{N} \frac{1}{\lambda_s} \left(\frac{\psi_{s,i}^2}{\pi_i} - \frac{\psi_{s,i}\psi_{s,j}}{\sqrt{\pi_i \pi_j}} \right), \tag{5.22}$$

seems much easier to calculate. From the obvious inequality $\lambda_2 \leq \lambda_r$, it follows that the first-passage times are asymptotically bounded by the spectral gap, namely $\lambda_2 = 1 - \mu_2$.

The matrix of first-hitting times is not symmetric, $H_{ij} \neq H_{ji}$, even for a regular graph. However, a deeper *triangle symmetry property* (see Fig. 5.1) has been observed in Coopersmith et al. (1993) for random walks defined by the transition operator (4.3). Namely, for every three nodes in the graph, the consecutive sums of the first-hitting times in the clockwise and in the counterclockwise directions are equal,

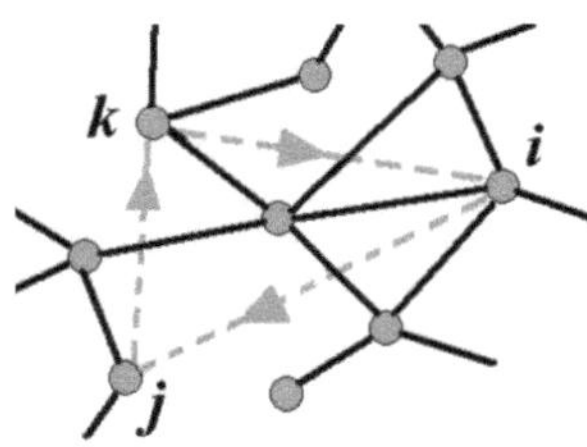

Fig. 5.1 The triangle symmetry of the first-hitting times: the sum of first-hitting times calculated for random walks defined by (4.3) visiting any three nodes i, j, and k, equals to the sum of the first-hitting times in the reversing direction

$$H_{ij} + H_{jk} + H_{ki} = H_{ik} + H_{kj} + H_{ji}. \tag{5.23}$$

We can now use the first-hitting times in order to quantify the accessibility of nodes and subgraphs for random walkers.

From the spectral representation (5.22), we may conclude that the average of the first-hitting times with respect to its first index is nothing else, but the first-passage time to the node (Lovász 1993),

$$\|\mathbf{e}_i\|_T^2 = \sum_{j \in V} \pi_j \, H_{ji}. \tag{5.24}$$

The average of the first-hitting times with respect to its second index is called the *random target access time* (Lovász 1993). It quantifies the expected number of steps required for a random walker to reach a randomly chosen node in the graph (a target). In contrast to (5.24), the random target access time $\mathfrak{T}_G$ is independent of the starting node $i \in V$ being a global spectral characteristic of the graph,

$$
\begin{aligned}
\mathfrak{T}_G &= \sum_{j \in V} \pi_j \, H_{ij} \\
&= \sum_{k=2}^{N} \lambda_k^{-1}.
\end{aligned}
\tag{5.25}
$$

The latter equation expresses the so called *random target identity* (Lovász 1993).

The scalar product $\left(\mathbf{e}_i, \mathbf{e}_j \right)_T$ estimates the expected overlap of random paths toward the nodes i and j starting from a node randomly chosen in accordance with the stationary distribution of random walks π. The normalized expected overlap of random paths given by the cosine of an angle (5.16) calculated in the $(N - 1)$–dimensional Euclidean space associated to random walks has the structure of Pearson's coefficient of linear correlations that reveals it's natural statistical interpretation. If the cosine of (5.16) is close to 1, the expected random paths toward the both nodes are mostly identical. The value of cosine is close to -1 if the walkers share the same random paths but in the opposite direction. Finally, the correlation coefficient equals 0 if the expected random paths toward the nodes do not overlap at all.

Several models were developed to study the mean first-passage time taken by a walker to move from an arbitrary source to a target in complex media. For instance, such situations were usually encountered in predatory animals and biological cells (Shlesinger 2007).

5.5   Probabilistic Embedding of Simple Graphs

In order to illustrate the approach, we have shown the probabilistic images of a chain (1D-lattice of $N = 100$ nodes) (see Fig. 5.2a,b), a polyhedron (a cycle of $N = 50$ nodes) (see Fig. 5.2c,d), and a 2D-lattice $\mathcal{L}_2$ containing 10^2 nodes (see Fig. 5.3).

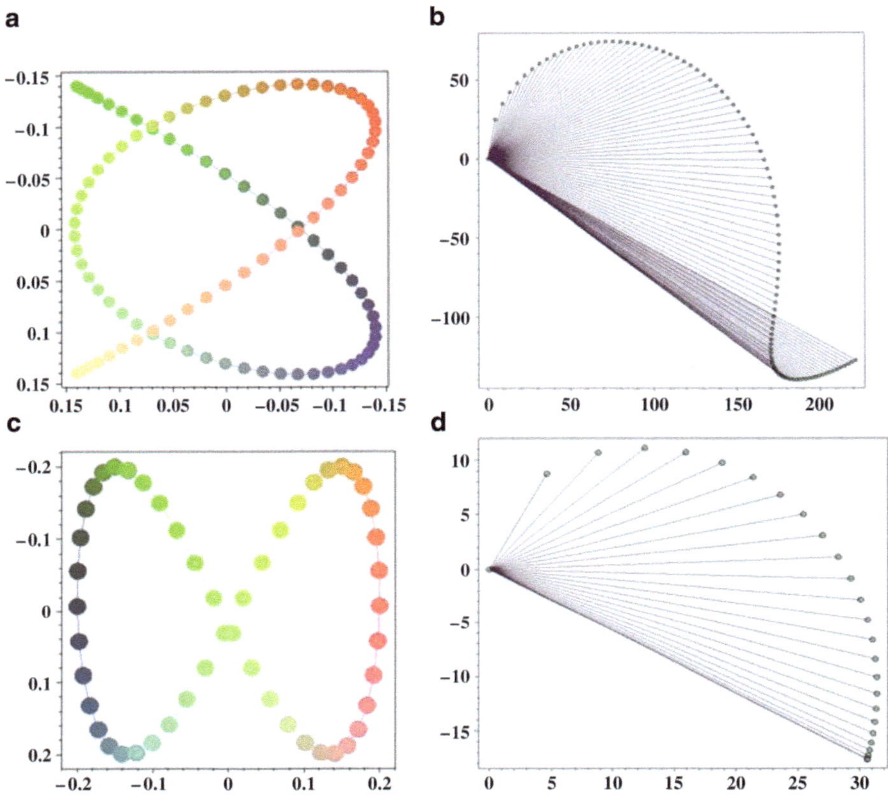

Fig. 5.2 Probabilistic images of a chain ($N = 100$) and a polyhedron ($N = 50$)

We suppose that all edges of these simple graphs have a unit weight, so that the respective affinity matrices are just the adjacency matrices of the graphs.

Random walks defined on the above networks embed them into the $(N - 1)$-dimensional locus of Euclidean space, in which all nodes acquire certain norms quantified by the first-passage times to them from randomly chosen nodes. Indeed, the structure of $(N - 1)$-dimensional vector spaces induced by random walks cannot be represented visually. In order to obtain a 3D visual representation of these graphs, we have calculated their three major eigenvectors $\{\psi_2, \psi_3, \psi_4\}$, belonging to the largest eigenvalues $\mu_k < 1$ of the symmetric transition operators (4.38). The $(\mathbf{e}_1, \mathbf{e}_2, \mathbf{e}_3)$-coordinates of the node $x \in V$ of the graph in 3D space have been taken equal to the relevant xth-components of three eigenvectors $\{\psi_2, \psi_3, \psi_4\}$. The radii of balls representing nodes in Figs. 5.2 and 5.3 have been taken proportional to the degrees of nodes. In Figs. 5.2a,c and in Fig. 5.3a, we have presented the 3D images of the chain, polyhedron, and the lattice $\mathcal{L}_2$. The connections between nodes represent the actual connections between them in the real space.

If we choose one node of a graph as a point of reference, we can draw the two-dimensional projection of the $(N - 1)$-dimensional locus by arranging other nodes

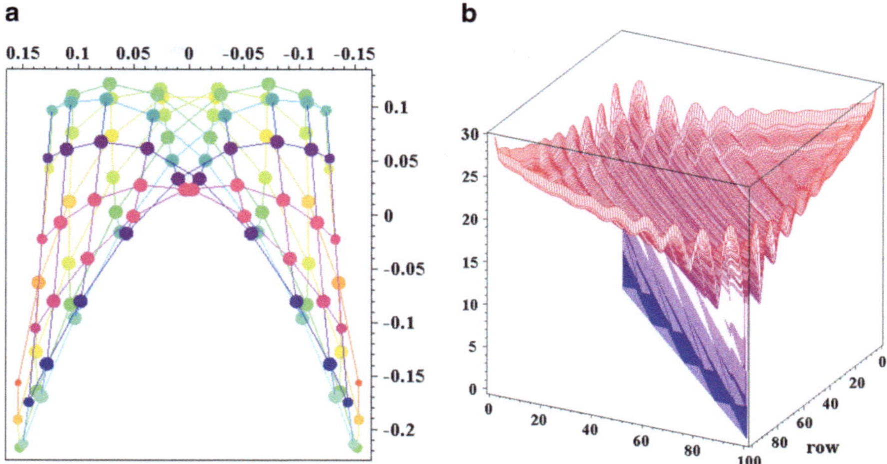

Fig. 5.3 (**a**) Probabilistic image of a 2D-lattice ($N = 10^2$). (**b**) The matrix plot of the probabilistic Euclidean distances (the commute times) between the nodes of the lattice $\mathcal{L}_2$

at the distances calculated accordingly to (5.19) and under the angles (5.16) they are with respect to the chosen reference node. The examples are given in Fig. 5.2b,d.

In particular, in Fig. 5.2b, we have presented the two-dimensional projection of the 99-dimensional Euclidean locus for a chain with respect to the marginal left node. The probabilistic Euclidean distance measured by the commute time (5.19) from the left end node is increasing node by node approximately as $\propto \sqrt{N}$ from 18.166 random steps for the nearest neighbor node to 180.748 random steps for the node at the opposite end of the chain ($x = 100$). In Fig. 5.2d, the similar diagram is represented for the polyhedron. It is worth to mention that the symmetry of the polyhedron can also be seen in the image of its 49-dimensional probabilistic locus. Chosen a node of the polyhedron as the reference node, the commute time with its nearest neighbors equals 9.899 random steps, while it takes in average 35.355 random steps in order to commute with the node on the circle diametrically opposite to the origin.

In Fig. 5.3b, we have shown the matrix plot of the probabilistic Euclidean distances (the commute times) between all nodes of the lattice $\mathcal{L}_2$. The distances on the diagonal $d(x, x) = 0$, and vary harmonically from 15 to 30 random steps for different pairs of nodes.

5.6   Group Generalized Inverse of the Laplace Operator for Directed Graphs

While the elements of the transition matrix **T** inform us on whether or not it is possible to move between any pair of nodes in the graph in a single step, an infinite power $\lim_{k\to\infty} \mathbf{T}^k$ must be considered to estimate the eventual movement between them. It was shown in Meyer (1975, 1982) that the method of group generalized inverse involving calculation of all powers of the transition matrix might be applied for analyzing every Markov chain, regardless of the structure, since the corresponding Laplace operator is always a member of a multiplicative matrix group.

Contrary to undirected graphs, the Markov chains defined on the directed graphs can have absorbing states correspondent to the nodes which are impossible to leave while in a directed walk respecting the direction of edges. Let us define a matrix which entries N_{ij} correspond to the average number of times a random walker is observed in the node j starting in i. It is clear that

$$\mathbf{N} = \lim_{t\to\infty} \sum_{s=0}^{t} \mathbf{T}^s. \tag{5.26}$$

Following Meyer (1975, 1982) and Campbell et al. (1976), we establish a relation between $\mathbf{N}$ and the group diffusion inverse $\mathbf{L}^\sharp$. The transition matrix $\mathbf{T}$ is unitary equivalent, by way of a permutation $\mathbf{\Pi}$, to a matrix

$$\mathbf{T} = \mathbf{\Pi}^{-1} \begin{pmatrix} \mathbf{1} & 0 \\ 0 & \mathbf{K} \end{pmatrix} \mathbf{\Pi}. \tag{5.27}$$

where $\lim_{t\to\infty} \mathbf{K}^t = 0$ and $\mathbf{1}$ is the block correspondent to the absorbing states. On the one hand, the matrix $\mathbf{1} - \mathbf{K}$ is nonsingular, as all eigenvalues of $\mathbf{K}$ are less than 1, and thus

$$\mathbf{N} = \mathbf{\Pi}^{-1} \begin{pmatrix} \lim\limits_{t\to\infty} \sum\limits_{s=0}^{t} \mathbf{1} & 0 \\ \mathbf{M} & (\mathbf{1} - \mathbf{K})^{-1} \end{pmatrix} \mathbf{\Pi}, \tag{5.28}$$

in which $\mathbf{M}$ is some matrix. On the other hand, it is clear that the matrix

$$\mathbf{L}^\sharp = \mathbf{\Pi}^{-1} \begin{pmatrix} 0 & 0 \\ 0 & (\mathbf{1} - \mathbf{K})^{-1} \end{pmatrix} \mathbf{\Pi} \tag{5.29}$$

satisfies all conditions (8.51), as being the group inverse of $\mathbf{L}$.

Therefore, the elements $L_{ij}^\sharp$ of the matrix (5.29) where the i and j are both non-absorbing states, might be interpreted as the *sojourn time* at j, the expected number of times the random walk is in j when being initially in i (Meyer 1982). Let us denote by $|x_1\rangle$ and $|y_1\rangle$ the right and left column eigenvectors of $\mathbf{L}$ belonging to the eigenvalue $\lambda_1 = 0$,

$$\mathbf{L} |x_1\rangle = 0 = \mathbf{L}^\top |y_1\rangle. \tag{5.30}$$

It is clear that

$$\frac{|y_1\rangle \langle x_1|}{\langle y_1|x_1\rangle} = \lim_{t\to\infty} \mathbf{T}^t$$

$$= \mathbf{\Pi}^{-1} \begin{pmatrix} 1 & 0 \\ 0 & \mathbf{0} \end{pmatrix} \mathbf{\Pi} \tag{5.31}$$

$$= \mathbf{1} - \mathbf{L}\mathbf{L}^{\sharp}$$

is the matrix of orthogonal projection on the one-dimensional null space of $\mathbf{L}$. Finally, since we might represent

$$\lim_{t\to\infty} \mathbf{T}^t = \frac{1}{t} \sum_{s=0}^{t-1} \mathbf{T}^s$$

$$= \mathbf{1} - \mathbf{L}\mathbf{L}^{\sharp} + \frac{(\mathbf{1} - \mathbf{T}^t)\mathbf{L}^{\sharp}}{t},$$

and because of $\lim_{t\to\infty} \|\mathbf{T}^t\| = 1$, we also conclude that

$$\lim_{t\to\infty} \frac{1}{t} \sum_{s=0}^{t-1} \mathbf{T}^s = \mathbf{1} - \mathbf{L}\mathbf{L}^{\sharp}. \tag{5.32}$$

The spectral representation of the matrix $\mathbf{1} - \mathbf{L}\mathbf{L}^{\sharp}$ can be readily calculated as

$$\left(\mathbf{1} - \mathbf{L}\mathbf{L}^{\sharp}\right)_{ij} = \left(\mathbf{1} - \sum_{k=2}^{N} |x_k\rangle \langle y_k|\right)_{ij} = (|x_1\rangle \langle y_1|)_{ij} = \pi_i, \tag{5.33}$$

the rows of (5.33) are all equal to the corresponding components of the stationary distribution of random walks. Should j is an absorbing state, then $\left(\mathbf{1} - \mathbf{L}\mathbf{L}^{\sharp}\right)_{ij}$ is the probability of being absorbed at j when initially in i.

Then, we can define the quantities analogous to the first passage time (5.18),

$$\mathcal{F}_i = \left(\mathbf{L}^{\sharp}\right)_{ii} / \pi_i, \tag{5.34}$$

the first-hitting times (5.20),

$$H_{ij} = \mathcal{F}_i - \left(\mathbf{L}^{\sharp}\right)_{ij} / \sqrt{\pi_i \pi_j}, \tag{5.35}$$

and the commute times (5.19), $K_{ij} = H_{ij} + H_{ji}$, for general Markov chains.

5.7 Summary

The Laplace operator possesses a group generalized inverse that can be used in order to define an Euclidean space metric on any finite connected undirected graph. Each node of the graph is characterized by a vector which squared norm is nothing else but the first passage time to the node, the expected number of steps required to reach the node for the first time starting from a node randomly chosen among all nodes of the graph accordingly to the stationary distribution of random walks. The Euclidean distance between any two nodes of the graph is given by the commute time of random walks between them, the expected number of steps required for a random walker starting at the first node visits the second one for the first time and then returns back to the first node (again, for the first time). These characteristic times describing the first encounter properties of the random walk remains finite even for a directed graph, although it lacks the Euclidean space structure.

Chapter 6
Random Walks and Electric Resistance Networks

In the present chapter, an electrical network is considered as an interconnection of resistors. We demonstrate that random walks defined on connected undirected graphs have a profound connection to electric resistor networks (Doyle and Snell 1984; Tetali 1991; Chandra et al. 1996; Bollobas 1998). In the present chapter, we discuss the effective resistance of electrical networks, the relation between the shortest path (geodesic) distance and the effective resistance distance, Kirchhoff and Wiener indexes of a graph.

6.1 Electric Resistance Network and its Probabilistic Interpretation

Currents through a resistor network spanned by a finite connected graph $G(V, E)$ of N nodes connected by a number of resistors

$$r_{ij} = r_{ji} > 0 \quad \text{if } i \sim j,$$
$$r_{ij} = r_{ji} = \infty \text{ if } i \nsim j$$

can be described by the *Kirchhoff circuit law*,

$$\mathbf{L}_C \mid U \rangle = \mid I \rangle, \ (\mathbf{L}_C)_{ij} = c_i - c_{ij},$$
$$c_{ij} = r_{ij}^{-1}, \quad c_i = \sum_{j \in V} c_{ij}, \tag{6.1}$$

in which $\mid U \rangle = (U_1, U_2, \ldots, U_N)^\top$ is the vector of electric potentials, and $\mid I \rangle = (I_1, I_2, \ldots, I_N)^\top$ is the vector of net currents.

P. Blanchard and D. Volchenkov, *Random Walks and Diffusions on Graphs and Databases*, Springer Series in Synergetics 10, DOI 10.1007/978-3-642-19592-1_6,
© Springer-Verlag Berlin Heidelberg 2011

Following Doyle and Snell (1984), we define a random walk on the electric resistor network (6.1), with the transition matrix given by

$$T_{ij} = \frac{c_{ij}}{c_i}. \tag{6.2}$$

We suppose that the graph $G(V, E)$ is connected and therefore the Markov chain (6.2) is ergodic. The stationary distribution of random walks is then given by

$$\pi_i = \frac{c_i}{\sum_{j \in V} c_j} \tag{6.3}$$

and satisfies the time reversibility property

$$\pi_i T_{ij} = \pi_j T_{ji}, \quad \text{since} \quad c_i T_{ij} = c_j T_{ji}. \tag{6.4}$$

Let us consider the left and right eigenvectors of the Laplace operator

$$\mathbf{L} = \mathbf{1} - \mathbf{T}$$

corresponding to the minimal eigenvalue $\lambda_1 = 0$,

$$\mathbf{L} \mid x_1 \rangle = 0 = \mathbf{L}^\top \mid y_1 \rangle. \tag{6.5}$$

By applying the relation (6.4) to (6.5, the right equality), we arrive at the equation

$$\frac{y_{1,i}}{c_i} = \sum_{j \in V} T_{ij} \frac{y_{1,j}}{c_j}$$

equivalent to (6.5, the left equality). Thus,

$$x_{1,i} = \frac{y_{1,i}}{c_i}. \tag{6.6}$$

Given two reference nodes $a, b \in V$, we can show (see Doyle and Snell 1984 for details) that the harmonic function x_1 satisfying (6.5, the left equality) supplied with the boundary conditions

$$x_{1,a} = 1, \quad x_{1,b} = 0 \tag{6.7}$$

determines the probability that starting at $i \neq a, b$ the node a is reached before b.

In a similar way, the function y_1 satisfying (6.5, the right equality) supplied with the boundary condition

$$y_{1,a} = c_a, \quad y_{1,b} = 0 \tag{6.8}$$

determines the expected number of visits to node $i \neq a, b$, starting at a, before reaching b. Indeed, every entrance to $i \neq a, b$ must come from some other node $j \in V$, so that (6.5) is satisfied.

If we apply a one-volt voltage bias between the nodes $a, b \in V$ (say, $U_a = 1$ and $U_b = 0$), we may ask a natural question: What are the voltages U_i and currents I_{ij} in the circuit?

By Ohm's law, the currents thorough the edges are determined by the voltages by

$$I_{ij} = (U_i - U_j)c_{ij}. \tag{6.9}$$

Furthermore, since by Kirchhoff's current law $\sum_{j \in V} I_{ij} = 0$, it is clear that

$$\sum_{j \in V}(U_i - U_j)c_{ij} = 0$$

for all $i \neq a, b$ and, finally,

$$U_i = \sum_{j \in V} U_j \frac{c_{ij}}{c_i} = \sum_{j \in V} T_{ij} U_j, \tag{6.10}$$

i.e., the voltage U_i is harmonic at all nodes $i \neq a, b$. Consequently, the probabilities $x_{1,i}$ satisfying (6.5), with the boundary conditions (6.7), give the probabilistic interpretation of *voltages*.

Turning to the function y_1, we can conclude from (6.6) that $y_{1,i}$ is nothing else, but *charge* at the node $i \in V$, while c_i is its capacitance (Kelly 1979).

Then, it is easy to see that the relation (6.9) can be written as a skew-symmetric function

$$\begin{aligned} I_{ij} &= \left(\frac{y_{1,i}}{c_i} - \frac{y_{1,j}}{c_j}\right) c_{ij} \\ &= x_{1,i} T_{ij} - x_{1,j} T_{ji}. \end{aligned} \tag{6.11}$$

The term $x_{1,i} T_{ij}$ equals the expected number of times the random walker goes from i to its immediate neighbor j, and $x_{1,j} T_{ji}$ is that in the other way round. Therefore, the *current I_{ij}* is the expected number of times the walker passes along the edge $(i, j) \in E$.

6.2   Dissipation and Effective Resistance in Electric Resistance Networks

Let us define the *dissipation* of a current (6.11) in the network by

$$\begin{aligned} D(I) &= \sum_{i,j \in V} \frac{r_{ij} I_{ij}^2}{2} \\ &= \tfrac{1}{2} \sum_{(i,j) \in E} \frac{1}{r_{ij}} \left(\frac{y_{1,i}}{c_i} - \frac{y_{1,j}}{c_j}\right)^2. \end{aligned} \tag{6.12}$$

The reason for the one-half in the above expression is that we are counting each edge twice, while calculating the sum.

Given an electric current from a to b of amount one, the *effective resistance* $\mathfrak{R}(a,b)$ of a network as the potential difference between a and b,

$$\mathfrak{R}(a,b) = \{U_a - U_b : I_{ab} = 1\}. \tag{6.13}$$

Let us consider the potentials $U_a = \mathfrak{R}(a,b)$ and $U_b = 0$ realizing the electrical flow

$$I_a = \sum_{(a,i)\in E} I_{ai} = 1.$$

We may conclude that

$$\begin{aligned}
\mathfrak{R}(a,b) &= I_a\,(U_a - U_b)\\
&= \sum_{(a,i)\in E,i\neq b} I_{ai}\,(U_a - U_b)\\
&= \frac{1}{2}\sum_{s\in V,s\neq a,b}\ \sum_{(s,i)\in E,i\neq a,b} I_{si}U_s\\
&= \frac{1}{2}\left(\sum_{s\in V,s\neq a,b}\ \sum_{(s,i)\in E,i\neq a,b} I_{si}U_i - \sum_{i\in V,i\neq a,b}\ \sum_{(s,i)\in E} I_{si}U_s\right)\\
&= \frac{1}{2}\sum_{(l,j)\in E,l,j\neq a,b} I_{lj}\,(U_l - U_j)\\
&= \frac{1}{2}\sum_{(l,j)\in E,l,j\neq a,b} I_{lj}^2\,r_{lj}\\
&= D(I_{ab} = 1).
\end{aligned} \tag{6.14}$$

Since the dissipation (6.12) is a continuous function of the potential $y_{1,i}/c_i$, the minimum of $D(I)$ is attained at the potential satisfying

$$\sum_{(i,j)\in E} \frac{1}{r_{ij}}\left(\frac{y_{1,i}}{c_i} - \frac{y_{1,j}}{c_j}\right) = 0 \tag{6.15}$$

that gives a proper electric current conforming to the Kirchoff circuit law (6.1) (Bollobas 1998). Thus, the potentials in (6.14) automatically minimize the dissipation of a current $D(I)$, and we may conclude following (Jorgensen and Pearse 2008, 2009) that the effective resistance between a and b is also given by

$$\mathfrak{R}(a,b) = \inf\{D(I) : I_{ab} = 1\}. \tag{6.16}$$

6.3 Effective Resistance is Bounded Above by the Shortest Path Distance

Let

$$W_\ell(a,b) = \{a = v_1, v_2, \ldots, v_\ell = b\}$$

be a walk of length ℓ in G, from a to b. With the use of the Cauchy-Schwarz inequality, we can obtain

$$\mathfrak{R}(a,b)^2 = |U_a - U_b|^2$$

$$= \inf \left| \sum_{v_k \in W_\ell(a,b)} \left(U_{v_k} - U_{v_{k-1}}\right) \right|^2$$

$$= \inf \left| \sum_{v_k \in W_\ell(a,b)} r_{v_k v_{k-1}} I_{v_k v_{k-1}} \right|^2$$

$$\leq \inf \left(\sum_{v_k \in W_\ell(a,b)} r_{v_k v_{k-1}} \right) \cdot \inf \left(\sum_{v_k \in W_\ell(a,b)} D\left(I_{v_k v_{k-1}}\right) \right)$$

$$= \inf r\left(W_\ell(a,b)\right) \cdot \mathfrak{R}(a,b) \tag{6.17}$$

where inf is taken over all possible walks connecting a and b in the network. From the latter equation, it follows that

$$\mathfrak{R}(a,b) \leq \inf r\left(W_\ell(a,b)\right). \tag{6.18}$$

Let us note that $\inf r\left(W_\ell(a,b)\right)$ defined in (6.17) is nothing else but the shortest path (geodesic) distance between a and b in the weighted graph G. In particular, if for all $(i,j) \in E$ we have $r_{ij} = 1$,

$$\inf r\left(W_\ell(a,b)\right) \equiv \mathrm{dist}_G(a,b)$$

where $\mathrm{dist}_G(a,b)$ is the standard geodesic distance on graphs that is evaluated as the minimal number of edges in a walk from a to b in G. We conclude that the effective resistance is bounded above by the shortest path distance (Jorgensen and Pearse 2008, 2009). It is important to note that the inequality (6.18) turns into equality only if the electric resistance network forms a tree, in which any two nodes are connected by the only possible path.

6.4 Kirchhoff and Wiener Indexes of a Graph

The *Kirchhoff index* of a graph, $\mathcal{K}_G$, was introduced in Klein and Randić (1993) as the sum of effective resistances between all pairs of nodes in G,

$$\mathcal{K}_G \equiv \frac{1}{2} \sum_{i,j \in V} \mathfrak{R}(i,j) \leq \frac{1}{2} \sum_{i,j \in V} \mathrm{dist}_G(i,j) \equiv \mathcal{W}_G \tag{6.19}$$

where $\mathcal{W}_G$ is called the *Wiener index* of the graph (Wiener 1947).

Provided that all edges in the complete graph K_N of N nodes have a unit resistance, it is obvious that

$$\mathcal{K}_{K_N} = N - 1. \tag{6.20}$$

Then, the old result on electric networks known as the *Foster theorem* states that for any electric resistance network

$$\sum_{(i,j) \in E} \mathfrak{R}(i,j) = N - 1. \tag{6.21}$$

Plugging (6.20) and (6.21) back, into (6.19), we obtain that

$$\begin{aligned}
\mathcal{K}_{K_N} &= \sum_{(i,j) \in E} \mathfrak{R}(i,j) \\
&\leq \sum_{(i,j) \in E} \mathfrak{R}(i,j) + \sum_{(i,j) \notin E} \mathfrak{R}(i,j) \\
&= \mathcal{K}_G \leq \mathcal{W}_G.
\end{aligned} \tag{6.22}$$

and, finally,

$$\mathcal{K}_{K_N} \leq \mathcal{K}_G \leq \mathcal{W}_G,$$

where the left inequality turns into an equality if G is a complete graph, and the right inequality makes up an equality for trees.

6.5 Relation Between Effective Resistances and Commute Times

It was established in Tetali (1991) and Chandra et al. (1996) that the effective resistance $\mathfrak{R}(i,j)$ might be interpreted as the expected number of times a random walker visits all nodes of the network in a random round trip from i to j and back. In particular, a simple relation between commute times K_{ij} (5.19) of random walks and the effective resistance was found (Tetali 1991; Chandra et al. 1996),

$$K_{ij} = 2\mathfrak{R}(i,j) \sum_{i \in V} c_i. \tag{6.23}$$

It follows immediately from (6.23) that the effective resistance allows for the spectral representation (Chen and Zhang 2007):

$$\mathfrak{R}(i,j) = \sum_{s=2}^{N} \frac{1}{\lambda_s} \left(\frac{\psi_{s,i}}{\sqrt{\sum_{i \in V} c_i}} - \frac{\psi_{s,j}}{\sqrt{\sum_{j \in V} c_j}} \right)^2, \tag{6.24}$$

in which $\{\psi_s : V \to S_1^{N-1}\}$ are the eigenvectors of the normalized Laplace operator

$$\widehat{\mathbf{L}} = \mathbf{1} - \widehat{\mathbf{T}}$$

(the self-adjoint transition operator $\widehat{\mathbf{T}}$ is defined by (4.38)) ordered with respect to the eigenvalues

$$0 < \lambda_1 \leq \lambda_2 \leq \ldots \leq \lambda_N \leq 2.$$

According to the representation (6.24), the Kirchhoff index $\mathcal{K}_G$ may be related to the random target time $\mathfrak{T}_G$ (5.25) (Xiao and Gutman 2003a,b, 2004; Bapat et al. 2003),

$$\mathcal{K}_G = N \sum_{s=2}^{N} \frac{1}{\lambda_s} = N \cdot \mathfrak{T}_G. \tag{6.25}$$

6.6 Summary

Random walks defined on connected undirected graphs have a profound connection to electric resistor networks. The effective resistance between two nodes of a electric resistor network defined as the potential difference between them at a unit current is equal (up to a normalization) to the commute time of a random walk between them. The effective resistance distance is bounded above by the shortest path distance and equals the shortest path distance only if the graph forms a tree, in which any two nodes are connected by the only possible path.

Chapter 7
Random Walks and Diffusions on Directed Graphs and Interacting Networks

The algebraic approach for directed graphs has not been as well developed as for undirected graphs since it is not always possible to define a *unique* self-adjoint operator on directed graphs. In general, any node i in a directed graph $\mathbf{G}$ can have different number of in-neighbors and out-neighbors,

$$\deg_{\mathrm{in}}(i) \neq \deg_{\mathrm{out}}(i). \tag{7.1}$$

In particular, a node i is a *source* if $\deg_{\mathrm{in}}(i) = 0$, $\deg_{\mathrm{out}}(i) \neq 0$, and is a *sink* if $\deg_{\mathrm{out}}(i) = 0$, $\deg_{\mathrm{in}}(i) \neq 0$. If the graph has neither sources nor sinks, it is called *strongly connected*. A directed graph is called Eulerian if $\deg_{\mathrm{in}}(i) = \deg_{\mathrm{out}}(i)$ for $\forall i \in \mathbf{G}$.

In general, the local structure of directed graphs is fundamentally different from that of undirected graphs. In particular, the diameters of directed networks can essentially exceeds that one for the same networks regarded as undirected. A recent investigation in Bianconi et al. (2008) shows that directed networks often have very few short loops as compared to finite random graph models.

In undirected networks, the high density of short loops (high clustering coefficient) together with small graph diameter gives rise to the small-world effect (Watts and Strogatz 1998). In directed networks, the correlation between number of incoming and outgoing edges modulates the expected number of short loops. In particular, it has been demonstrated in Bianconi et al. (2008) that if the values $\deg_{\mathrm{in}}(i)$ and $\deg_{\mathrm{out}}(i)$ are not correlated, then the number of short loops is strongly reduced as compared to the case when the degrees are positively correlated.

7.1 Random Walks on Directed Graphs

Finite random walks are defined on a strongly connected directed graph $\mathbf{G}(V, \mathbf{E})$ as finite node sequences $\mathfrak{w} = \{v_0, \ldots, v_n\}$ (time forward) and $\mathfrak{w}' = \{v_{-n}, \ldots, v_0\}$ (time backward) such that each pair (v_{i-1}, v_i) of nodes adjacent either in $\mathfrak{w}$ or in $\mathfrak{w}'$ constitutes a directed edge $v_{i-1} \to v_i$ in $\mathbf{G}$.

P. Blanchard and D. Volchenkov, *Random Walks and Diffusions on Graphs and Databases*, Springer Series in Synergetics 10, DOI 10.1007/978-3-642-19592-1_7, © Springer-Verlag Berlin Heidelberg 2011

7.1.1 A Time Forward Random Walk

A *time forward random walk* is defined by the transition probability matrix (Chung 2005) P_{ij} for each pair of nodes $i, j \in \mathbf{G}$ by

$$P_{ij} = \begin{cases} 1/\deg_{\text{out}}(i), & i \to j, \\ 0, & \text{otherwise}, \end{cases} \tag{7.2}$$

which satisfies the probability conservation property:

$$\sum_{j,\, i \to j} P_{ij} = 1. \tag{7.3}$$

The definition (7.2) can be naturally extended for weighted graphs (Chung 2005) with $w_{ij} > 0$,

$$P_{ij} = \frac{w_{ij}}{\sum_k w_{ik}}. \tag{7.4}$$

Matrices (7.2) and (7.4) are real, but not symmetric and therefore have complex conjugated pairs of eigenvalues. For each pair of nodes $i, j \in \mathbf{G}$, the *forward* transition probability is given by $p_{ij}^{(t)} = (\mathbf{P}^t)_{ij}$ that is equal zero, if $\mathbf{G}$ contains no a directed path from i to j.

7.1.2 Backward Time Random Walks

Backward time random walks are defined on the strongly connected directed graph $\mathbf{G}$ by the stochastic transition matrix

$$P_{ij}^{\star} = \begin{cases} 1/\deg_{\text{in}}(i), & j \to i, \\ 0, & \text{otherwise}, \end{cases} \tag{7.5}$$

satisfying *another* probability conservation property

$$\sum_{i,\, i \to j} P_{ij}^{\star} = 1. \tag{7.6}$$

It describes random walks unfolding backward in time: should a random walker arrives at $t = 0$ at a node i, then

$$p_{ij}^{(-t)} = ((\mathbf{P}^{\star})^t)_{ij} \tag{7.7}$$

defines the probability that t steps *before* the walker had been at j. The matrix element (7.7) is zero, provided there is no a directed path from j to i in $\mathbf{G}$.

It is well known that the evolution of densities $f \in \mathbb{R}^N$ such that $\sum_{v \in V} f(v) = 1$ in systems for which the dynamics are deterministic may be studied by the use of the linear Perron-Frobenius and *Koopman operators* (Mackey 1991). Strongly connected directed graphs $\mathbf{G}(V, \mathbf{E})$ can be interpreted as the discrete time dynamical systems specified by the dynamical law $\mathcal{S} : V \to V$,

$$\mathcal{S}(A) = \{w \in V | v \in A, v \to w\}. \tag{7.8}$$

Therefore, the transition operators of random walks defined above can be readily related to the Perron-Frobenius and Koopman operators (Koopman 1931) (the reference has been given in Mackey 1991). In particular, the Perron-Frobenius operator $\mathbf{P}^t$, $t \in \mathbb{Z}_+$, transports the density function f, supported on the set A, forward in time to a function supported on some subset of $\mathcal{S}_t(A)$,

$$\sum_{v \in A} \mathbf{P}^t f(v) = \sum_{\mathcal{S}_t^{-1}(A)} f(v). \tag{7.9}$$

The Koopman operator adjoint to the Perron-Frobenius one is thought of as transporting the density f supported on A backward in time to a density supported on a subset of $\mathcal{S}_t^{-1}(A)$:

$$(\mathbf{P}^\star)^t f(v) = f\left(\mathcal{S}_t(v)\right). \tag{7.10}$$

7.1.3   Stationary Distributions of Random Walks on Directed Graphs

The stationary distribution for general directed graphs is not so easy to describe. Even if π exists for a given directed graph $\mathbf{G}$, usually it can be evaluated only numerically in polynomial time being very far from a uniform one since the probability that some nodes could be visited may be exponentially small in the number of edges (Lovász and Winkler 1995).

Stationary distributions on aperiodic general directed graphs are typically non-local as the value of π_i might depend on the entire subgraph (the number of spanning arborescences of $\mathbf{G}$ rooted at i (Lovász and Winkler 1995)), but not on the local connectivity property of a node itself like it was in undirected graphs. Furthermore, if the greatest common divisor of its cycle lengths in $\mathbf{G}$ exceeds 1, then the transition probability matrices (7.2) and (7.5) can have several eigenvectors belonging to the largest eigenvalue 1.

If G is strongly connected and *aperiodic*, the random walk converges (Lovász and Winkler 1995; Chung 2005; Bjorner et al. 1991; Bjorner and Lovász 1992) to the single stationary distribution π,

$$\pi \mathbf{P} = \pi. \tag{7.11}$$

If the graph $\mathbf{G}$ is periodic, then the transition probability matrix $\mathbf{P}$ can have more than one eigenvalue with absolute value 1 (Chung 1997). The components of Perron's vector (7.11) can be normalized in such a way that $\sum_i \pi_i = 1$, moreover there is a bound for the ratio of their maximal and minimal values (Chung 2005),

$$\frac{\max_i \pi_i}{\min_i \pi_i} \leq \max_i \deg_{\text{out}}(i)^{\text{diam}(\mathbf{G})}, \tag{7.12}$$

where $\text{diam}(\mathbf{G})$ denotes the diameter of the strongly connected graph $\mathbf{G}$. In particular, it follows from (7.12) that the Perron vector π for random walks defined on a strongly connected directed graph can have coordinates with exponentially small values. Any non-negative function $F_{i \to j}$ defined on the set of graph edges E is called a circulation if it enjoys the Kirchhoff's law at each node $i \in \mathbf{G}$,

$$\sum_{s,\, \{s \to i\}} F_{s \to i} = \sum_{j,\, \{i \to j\}} F_{i \to j}. \tag{7.13}$$

For a directed graph G, the Perron eigenvector π satisfies

$$F_{i \to j} = \pi_i P_{ij} \tag{7.14}$$

for any edge $i \to j$ (Chung 2005).

7.2   Laplace Operator Defined on Aperiodic Strongly Connected Directed Graphs

Given an aperiodic strongly connected graph $\mathbf{G}$, the Laplace operator can be defined (Chung 2005) as

$$\widehat{L}_{ij} = \delta_{ij} - \frac{1}{2} \left(\pi^{1/2} \mathbf{P} \pi^{-1/2} + \pi^{-1/2} \mathbf{P}^{\top} \pi^{1/2} \right)_{ij}, \tag{7.15}$$

where π is the diagonal matrix of the stationary distribution of random walks. The matrix (7.15) is symmetric and has therefore real non-negative eigenvalues, $0 = \lambda_1 < \lambda_2 \leq \ldots \leq \lambda_N$, and real eigenvectors.

It was proved in (Butler 2007) that the Laplace operator (7.15) defined on the aperiodic strongly connected graph $\mathbf{G}$ is equivalent to the Laplace operator defined on a symmetric undirected weighted graph G_w on the same node set with weights defined by

$$\begin{aligned} w_{ij} &= F_{i \to j} + F_{j \to i} \\ &\equiv \pi_i P_{ij} + \pi_j P_{ji}. \end{aligned} \tag{7.16}$$

Suppose that the transition probability matrix (7.2) has eigenvalues $\{\mu_i\}$, then it can be proved (Chung 2005) that

$$\min_{i \geq 2} (1 - |\mu_i|) \leq \lambda_2 \leq \min_{i \geq 2} (1 - \Re(\mu_i)), \tag{7.17}$$

where $\Re(z)$ denotes the real part of $z \in \mathbb{C}$.

Some results on spectral properties of Laplace operators can be translated to directed graphs. For instance, the Cheeger inequality for a directed graph $\mathbf{G}$ has been established in Chung (2005). Given a subset of nodes $\Gamma \subset \mathbf{G}$, the out-boundary of Γ is

$$\partial \Gamma = \{i \rightarrow j, \ i \in \Gamma, j \in \mathbf{G} \setminus \Gamma\}. \tag{7.18}$$

The circulation through the out-boundary is defined by

$$F_{\partial \Gamma} = \sum_{i \in \Gamma, \ j \in \mathbf{G} \setminus \Gamma} F_{i \rightarrow j}, \tag{7.19}$$

and let

$$F_\Gamma \equiv \sum_{i \in \Gamma} F_i \equiv \sum_{j, \ j \rightarrow i} F_{j \rightarrow i}. \tag{7.20}$$

For a strongly connected graph $\mathbf{G}$ with stationary distribution π, the Cheeger constant can be defined as

$$h(\mathbf{G}) = \inf_{\Gamma \subset \mathbf{G}} \frac{F_{\partial \Gamma}}{\min_\Gamma (F_\Gamma, F_{\mathbf{G} \setminus \Gamma})}. \tag{7.21}$$

Given λ, the first nontrivial eigenvalue of Laplace operator (7.15), the Cheeger constant (7.21) satisfies the following inequality (Chung 2005):

$$2h(\mathbf{G}) \geq \lambda \geq \frac{h^2(\mathbf{G})}{2}. \tag{7.22}$$

For a strongly connected graph $\mathbf{G}$, the diameter $\mathrm{diam}(\mathbf{G})$ satisfies

$$\mathrm{diam}(\mathbf{G}) \leq 1 + \left\lfloor 2\frac{\min_i \ln(\pi_i^{-1})}{\ln \frac{2}{2-\lambda}} \right\rfloor, \tag{7.23}$$

where λ is the first non-trivial eigenvalue of the Laplace operator (7.15) and π is the Perron eigenvector of the random walk defined on $\mathbf{G}$ with the transition probability matrix (7.2).

7.2.1   Bi-orthogonal Decomposition of Random Walks Defined on Strongly Connected Directed Graphs

7.2.1.1   Dynamically Conjugated Operators of Random Walks

Let us choose the natural counting measure μ_0 defined on a strongly connected directed graph $\mathbf{G}$ specified by its non-symmetric adjacency matrix such that $\deg_{\text{in}}(i) \neq 0$ and $\deg_{\text{out}}(i) \neq 0$ for all nodes $i \in \mathbf{G}$ and consider two random walks operators. The first operator represented by the matrix

$$\mathbf{P} = \mathbf{D}_{\text{out}}^{-1}\mathbf{A_G}, \tag{7.24}$$

in which $\mathbf{D}_{\text{out}}$ is a diagonal matrix with entries $\deg_{\text{out}}(i)$, describes the time forward random walks of the nearest neighbor type defined on $\mathbf{G}$. Given a time forward node sequence $\mathfrak{w}$ rooted at $i \in \mathbf{G}$, the matrix element P_{ij} gives the probability that $j \in \mathbf{G}$ is the node next to i in $\mathfrak{w}$. Another operator is a dynamically conjugated operator to (7.24),

$$\begin{aligned}
\mathbf{P}^\star &= \mathbf{D}_{\text{in}}^{-1}\mathbf{A_G^\top} \\
&= \mathbf{D}_{\text{in}}^{-1}\mathbf{P}^\top\mathbf{D}_{\text{out}},
\end{aligned} \tag{7.25}$$

where $\mathbf{D}_{\text{in}} = \text{diag}(\deg_{\text{in}}(i))$. It describes the time backward random walks. It is worth to mention that being defined on undirected graphs $\mathbf{P}^\star \equiv \mathbf{P}$, since $\deg_{\text{in}}(i) = \deg_{\text{out}}(i)$ for all nodes $i \in \mathbf{G}$ and $\mathbf{A}_G = \mathbf{A}_G^\top$. While on directed graphs, $\mathbf{P}^\star$ is related to $\mathbf{P}$ by the transformation

$$\mathbf{P} = \mathbf{D}_{\text{out}}^{-1}\left(\mathbf{P}^\star\right)^\top \mathbf{D}_{\text{in}}, \tag{7.26}$$

so that these operators are not adjoint, in general $\mathbf{P}^\top \neq \mathbf{P}^\star$.

7.2.1.2   Measures Associated to Random Walks on Directed Graphs

The measure associated to random walks defined on undirected graphs was specified by (4.37). Correspondingly, we can define two different measures

$$\mu_+ = \sum_j \deg_{\text{out}}(j)\delta(j), \quad \mu_- = \sum_j \deg_{\text{in}}(j)\delta(j) \tag{7.27}$$

associated with the *out-* and *in*-degrees of nodes of the directed graph. In accordance to (7.27), we also define two Hilbert spaces $\mathcal{H}_+$ and $\mathcal{H}_-$ associated with the spaces of square summable functions, $\ell^2(\mu_+)$ and $\ell^2(\mu_-)$, by setting the norms as

$$\|x\|_{\mathcal{H}_\pm} = \sqrt{\langle x, x\rangle}_{\mathcal{H}_\pm},$$

where $\langle \cdot, \cdot \rangle_{\mathcal{H}_{\pm}}$ denotes the inner products with respect to measures (7.27). Then a function $f(j)$ defined on the set of graph nodes is $f_{\mathcal{H}_-}(j) \in \mathcal{H}_-$ if transformed by

$$f_{\mathcal{H}_-}(j) \rightarrow \mathbf{J}_- f(j) \equiv \mu_{-j}^{-1/2} f(j) \tag{7.28}$$

and is $f_{\mathcal{H}_+}(j) \in \mathcal{H}_+$ while being transformed accordingly to

$$f_{\mathcal{H}_+}(j) \rightarrow \mathbf{J}_+ f(j) \equiv \mu_{+j}^{-1/2} f(j). \tag{7.29}$$

The obvious advantage of the measures (7.27) against the natural counting measure μ_0 is that the matrices of the transition operators $\mathbf{P}$ and $\mathbf{P}^\star$ being transformed

$$\mathbf{P}_\mu = \mathbf{J}_+^{-1} \mathbf{P} \mathbf{J}_-, \quad \mathbf{P}_\mu^\star = \mathbf{J}_-^{-1} \mathbf{P}^\star \mathbf{J}_+, \tag{7.30}$$

as the measures change, become adjoint,

$$\begin{aligned}
\left(\mathbf{P}_\mu\right)_{ij} &= \frac{A_{\mathbf{G}ij}}{\sqrt{\deg_{\text{out}}(i)}\sqrt{\deg_{\text{in}}(j)}}, \\
\left(\mathbf{P}_\mu^\star\right)_{ij} &\equiv \left(P_\mu^\top\right)_{ij} = \frac{A_{\mathbf{G}ij}^\top}{\sqrt{\deg_{\text{in}}(i)}\sqrt{\deg_{\text{out}}(j)}}.
\end{aligned} \tag{7.31}$$

It is also important to note that

$$\mathbf{P}_\mu : \mathcal{H}_- \rightarrow \mathcal{H}_+ \quad \text{and} \quad \mathbf{P}_\mu^\star : \mathcal{H}_+ \rightarrow \mathcal{H}_-.$$

7.2.1.3 Bi-orthogonal Decomposition of Random Walks on Directed Graphs

The singular value dyadic expansion (*biorthogonal decomposition*, Aubry et al. 1991; Aubry 1991) for the operator,

$$\mathbf{P}_\mu = \sum_{k=1}^{N} \Lambda_k \varphi_{ki} \psi_{ki} \equiv \sum_{k=1}^{N} \Lambda_k |\varphi_k\rangle \langle \psi_k|, \tag{7.32}$$

where $0 \leq \Lambda_1 \leq \ldots \Lambda_N$ and the functions $\varphi_k \in \mathcal{H}_+$ and $\psi_k \in \mathcal{H}_-$ are related by the *Karhunen-Loève dispersion* (Karhunen 1944; Loève 1955),

$$\mathbf{P}_\mu \varphi_k = \Lambda_k \psi_k \tag{7.33}$$

satisfying the orthogonality condition:

$$\langle \varphi_k, \varphi_s \rangle_{\mathcal{H}_+} = \langle \psi_k, \psi_s \rangle_{\mathcal{H}_-} = \delta_{ks}. \tag{7.34}$$

Since the operators $\mathbf{P}_\mu$ and $\mathbf{P}_\mu^\top$ act between the alternative Hilbert spaces, not just one equation has to be solved in order to determine the eigenvectors φ_k and ψ_k (Aubry and Lima 1993). Instead, the two equations have to be solved,

$$\begin{cases} \mathbf{P}_\mu \varphi = \Lambda \psi, \\ \mathbf{P}_\mu^\top \psi = \Lambda \varphi, \end{cases} \tag{7.35}$$

or, equivalently,

$$\begin{pmatrix} 0 & \mathbf{P}_\mu \\ \mathbf{P}_\mu^\top & 0 \end{pmatrix} \begin{pmatrix} \varphi \\ \psi \end{pmatrix} = \Lambda \begin{pmatrix} \varphi \\ \psi \end{pmatrix}. \tag{7.36}$$

The latter equation allows for a graph-theoretical interpretation. The block anti-diagonal operator matrix in the left hand side of (7.36) describes random walks defined on a bipartite graph. Bipartite graphs contain two disjoint sets of nodes such that no edge has both end-points in the same set. However, in (7.36), both sets are formed by one and the same nodes of the original graph $\mathbf{G}$ on which two different random walk processes specified by the operators $\mathbf{P}_\mu$ and $\mathbf{P}_\mu^\top$ are defined. It is obvious that any solution of the equation (7.36) is also a solution of the system

$$\mathbf{U}\varphi = \Lambda^2 \varphi, \quad \mathbf{V}\psi = \Lambda^2 \psi, \tag{7.37}$$

in which $\mathbf{U} \equiv \mathbf{P}_\mu^\top \mathbf{P}_\mu$ and $\mathbf{V} \equiv \mathbf{P}_\mu \mathbf{P}_\mu^\top$, although the converse is not necessarily true.

The self-adjoint nonnegative operators $\mathbf{U} : \mathcal{H}_- \to \mathcal{H}_-$ and $\mathbf{V} : \mathcal{H}_+ \to \mathcal{H}_+$ share one and the same set of eigenvalues $\Lambda^2 \in [0, 1]$, and the orthonormal functions $\{\varphi_k\}$ and $\{\psi_k\}$ constitute the orthonormal basis for the Hilbert spaces $\mathcal{H}_+$ and $\mathcal{H}_-$ respectively. The Hilbert-Schmidt norm of both operators,

$$\mathrm{Tr}\left(\mathbf{P}_\mu^\top \mathbf{P}_\mu\right) = \mathrm{Tr}\left(\mathbf{P}_\mu \mathbf{P}_\mu^\top\right) = \sum_{k=1}^{N} \Lambda_k^2 \tag{7.38}$$

is the global characteristic of the directed graph. Provided the random walks are defined on a strongly connected directed graph $\mathbf{G}$, let us consider the functions $\rho^{(t)}(k) \in [0, 1] \times \mathbb{Z}_+$ representing the probability for finding a random walker at the node k, at time t. A random walker started at k can reach the destination k' through either nodes and all paths are combined in superposition. With the use of (7.28) and (7.29), these functions take the following forms:

$$(\rho_k^{(t)})'_{\mathcal{H}_-} = \mu_-^{-1/2} \rho^{(t)}(k)$$

and

$$(\rho_k^{(t)})_{\mathcal{H}_+} = \mu_+^{-1/2} \rho^{(t)}(k).$$

Then, the self-adjoint operators $\mathbf{U} : \mathcal{H}_- \to \mathcal{H}_-$ and $\mathbf{V} : \mathcal{H}_+ \to \mathcal{H}_+$ with the matrix elements

$$U_{kk'} = \frac{1}{\sqrt{\deg_{\mathrm{out}}(k)\,\deg_{\mathrm{in}}(k')}} \sum_{i \in \mathbf{G}} \frac{A_{\mathbf{G}ik}^{\top} A_{\mathbf{G}ik'}}{\sqrt{\deg_{\mathrm{out}}(i)\,\deg_{\mathrm{in}}(i)}},$$
$$V_{k'k} = \frac{1}{\sqrt{\deg_{\mathrm{out}}(k')\,\deg_{\mathrm{in}}(k)}} \sum_{i \in \mathbf{G}} \frac{A_{\mathbf{G}k'i} A_{\mathbf{G}ki}^{\top}}{\sqrt{\deg_{\mathrm{out}}(i)\,\deg_{\mathrm{in}}(i)}} \tag{7.39}$$

define the dynamical system

$$\begin{cases} \sum_{k \in \mathbf{G}} \left(\rho_k^{(t)} \right)_{\mathcal{H}_-} U_{kk'} = \left(\rho_{k'}^{(t+2)} \right)_{\mathcal{H}_-}, \\ \sum_{k' \in \mathbf{G}} \left(\rho_{k'}^{(t)} \right)_{\mathcal{H}_+} V_{k'k} = \left(\rho_k^{(t+2)} \right)_{\mathcal{H}_+}. \end{cases} \tag{7.40}$$

Following Aubry and Lima (1993), we can define a "symmetry" $(\mathbf{S}, \widetilde{\mathfrak{S}})$ by the action of two operators $\mathfrak{S} : \mathcal{H}_- \to \mathcal{H}_-$ and $\widetilde{\mathfrak{S}} : \mathcal{H}_+ \to \mathcal{H}_+$ such that

$$\mathbf{P}_\mu \mathfrak{S} = \widetilde{\mathfrak{S}} \mathbf{P}_\mu, \quad \mathbf{P}_\mu \mathfrak{S}^{\top} = \widetilde{\mathfrak{S}}^{\top} \mathbf{P}_\mu, \tag{7.41}$$

The commutation of $\mathbf{P}_\mu$ with the "symmetry" $(\mathfrak{S}, \widetilde{\mathfrak{S}})$ implies that $\mathbf{U}$ commutes with the partial symmetry $\mathfrak{S}$,

$$\mathbf{U}\mathfrak{S} = \mathfrak{S}\mathbf{U}, \tag{7.42}$$

and $\mathbf{V}$ – with the symmetry $\widetilde{\mathfrak{S}}$,

$$\mathbf{V}\widetilde{\mathfrak{S}} = \widetilde{\mathfrak{S}}\mathbf{V}. \tag{7.43}$$

If $\mathfrak{S}$ and $\widetilde{\mathfrak{S}}$ constitute the unitary representations of the finite symmetry group, its irreducible representations can be used in order to classify structures of the directed graph $\mathbf{G}$.

7.3   Spectral Analysis of Self-adjoint Operators Defined on Directed Graphs

The self-adjoint operators $\mathbf{U}$ and $\mathbf{V}$ describe correlations between flows of random walkers entering and leaving nodes in a directed graph. In the framework of spectral approach, these correlations are labeled by the eigenvalues Λ_k^2, and those belonging to the largest eigenvalues are essential for coherence of traffic through the different components of the transport network. The approach which we propose below for the analysis of coherent structures that arise in strongly connected directed graphs is similar to that one used in purpose of the spatiotemporal analysis of complex signals in Aubry et al. (1991); Aubry and Lima (1993) and refers to the Karhuen-Loéve decomposition in classical signal analysis (Karhunen 1944).

In the framework of bi-orthogonal decomposition discussed in Aubry et al. (1991), the complex spatio-temporal signal has been decomposed into orthogonal temporal modes called chronos and orthogonal spatial modes called topos. Then the

spectral analysis of the phase-space of the dynamics and the spatio-temporal intermittency in particular has naturally led to the notions of "energies" and "entropies"
(temporal, spatial, and global) of signals. Each spatial mode has been associated
with an instantaneous coherent structure which has a temporal evolution directly
given by its corresponding temporal mode. In view of that the thermodynamic-
like quantities had been used in order to describe the complicated spatio-temporal
behavior of complex systems. In the present subsection, we show that a somewhat
similar approach can also be applied to the spectral analysis of directed graphs.
Namely, that the morphological structure of directed graphs can be related to the
quantities extracted from bi-orthogonal decomposition of random walks defined
on them. Furthermore, in such a context, the temporal modes and spatial modes
introduced in Aubry et al. (1991) are the eigenfunctions of correlation operators $\mathbf{U}$
and $\mathbf{V}$.

Although our approach can be viewed as a version of the signal analysis, it
is fundamentally different from that in principle. Transition probability operators
satisfy the probability conservation property by definition. In general, this is not the
case for the spatio-temporal signals generated by complex systems. All coherent
segments of a directed graph participate independently to the Hilbert-Schmidt norm
(7.38) of the self-adjoint operators $\mathbf{U}$ and $\mathbf{V}$,

$$\mathfrak{E}(\mathbf{G}) = \sum_{k=1}^{N} \Lambda_k^2. \tag{7.44}$$

Borrowing the terminology from the theory of signals and Aubry et al. (1991), we
can call this quantity energy, the only additive characteristic of the directed graph
$\mathbf{G}$. While introducing the projection operators (in Dirac's notation) by

$$\mathbb{P}_k^{(+)} = |\psi_k\rangle\langle\psi_k|, \quad \mathbb{P}_k^{(-)} = |\varphi_k\rangle\langle\varphi_k|, \tag{7.45}$$

we can decompose the energy (7.44) into two components related to the Hilbert
spaces $\mathcal{H}_+$ and $\mathcal{H}_-$:

$$\begin{aligned}
\mathfrak{E}(\mathbf{G}) &= \mathfrak{E}^{(+)}(\mathbf{G}) + \mathfrak{E}^{(-)}(\mathbf{G}) \\
&= \sum_{k=1}^{N} \Lambda_k^2 \mathbb{P}_k^{(+)} + \sum_{k=1}^{N} \Lambda_k^2 \mathbb{P}_k^{(-)}.
\end{aligned} \tag{7.46}$$

Using (7.44) as a normalizing factor, we can introduce the relative energy for each
coherent structure of the directed graph by

$$\mathfrak{e}_k = \frac{\Lambda_k^2}{\mathfrak{E}(\mathbf{G})} \tag{7.47}$$

and following Aubry et al. (1991) define the global entropy of coherent structures
in the graph $\mathbf{G}$ as

$$\mathfrak{H}(\mathbf{G}) = -\frac{1}{\ln N} \sum_{k=1}^{N} \mathfrak{e}_k \ln \mathfrak{e}_k, \tag{7.48}$$

which is independent on the graph order N due to the presence of normalizing factor $1/\ln N$, and therefore can be used in order to compare different directed graphs. The global entropy of the graph $\mathbf{G}$ is zero if all its nodes belong to one and the same coherent structure (i.e., only one eigenvalue $\Lambda_k^2 \neq 0$). In the opposite case, $\mathfrak{H}(\mathbf{G}) \to 1$ if most of the eigenvalues Λ_k^2 are equal (multiple). The relative energy (7.47) can also be decomposed into the $\mathcal{H}_+$- and $\mathcal{H}_-$-components:

$$\mathfrak{e}_k^{(\pm)} = \frac{\Lambda_k^2 \mathbb{P}_k^{(\pm)}}{\mathfrak{E}^{(\pm)}(\mathbf{G})}, \tag{7.49}$$

and then the partial entropies can be defined as

$$\mathfrak{H}^{(\pm)}(\mathbf{G}) = -\frac{1}{\ln N} \sum_{k=1}^{N} \mathfrak{e}_k^{(\pm)} \ln \mathfrak{e}^{(\pm)}{}_k. \tag{7.50}$$

To conclude, we have seen that any strongly connected directed graph $\mathbf{G}$ can be considered as a bipartite graph with respect to the in– and out– connectivity of nodes. The bi-orthogonal decomposition of random walks is then used in order to define the self-adjoint operators on directed graphs describing correlations between flows of random walkers which reach at and depart from graph nodes. These self-adjoint operators share the non-negative real spectrum of eigenvalues, but different orthonormal sets of eigenvectors. The standard principal component analysis can be applied also to directed graphs. The global characteristics of the directed graph and its components can be obtained from the spectral properties of the self-adjoint operators.

7.4   Self-adjoint Operators Defined on Interacting Networks

The biorthogonal decomposition can also be implemented in order to determine coherent segments of two or more interacting networks defined on one and the same N–set of nodes. Given two different strongly connected weighted directed graphs $\mathbf{G}_1$ and $\mathbf{G}_2$ specified on the same set of N nodes by the non-symmetric adjacency matrices $\mathbf{A}^{(1)}$, $\mathbf{A}^{(2)}$, which entries are the edge weights, $w_{ij}^{(1,2)} \geq 0$, then the four following transition operators of random walks can be defined on both networks as

$$\mathbf{P}^{(\alpha)} = \left(\mathbf{D}_{\text{out}}^{(\alpha)}\right)^{-1} \mathbf{A}^{(\alpha)}, \quad \left(\mathbf{P}^{(\alpha)}\right)^{\star} = \left(\mathbf{D}_{\text{in}}^{(\alpha)}\right)^{-1} \mathbf{A}^{(\alpha)}, \quad \alpha = 1, 2, \tag{7.51}$$

where $\mathbf{D}_{\text{out/in}}$ are the diagonal matrices associated to the following entries:

$$\deg_{\text{out}}^{(\alpha)}(j) = \sum_{i, j \to i} w_{ji}^{(\alpha)}, \quad \deg_{\text{in}}^{(\alpha)}(j) = \sum_{i, i \to j} w_{ij}^{(\alpha)}, \quad \alpha = 1, 2. \tag{7.52}$$

We can therefore define four different measures,

$$\mu_-^{(1)} = \sum_j \deg_{\text{out}}^{(1)}(j)\delta(j), \; \mu_+^{(1)} = \sum_j \deg_{\text{in}}^{(1)}(j)\delta(j),$$

$$\mu_-^{(2)} = \sum_j \deg_{\text{out}}^{(2)}(j)\delta(j), \; \mu_+^{(2)} = \sum_j \deg_{\text{in}}^{(2)}(j)\delta(j). \tag{7.53}$$

and four Hilbert spaces $\mathcal{H}_{\pm}^{(\alpha)}$ associated with the spaces of square summable functions, $\ell^2\left(\mu_{\pm}^{(\alpha)}\right)$, $\alpha = 1, 2$. Then the transitions operators

$$\mathbf{P}_{\mu}^{(\alpha)} : \mathcal{H}_-^{(\alpha)} \to \mathcal{H}_-^{(\alpha)},$$

$$\left(\mathbf{P}_{\mu}^{(\alpha)}\right)^{\star} : \mathcal{H}_+^{(\alpha)} \to \mathcal{H}_+^{(\alpha)}$$

adjoint with respect to the measures $\mu_{\pm}^{(\alpha)}$ are defined by the following matrices:

$$\left(\mathbf{P}_{\mu}^{(\alpha)}\right)_{ij} = \frac{A_{\mathbf{G}}^{(\alpha)}{}_{ij}}{\sqrt{k_{\text{out}}^{(\alpha)}(i)}\sqrt{k_{\text{in}}^{(\alpha)}(j)}},$$

$$\left(\mathbf{P}_{\mu}^{(\alpha)}\right)_{ij}^{\star} = \frac{A_{\mathbf{G}}^{(\alpha)\top}{}_{ij}}{\sqrt{k_{\text{in}}^{(\alpha)}(i)}\sqrt{k_{\text{out}}^{(\alpha)}(j)}}. \tag{7.54}$$

The spectral analysis of the above operators requires the solution of the four following equations:

$$\begin{cases} \mathbf{P}_{\mu}^{(\alpha)}\boldsymbol{\varphi}^{(\alpha)} &= \Lambda^{(\alpha)}\boldsymbol{\psi}^{(\alpha)}, \\ \mathbf{P}_{\mu}^{(\alpha)\top}\boldsymbol{\psi}^{(\alpha)} &= \Lambda^{(\alpha)}\boldsymbol{\varphi}^{(\alpha)}, \end{cases} \tag{7.55}$$

where $\alpha = 1, 2$.

Any solution $\{\boldsymbol{\varphi}^{(\alpha)}, \boldsymbol{\psi}^{(\alpha)}\}$ of the system (7.55), up to possible partial isometries,

$$\mathbf{G}^{(\alpha)}\boldsymbol{\varphi}^{(\alpha)} = \boldsymbol{\psi}^{(\alpha)}, \tag{7.56}$$

also satisfies the system

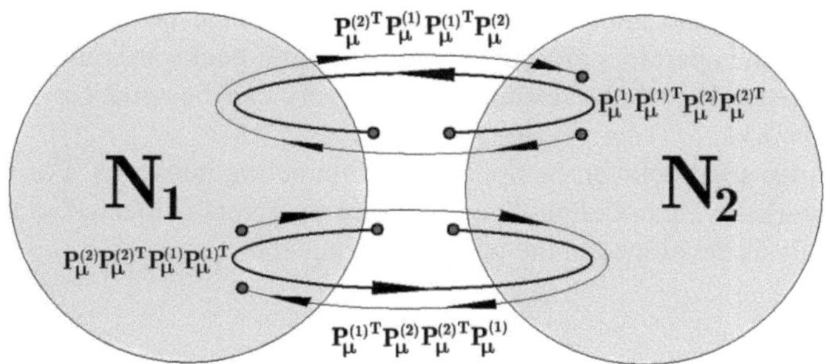

Fig. 7.1 Self-adjoint operators for two interacting networks sharing the same set of nodes

$$\mathbf{P}_{\mu}^{(2)\top}\mathbf{P}_{\mu}^{(1)}\mathbf{P}_{\mu}^{(1)\top}\mathbf{P}_{\mu}^{(2)}\,\boldsymbol{\psi}^{(2)} = \left(\Lambda^{(1)}\Lambda^{(2)}\right)^{2}\boldsymbol{\psi}^{(2)},$$
$$\mathbf{P}_{\mu}^{(1)\top}\mathbf{P}_{\mu}^{(2)}\mathbf{P}_{\mu}^{(2)\top}\mathbf{P}_{\mu}^{(1)}\,\boldsymbol{\psi}^{(1)} = \left(\Lambda^{(1)}\Lambda^{(2)}\right)^{2}\boldsymbol{\psi}^{(1)},$$
$$\mathbf{P}_{\mu}^{(1)}\mathbf{P}_{\mu}^{(1)\top}\mathbf{P}_{\mu}^{(2)}\mathbf{P}_{\mu}^{(2)\top}\,\boldsymbol{\varphi}^{(1)} = \left(\Lambda^{(1)}\Lambda^{(2)}\right)^{2}\boldsymbol{\varphi}^{(1)},\qquad(7.57)$$
$$\mathbf{P}_{\mu}^{(2)}\mathbf{P}_{\mu}^{(2)\top}\mathbf{P}_{\mu}^{(1)}\mathbf{P}_{\mu}^{(1)\top}\,\boldsymbol{\varphi}^{(2)} = \left(\Lambda^{(1)}\Lambda^{(2)}\right)^{2}\boldsymbol{\varphi}^{(2)},$$

in which $\left(\Lambda^{(1)}\Lambda^{(2)}\right)^{2} \in [0,1]$. Operators in the l.h.s of the system (7.57) describe correlations between flows of random walkers which go through nodes following the links of either networks. Their spectrum can also be investigated by the methods discussed in the previous subsection.

It is convenient to represent the self-adjoint operators from the l.h.s. of (7.57) by the closed directed paths shown in the diagram in Fig. 7.1. Being in the self-adjoint products of transition operators, $\mathbf{P}_{\mu}^{(\alpha)}$ corresponds to the flows of random walkers which start from either networks, and $\mathbf{P}_{\mu}^{(\alpha)\top}$ is for those which arrive at the network α. From Fig. 7.1, it is clear that the self-adjoint operators in (7.57) represent all possible closed trajectories visiting both networks $\mathbf{N}_1$ and $\mathbf{N}_2$.

In general, given a complex system consisting of $n > 1$ interacting networks operating on the same set of nodes, we can define 2^n self-adjoint operators related to the different modes of random walks. Then the set of network nodes can be separated into a number of essentially correlated segments with respect to each of self-adjoint operators.

7.5   Summary

Spectral methods have not been applied to directed networks because it is not always possible to define uniquely a linear self-adjoint operator on them.

We have implemented the method of bi-orthogonal decomposition on strongly connected directed graphs in order to define two self-adjoint operators - one for the in-component and another for the out-component of the graph. They can be interpreted as the operators of time-forward and time-backward random walks. The spectral properties of these self-adjoint operators can be used for analyzing the directed networks.

The similar approach can be applied for interacting networks. For $n > 1$ interacting networks, we can define 2^n self-adjoint operators subjected to the standard spectral analysis developed in the preceding chapters.

Chapter 8
Structural Analysis of Networks and Databases

The structural analysis is an important problem motivating many studies of real – world networks. For instance, the efficient partitioning of a transport network is one of the most effective logistics optimization tools and the biggest opportunity to significantly reduce transportation costs. The structural analysis is also of importance when dealing with extremely large graphs, when we need to cluster the vertices into logical components for storage (to improve virtual memory performance) or for drawing purposes (to collapse dense subgraphs into single nodes in order to reduce cluttering). Finally, the structural analysis is important for accessing large databases. In the present chapter, we apply the methods related to random walks for analyzing of urban structures, evolution of languages, and musical compositions.

In particular, we demonstrate that random walks and diffusions defined on spatial city graphs might spot hidden areas of geographical isolation in the urban landscape going downhill. First passage time to a place correlates with assessed value of land in that. The method accounting the average number of random turns at junctions on the way to reach any particular place in the city from various starting points could be used to identify isolated neighborhoods in big cities with a complex web of roads, walkways and public transport systems.

We show how a Markov chain analysis of a network generated by the matrix of lexical distances allows for representing complex relationships between different languages in a language family geometrically, in terms of distances and angles. We test the fully automated method for construction of language taxonomy on a sample of fifty languages of the Indo-European language group and applied to a sample of fifty languages of the Austronesian language group. The Anatolian and Kurgan hypotheses of the Indo-European origin and the 'express train' model of the Polynesian origin are thoroughly discussed.

We also present our study of entropy, redundancy, complexity, and first passage times to notes for the musical dice games generated by the transition matrices between pitches encoded for 804 musical pieces of 29 composers. The successful understanding of tonal music calls for an experienced listener, as entropy dominates over redundancy in musical messages. First passage times to notes resolve tonality

P. Blanchard and D. Volchenkov, *Random Walks and Diffusions on Graphs and Databases*, Springer Series in Synergetics 10, DOI 10.1007/978-3-642-19592-1_8,
© Springer-Verlag Berlin Heidelberg 2011

and feature a composer. We also discuss the possible distances in space of musical dice games and introduced the geodesic distance based on the Riemann structure associated to the probability vectors (rows of the transition matrices).

8.1 Structure and Function in Complex Networks and Databases

The relations between structure and function have been at the focus of attention of complex network research (Newman 2003a) for the last decade. The availability of computers and communication networks made possible to analyze extremely large networks with millions or even billions of vertices. Thus, the analysis of large scale networks has called for a statistical description of the structure rather than for a direct investigation of network topology influencing the behavior of networked systems. There are many excellent reviews devoted to complex networks theory such as Albert and Barabási (2002), Newman (2003a), Dorogovtsev and Mendes (2002), Boccaletti et al. (2006) and Costa et al. (2007), to mention just a few of them. The many efforts have been devoted to understanding synchronization phenomena (Arenas et al. 2008) and critical phenomena (Dorogovtsev et al. 2008) in complex networks; extensive numerical work as well as many analytical approaches to these problems have been presented with applications ranging from biological systems to computer science. Studies of the statistical properties of many large real-world networks have revealed their highly inhomogeneous, hierarchical structure. The typical degree distribution $P(k)$ that indicates the probability of a node chosen uniformly at random to have precisely k neighbors decays very fast significantly deviating from the Poisson distribution expected for a random graph of Erdös-Rényi and often exhibits the property of scale invariance. The massive and comparative analysis of large networks from different fields have been performed, in which the real-world networks were compared with large scale-free random graphs, which almost certainly have the structure of a random tree, while the probability to observe a cycle vanishes as the order of the graphs tends to infinity (Dorogovtsev 2010). In particular, it has been found that in contrast to random scale-free graphs real-world networks might be characterized by correlations in the node degrees, relatively short paths between any two nodes (small-world property), by the specific motifs, short cycles and communities of nodes that are linked together in densely connected groups.

A number of different parameters has been proposed in so far in a wide range of studies to assess the various properties of complex networks. Some of these measures can be computed directly from the graph adjacency matrix, such as *likelihood* (Li et al. 2004), *assortativity* (Newman 2002), *clustering* (Newman 2003b), *degree centrality* (Nieminen 1974; Freeman 1979; Wasserman and Faust 1994), *betweenness centrality* (Freeman 1979), *link value* (Tangmunarunkit et al. 2001), *structural similarity* (Leicht et al. 2006), *distance* (counting the number of paths between vertices) (Girvan and Newman 2002). Other measures (concerned with the networks embedded into Euclidean space) involve the lengths of links or the true Euclidean distances between nodes: *closeness centrality* (Wasserman and Faust 1994; Sabidussi 1966), *straightness centrality* (Vragovic 2005; Latora and Marchiori 2001), *expansion* (Tangmunarunkit et al. 2001), *information centrality* and *graph efficiency* (Latora and Marchiori 2001, 2005), the modularity parameter (Newman 2006). A good summary on the several centrality measures can be found in Crucitti et al. (2006) and in Mahadevan et al. (2006), for the Internet related measures. The list of available measures is still far from being complete, as the new measures appear day by day, together with any forthcoming network model. It is also worth to add some spectral measures (concerned the eigenvalues of graph adjacency matrix), such as *subgraph centralization* (Cvetkovic et al. 1997), *subgraph centrality* (Estrada and Rodríguez-Velázquez 2005a), *network bipartivity* (Estrada and Rodríguez-Velázquez 2005b) and many others. The reason for such an excessive proliferation of heuristic parameters is obvious: the most of real-world networks modeled by and compared to infinite random scale-free graphs are in fact of a *mesoscopic scale*, as ranging from just several tens to several thousands of nodes. Biological systems, urban street networks, neural networks, social interacting species – they all are of the mesoscopic scale. The coupling architecture in them is essentially not random, as having important consequences on the network functional robustness and response to external perturbations, and thus strongly deviate from the structural properties of infinitely large scale-free random graphs.

Mesoscopic and macroscopic complex networks have in common that they both contain a relatively large number of nodes. The structure of mesoscopic networks calls for the alternative methods of analysis, as being affected by the rather strong fluctuations around the average structural properties established by complex network theory. We discuss them in the forthcoming sections.

8.2 Graph Cut Problems

In many networks, their individual components are sparsely connected by only few "bridges" between them.

Extensive literature exists on clustering and partitioning of graphs in two or more almost disjoint parts. Several different flavors of graph partitioning can be implemented depending on the desired objective function. The smallest set of edges to cut (the minimum cut set) that will disconnect a graph can be efficiently found using network flow methods described by Ahuja et al. (1993).

A better partition criterion seeks a small cut that partitions the vertices into roughly equal pieces. The basic approach to dealing with graph partitioning or max-cut problems is to construct an initial partition of the vertices (either randomly or according to some problem-specific strategy) and then sweep through the vertices, deciding whether the size of the cut would increase or decrease if we moved this vertex over to the other side.

Graph cut problems are often NP-hard (Nondeterministic Polynomial time hard). Let us assume that $G(V, E)$ is a connected undirected weighted graph with a symmetric affinity matrix $w_{ij} = w_{ji}$, if $i \sim j$, and $w_{ij} = 0$ otherwise. The degree $\deg(i)$ of a node is defined as the sum of all weights of incident edges.

In order to achieve a balanced partition of G in two components, we can use the Cheeger constant (Cheeger 1969). For the optimal partition $G = \Gamma \bigcup \{G \setminus \Gamma\}$ (at which the minimum conductance between two partitions is achieved) the value of the relevant *Cheeger ratio* coincides with Cheeger's constant,

$$h_G = \min_{\Gamma \subset G} \frac{|\partial \Gamma|}{\min\left(m(\Gamma), m(G \setminus \Gamma)\right)} \tag{8.1}$$

where the volume of the subgraph Γ is denoted by $m(\Gamma)$ and the volume of the edge boundary $\partial \Gamma$ connecting the subgraph Γ to its complement $G \setminus \Gamma$ is defined by

$$|\partial \Gamma| = \sum_{i \in \Gamma,\, j \in G \setminus \Gamma} w_{ij}. \tag{8.2}$$

Another objective function known as the normalized cut is defined by

$$n(G) = \min_{\emptyset \subset \Gamma \subset G} \left(\frac{|\partial \Gamma|}{m(\Gamma)} + \frac{|\partial \Gamma|}{m(G \setminus \Gamma)} \right) \tag{8.3}$$

has been proposed by Shi and Malik (2000).

It is obvious that both objective functions, (8.2) and (8.3), are closely related, since for any pair of positive numbers $a, b > 0$ it is always true that

$$\min(a, b) \leq \left(\frac{1}{a} + \frac{1}{b} \right)^{-1} \leq 2 \min(a, b).$$

In particular, both cut problems are NP-hard, because of a direct checking of all 2^{N-1} possible subsets of G is practically impossible especially if the graph is large, so that the graph segmentation problem needs to be approximated by computationally feasible methods.

A spectral heuristic can be implemented in order to detect bottlenecks and weakly connected subgraphs.

8.2.1 *Weakly Connected Graph Components*

A good heuristic can be obtained by writing the objective function as a quadratic form (by the Rayleigh quotient) and relaxing the discrete optimization problem to a continuous one which then can be solved using the standard methods.

Let us suppose that the connected undirected weighted graph G contains two components, G_1 and G_2, with just a few edges

$$e(G_1, G_2) = \{i \sim j, \ i \in G_1, \ j \in G_2\}$$

linking them. Transport property of a network spanned by the graph G can be characterized by the following reciprocal probabilities,

$$\begin{aligned}
P_{G_1 G_2} &= \Pr[G_1 \to G_2 | G_1] = m^{-1}(G_1) \sum_{i \in G_1, \ j \in G_2} w_{ij}, \\
P_{G_2 G_1} &= \Pr[G_2 \to G_1 | G_2] = m^{-1}(G_2) \sum_{i \in G_2, \ j \in G_1} w_{ij},
\end{aligned} \tag{8.4}$$

of that a random walker located in one of the components alternates it in the next step.

In a probabilistic setting, we say that two components, G_1 and G_2, of the graph $G = G_1 \cup G_2$ are the weakly connected components if the probability

$$\Pr[G_1 \leftrightarrow G_2] = \left(m^{-1}(G_1) + m^{-1}(G_2)\right) \sum_{i \in G_1, \ j \in G_2} w_{ij}, \tag{8.5}$$

of random traffic between them in one-step is ever minimal among all possible bisections of the graph G.

It is important to mention that the minimization of the inter-subgraph random traffic probability can be promptly reformulated as a discrete optimization problem (von Luxburg et al. 2004). Let us define the indicator vector χ_i for the subgraphs G_1 and G_2 as

$$\chi_i = \begin{cases} 1, \ i \in G_1, \\ -1, \ i \in G_2 \end{cases} \tag{8.6}$$

and note that $\chi^\top \chi = |V_G|$. Then, it can be readily obtained that

$$\begin{aligned}
\chi^\top \mathbf{L} \chi &= \sum_{[i \sim j]} w_{ij} (\chi_i - \chi_j)^2 = 4 \, |e(G_1, G_2)|, \\
\chi^\top \mathbf{D} \chi &= \sum_{i \in G} w_{ij} = \mu(G_1) + \mu(G_2) = \mu(G),
\end{aligned} \tag{8.7}$$

where $\mathbf{D} = \operatorname{diag}\{\deg(i), i \in V\}$ is the diagonal matrix of vertex degrees and $\mathbf{L} = \mathbf{D} - \mathbf{w}$ is the canonical Laplace operator defined on the weighted graph G. If we then denote by $\mathbf{e}$ the column vector of all ones, we can write

$$\begin{aligned}
\chi^\top \mathbf{D} \mathbf{e} &= \sum_{i \in G_1} \deg(i) - \sum_{i \in G_2} \deg(i) \\
&= \mu(G_1) - \mu(G_2),
\end{aligned} \tag{8.8}$$

so that $\chi^\top \mathbf{D} \mathbf{e} = 0$ if and only if the components G_1 and G_2 are the volume-balanced components. It is then obvious that

$$\min_{G_1 \cup G_2 = G} \Pr[G_1 \leftrightarrow G_2] = \frac{\mu(G)}{4} \cdot \min_{f \mathbf{De} = 0} \frac{f^\top \mathbf{L} f}{f^\top \mathbf{D} f} \tag{8.9}$$

where the r.h.s. Rayleigh quotient is minimized over all vectors $f \in \{-1, 1\}^N$. If instead, we suppose that f can take real values, $f \in \mathbb{R}^N$ (von Luxburg et al. 2004), then standard linear algebra arguments show that the minimal value of (8.9),

$$\lambda_2 = \min_{f\mathbf{De}=0} \frac{f^\top \mathbf{L} f}{f^\top \mathbf{D} f}, \tag{8.10}$$

is achieved for the eigenvector f_2 (the *Fiedler eigenvector*, studied by Fiedler 1975) which belongs to the second smallest eigenvalue λ_2 of the generalized eigenvector problem,

$$\mathbf{L} f_2 = \lambda_2 \mathbf{D} f_2. \tag{8.11}$$

While the smallest eigenvalue of Laplace operator is always $\lambda_1 = 0$, the second smallest eigenvalue $\lambda_2 > 0$ if the graph G is connected. The normalized spectral clustering (von Luxburg et al. 2004) is given by the Fiedler eigenvector,

$$G_1 = \{i : f_2(i) > 0\}, \quad G_2 = \{i : f_2(i) \le 0\}. \tag{8.12}$$

In general, each nodal domain on which the components of the αth smallest eigenvector f_α do not change sign refers to a coherent flow of random walkers (characterized by its decay time $\tau_\alpha \simeq -1/\lambda_\alpha$) toward the domain of alternative sign. Nodal domains participating in the different diffusion eigenmodes as one and the same degrees of freedom can be considered as dynamically independent modules of a transport network (Volchenkov and Blanchard 2007b). It is known from (Davis et al. 2001) that the eigenvector f_α can have at most $\alpha + m_\alpha - 1$ strong nodal domains (the maximal connected induced subgraphs, on which the components of eigenvectors have a definite sign) where m_α is the multiplicity of the eigenvalue λ_α, but not less than 2 strong nodal domains (for $\alpha > 1$) (Biyikoglu et al. 2004). However, the actual number of nodal domains can be much smaller than the bound obtained in Biyikoglu et al. (2004). In the case of degenerate eigenvalues of the Laplace operator, the situation becomes even more difficult because this number may vary considerably depending upon which vector from the m_α-dimensional subspace belonging to the degenerate eigenvalue λ_α is chosen.

8.2.2 *Graph Partitioning Objectives as Trace Optimization Problems*

The generalized graph partitioning problem seeks to partition a weighted undirected graph G into n almost disjoint clusters $\Gamma_1, \ldots \Gamma_n$ such that $\bigcup_{i=1}^{n} \Gamma_n \subset G$ and either their properties share some common trait, or the graphs nodes belonging to them are close to each other according to some distance measure defined for the nodes of the graph.

A number of different graph partitioning strategies for undirected weighted graphs have been studied in connection with Object Recognition and Learning in Computer Vision, see Morris (2004).

Below, we generalize them into four basic cases:

- The Ratio Cut objective seeks to minimize the cut between clusters and the remaining vertices (Chan et al. 1994),

$$\mathrm{Rcut}(G) = \min_{\{\Gamma_v\}} \sum_{v=1}^{n} \frac{|\partial \Gamma_v|}{|V_{\Gamma_v}|},$$
(8.13)

where

$$|\partial \Gamma_v| = \sum_{i \in \Gamma_v,\ j \in G \setminus \Gamma_v} w_{ij}$$

is the size of boundary $\partial \Gamma_v$, $w_{ij} > 0$ is a symmetric matrix of edge weights, $|V_{\Gamma_v}|$ is the number of nodes in the subgraph Γ_v.

- The Normalized Cut objective (8.3) mentioned at the beginning of present section is one of the most popular graph partitioning objectives (Shi and Malik 2000; Yu and Shi 2003) that seeks to minimize the cut relative to the size $\mu(\Gamma_v)$ of a cluster Γ_v with respect to some counting measure μ instead of the number of its nodes $|V_{\Gamma_v}|$ used in (8.13).

  In particular, given the measure associated to random walks $m_0 = \sum_{i \in V} \deg(i)\delta_i$ then $\mu(\Gamma_v) = \sum_{i \in \Gamma_v} \deg(i)$,

$$\mathrm{Ncut}(G) = \min_{\{\Gamma_v\}} \sum_{v=1}^{n} \frac{|\partial \Gamma_v|}{\mu(\Gamma_v)}.$$
(8.14)

- The Ratio Association objective (also called average association) (Shi and Malik 2000) aims to maximize the size of a cluster,

$$|\Gamma_v| = \sum_{i,j \in \Gamma_v} w_{ij},$$

relative to the number of its nodes $|V_{\Gamma_v}|$,

$$\mathrm{RAssoc}(G) = \max_{\{\Gamma_v\}} \sum_{v=1}^{n} \frac{|\Gamma_v|}{|V_{\Gamma_v}|}.$$
(8.15)

- The Weighted Ratio Association objective (Dhillon et al. 2004a) generalizes (8.15) for the size $\mu(\Gamma_v)$ of the cluster with respect to the measure μ:

$$\mathrm{WRAssoc}(G) = \max_{\{\Gamma_v\}} \sum_{v=1}^{n} \frac{|\Gamma_v|}{\mu(\Gamma_v)}.$$
(8.16)

All graph partition objectives (8.14–8.16) can be formulated as trace maximization problems (Dhillon et al. 2004a).

We introduce partition indicator vectors $\{\chi_v\}_{v=1}^n \in \{0, 1\}^N$ by

$$\chi_v(i) = \begin{cases} 1, i \in \Gamma_v, \\ 0, i \notin \Gamma_v \end{cases} \tag{8.17}$$

and note that $\chi_v^\top \chi_v = |\Gamma_v|$.

The Ratio Association objective (8.15) can be reformulated as

$$\begin{aligned} \mathrm{RAssoc}(G) &= \max \left\{ \sum_{v=1}^n \frac{\chi_v^\top w \chi_v}{\chi_v^\top \chi_v} \right\} \\ &= \max_{\mathbf{X}^\top \mathbf{X}=\mathbf{1}} \ \mathrm{tr}(\mathbf{X}^\top w \mathbf{X}), \end{aligned} \tag{8.18}$$

where $\mathbf{w}$ is the graph affinity matrix, and

$$\mathbf{X} = \left[\frac{\chi_1}{\sqrt{|\Gamma_1|}}, \ldots, \frac{\chi_n}{\sqrt{|\Gamma_n|}} \right] \tag{8.19}$$

is the $n \times N$ rectangular orthogonal matrix $(\mathbf{X}^\top \mathbf{X} = \mathbf{1}_n)$ of normalized indicator vectors.

The Ratio Cut objective (8.13) can be rewritten in terms of the canonical Laplace operator defined on the weighted graph G,

$$\mathrm{Rcut}(G) = \max_{\mathbf{X}^\top \mathbf{X}=\mathbf{1}_n} \ \mathrm{tr}(\mathbf{X}^\top \mathbf{L} \mathbf{X}), \tag{8.20}$$

in which $\mathbf{L} = \mathbf{D} - \mathbf{w}$ and $\mathbf{D} = \mathrm{diag}\,\{\deg(i), i \in V\}$ is the diagonal matrix of vertex degrees, $\deg(i) = \sum_{i \sim j} w_{ij}$.

The Normalized Cut objective can be formulated as

$$\begin{aligned} \mathrm{Ncut}(G) &= \max \left\{ \sum_{v=1}^n \frac{\chi_v^\top w \chi_v}{\chi_v^\top D \chi_v} \right\} \\ &= \max_{\tilde{\mathbf{X}}^\top \tilde{\mathbf{X}}=\mathbf{1}_n} \ \mathrm{tr}(\tilde{\mathbf{X}}^\top \mathbf{D}^{-1/2} \mathbf{w} \mathbf{D}^{-1/2} \tilde{\mathbf{X}}), \end{aligned} \tag{8.21}$$

where $\tilde{\mathbf{X}}$ is the $n \times N$ rectangular matrix of normalized indicator vectors similar to (8.19),

$$\tilde{\mathbf{X}} = \left[\frac{\chi_1}{\sqrt{\mu(\Gamma_1)}}, \ldots, \frac{\chi_n}{\sqrt{|\mu(\Gamma_n)|}} \right]. \tag{8.22}$$

The Weighted Ratio Association is reduced to the trace maximization of

$$\mathrm{WRAssoc}(G) = \max_{\tilde{\mathbf{X}}^\top \tilde{\mathbf{X}}=\mathbf{1}_n} \ \mathrm{tr}(\tilde{\mathbf{X}}^\top \mathbf{D}^{-1/2} \mathbf{L} \mathbf{D}^{-1/2} \tilde{\mathbf{X}}). \tag{8.23}$$

The optimization of graph partitioning objectives gives us graph cuts balanced with respect to the numbers of nodes in the subgraphs Γ_i, their cumulative weights $|\Gamma_i|$, or their sizes $\mu(\Gamma_i)$ with respect to a certain measure assigned to nodes.

In the previous chapter, we have discussed that random walks defined on a connected undirected graph G set up the structure of $(N-1)$–dimensional Euclidean space such that for every pair of nodes $i \neq j$ we can introduce the positive symmetric distance $K(i, j) > 0$ (the commute time) (5.19). It is important to note that the probabilistic distance associated to random walks and the probabilistic angle (5.16) between them can be used as the measures of similarity between two nodes in the graph G in the purpose of the graph partitioning.

In the probabilistic space associated to random walks, each node of the graph G is represented by a certain vector $\mathbf{z}_i \in \mathbb{R}^{N-1}$. Then, we can assign each vector $\mathbf{z}_i$ to the cluster Γ_i whose center,

$$\mathbf{m}_i = \sum_{s=1}^{|\Gamma_i|} \frac{z_{is}}{|\Gamma_i|}, \tag{8.24}$$

(also called centroid) is nearest with respect to the graph random walks distance (5.19).

The objective we try to achieve is to minimize the total intra-cluster variance of the resulting partition $\mathfrak{P}$ of the graph G into n clusters, or, the squared error function (s.e.f.),

$$\mathrm{sef}(\mathfrak{P}) = \sum_{i=1}^{n} \sum_{s=1}^{|\Gamma_i|} |z_{is} - \mathbf{m}_i|^2. \tag{8.25}$$

Let $\mathbf{e} = (1, 1, \ldots, 1)^\top$ be a $(N - 1)$-dimensional vector of ones, and $\mathbf{Z} = (\mathbf{z}_i)$ be the $(N - 1) \times N$ matrix of node coordinates. Then it is clear that

$$\begin{aligned}
\mathrm{sef}(\mathfrak{P}) &= \sum_{i=1}^{n} |\mathbf{Z}_i - \mathbf{m}_i \mathbf{e}^\top|^2 \\
&= \sum_{i=1}^{n} |\mathbf{Z}_i \mathbf{P}_i|^2 ,
\end{aligned} \tag{8.26}$$

where

$$\mathbf{P}_i = \left(\mathbf{1}_{\Gamma_i} - \frac{\mathbf{e}\,\mathbf{e}^\top}{|\Gamma_i|} \right),$$

is the projection operator onto the cluster Γ_i. Since $\mathbf{P}_i^2 = \mathbf{P}_i$, we obtain

$$\begin{aligned}
\mathrm{sef}(\mathfrak{P}) &= \sum_{i=1}^{n} \mathrm{tr}(\mathbf{Z}_i^\top \mathbf{P}_i \mathbf{Z}_i) \\
&= \mathrm{tr}(\mathbf{Z}^\top \mathbf{Z}) - \mathrm{trace}(\mathbf{X}^\top \mathbf{Z}^\top \mathbf{Z} \mathbf{X}),
\end{aligned} \tag{8.27}$$

in which $\mathbf{X}$ is the $(N - 1) \times N$ orthogonal indicator matrix (8.19).

If we consider the elements of the $\mathbf{Z}^\top \mathbf{Z}$ matrix as measuring similarity between nodes, then it can be shown (Zha et al. 2001) that Euclidean distance leads to Euclidean inner-product similarity which can be replaced by a general Mercer kernel (Saitoh 1988; Wahba 1990) uniquely represented by a positive semi-definite matrix $K(i, j)$. If we relax the discrete structure of $\mathbf{X}$ and assume that $\mathbf{X}$ is an arbitrary orthonormal matrix, the minimization of sef($\mathfrak{P}$) function is reduced to the trace maximization problem,

$$\max_{\mathbf{X}^\top \mathbf{X} = \mathbf{1}_{N-1}} \mathrm{tr}(\mathbf{X}^\top \mathbf{Z}^\top \mathbf{Z} \mathbf{X}), \tag{8.28}$$

as it was the case for the graph partitioning according to the minimization of objective functions (8.13–8.16).

A standard result in linear algebra (proved by Fan 1949) provides a global solution to a related version of the trace optimization problems: Given a symmetric matrix $\mathbf{S}$ with eigenvalues $\lambda_1 \geq \ldots \geq \lambda_n \geq \ldots \geq \lambda_N$, and the matrix of corresponding eigenvectors, $[\mathbf{u}_1, \ldots, \mathbf{u}_N]$, the maximum of $\mathrm{tr}(\mathbf{Q}^\top \mathbf{S} \mathbf{Q})$ over all n-dimensional orthonormal matrices $\mathbf{Q}$ such that $\mathbf{Q}^\top \mathbf{Q} = \mathbf{1}_n$ is given by

$$\max_{\mathbf{Q}^\top \mathbf{Q} = \mathbf{1}_n} \mathrm{tr}(\mathbf{Q}^\top \mathbf{S} \mathbf{Q}) = \sum_{k=1}^{n} \lambda_k, \tag{8.29}$$

and the optimal n-dimensional orthonormal matrix

$$\mathbf{Q} = [\mathbf{u}_1, \ldots, \mathbf{u}_n]\mathbf{R} \tag{8.30}$$

where $\mathbf{R}$ is an arbitrary orthogonal $n \times n$ matrix (describing a rotation in $\mathbb{R}^n$).

The result (8.29, 8.30) relates the problem of network segmentations to the investigation of n primary eigenvectors of a symmetric matrix defined on the graph nodes (Golub and Van Loan 1996; Dhillon et al. 2004b). The eigenvectors $\mathbf{u}_{i>1}$ have both positive and negative entries, so that in general $[\mathbf{u}_1, \ldots, \mathbf{u}_n]$ differ substantially from the matrix of discrete cluster indicator vectors which have strictly positive entries.

It is important to note that even for not very large n it may be rather difficult to compute the appropriate $n \times n$ orthonormal transformation matrix $\mathbf{R}$ which would recover the necessary discrete cluster indicators structure. Furthermore, it can be shown that the post processing of eigenvectors into the cluster indicator vectors can be reduced to an optimization problem with $n(n-1)/2 - 1$ parameters (Ding and He 2004). Several methods have been proposed to obtain the partitions from the eigenvectors of various similarity matrices (see Dhillon et al. 2004a; Bach and Jordan 2004 for a review).

8.3 Markov Chains Estimate Land Value in Cities

In the present section, we report on the study of spatial graphs of several compact urban patterns: the city canal networks of Venice and Amsterdam, as well as the almost regular array of streets in Manhattan.

Multiple increases in urban population that had occurred in Europe at the beginning of the twentieth century made urban agglomerations suffer from the problems of urban decay such as wide-spread poverty, high unemployment, and rapid changes in the racial composition of neighborhoods. Riots and social revolutions happened in response to conditions of urban decay in many European countries established regimes affecting immigrants and certain population groups *de facto* alleviating the burden of the haphazard urbanization by increasing its deadly price. More than half of world's population, 3.3 billion people, is now living in cities, and the figure is about to double by 2030 (United Nations 2007). Urban sprawl in US covered 41,000 km^2 in 20 years, the area that equals that of the state of Switzerland (US Census Data). In Europe, urban sprawl has covered 8,000 km^2 in 10 years the area that equals the territory of Luxembourg (EEA 2006). Unsustainable pressure on resources causes the increasing loss of fertile lands through degradation and climate changes through the increasing carbon emissions warming the earth's atmosphere. City development planners will face great challenges in preventing cities from unlimited expansion. Global poverty is in flight becoming a primarily urban phenomenon in the developing world. two billion new urban settlers in the next 20 years will live in slums on no more than \$1 a day, adding to one billion already living there (Ravallion 2007). Faults in urban planning, poverty, redlining, immigration restrictions and clustering of minorities dispersed over the spatially isolated pockets of streets trigger urban decay, a process by which a city falls into a state of disrepair. The speed and scale of urban growth in that require urgent global actions to help cities prepare for growth and to avoid them of being the future epicenters of poverty and human suffering.

Sociologists think that isolation worsens an area's economic prospects by reducing opportunities for commerce, and engenders a sense of isolation in inhabitants, both of which can fuel poverty and crime. It is well known that many social variables demonstrate striking spatial distribution patterns, and therefore may be detected and predicted by a structural analysis. The proposed method could be used to identify isolated neighborhoods in big cities with a complex web of roads, walkways and public transport systems.

8.3.1   Spatial Networks of Urban Environments

In traditional urban researches, the dynamics of an urban pattern come from the landmasses, the physical aggregates of buildings delivering place for people and their activity. The relationships between certain components of the urban texture are often measured along streets and routes considered as edges of a planar graph, while the traffic end points and street junctions are treated as nodes. Such a primary graph representation of urban networks is grounded on relations between junctions through the segments of streets. The usual city map based on Euclidean geometry can be considered as an example of primary city graphs.

In space syntax theory (see Hillier and Hanson 1984; Hillier 1999), built environments are treated as systems of spaces of vision subjected to a configuration analysis. Being irrelevant to the physical distances, spatial graphs representing the urban environments are removed from the physical space. It has been demonstrated in multiple experiments that spatial perception shapes peoples understanding of how a place is organized and eventually determines the pattern of local movement (Hillier 1999). The aim of the space syntax study is to estimate the relative proximity between different locations and to associate these distances to the densities of human activity along the links connecting them (Hansen 1959; Wilson 1970; Batty 2004). The surprising accuracy of predictions of human behavior in cities based on the purely topological analysis of different urban street layouts within the space syntax approach attracts meticulous attention (Penn 2001).

The decomposition of urban spatial networks into the complete sets of intersecting open spaces can be based on a number of different principles. In Jiang and Claramunt (2004), while identifying a street over a plurality of routes on a city map, the named-street approach has been used, in which two different arcs of the primary city network were assigned to the same identification number (ID) provided they share the same street name.

In the present section, we take a "named-streets"-oriented point of view on the decomposition of urban spatial networks into the complete sets of intersecting open spaces following our previous works (Volchenkov and Blanchard 2007b,c). Being interested in the statistics of random walks defined on spatial networks of urban patterns, we assign an individual street ID code to each continuous segment of a street. The spatial graph of urban environment is then constructed by mapping all edges (segments of streets) of the city map shared the same street ID into nodes and all intersections among each pair of edges of the primary graph into the edges of the secondary graph connecting the corresponding nodes.

8.3.2 *Spectra of Cities*

If we take many, many random numbers from an interval of all real numbers symmetric with respect to a unit and calculate the sample mean in each case, then the distribution of these sample means will be approximately normal in shape and centered at 1 provided the size of samples was large. The probability density function of a normal distribution forms a symmetrical bell-shaped curve highest at the mean value indicating that in a random selection of the numbers around the mean (1) have a higher probability of being selected than those far away from the mean. Maximizing information entropy among all distributions with known mean and variance, the normal distribution arises in many areas of statistics.

It is interesting to compare the empirical distributions of eigenvalues of the normalized Laplace operator (4.39) defined on the spatial graphs of compact urban patterns – the spectra of cities – with the normal distribution centered at 1. In Fig. 8.1a, we have shown a probability-probability plot of the normal distribution (on the horizontal axis) against the empirical distribution of eigenvalues in the city spectra (the normal plot) of the city canal networks in Venice (96 canals) and Amsterdam (57 canals). A random sample of the normal distribution, having size

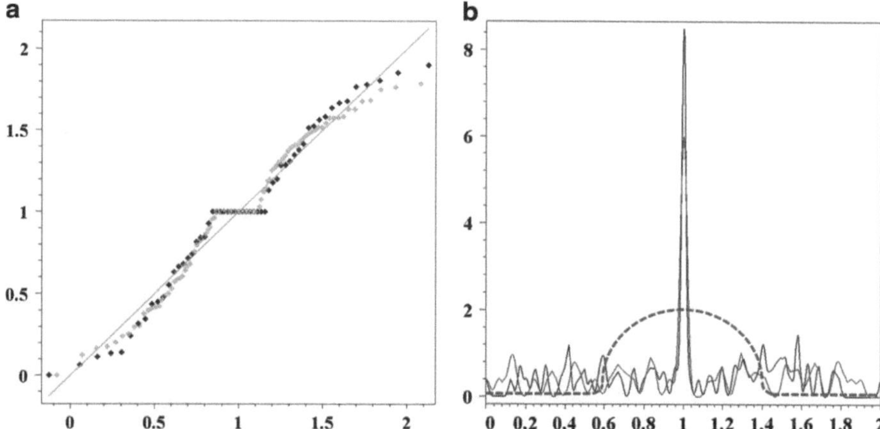

Fig. 8.1 (**a**) The probability-probability plot of the normal distribution (on the horizontal axis) against the empirical distribution of eigenvalues for the spectra of the city canal networks in Venice (dark diamonds) and Amsterdam (light diamonds). The diagonal line $y = x$ is set for a reference. (**b**) The spectral density distributions of eigenvalues of the normalized Laplace operators defined for the spatial networks of urban patterns of Venice and Amsterdam (given by solid lines). The dashed line representing the classical Wigner semicircle distribution typical for the eigenvalues of the normalized Laplace operator defined on random graphs (of Erdös and Rényi random graphs, according to Farkas et al. 2001, 2002, and of scale-free graphs, according to Chung et al. 2003) is set for a reference

equal to the number of eigenvalues in the spectrum has been be generated, sorted in ascending order, and plotted against the response of the empirical distribution of city eigenvalues. The spectra of canal maintained in the compact urban patterns of Venice and Amsterdam look also amazingly alike and are obviously tied to the normal distribution, although these canals had been founded in the dissimilar geographical regions and for the different purposes. While the Venetian canals mostly serve the function of transportation routs between the distinct districts of the gradually growing naval capital of the Mediterranean region, the concentric web of Amsterdam *gratchen* had been built in order to defend the city.

It is remarkable that the spectral density distributions of eigenvalues of the normalized Laplace operators defined for the spatial networks of urban patterns shown in Fig. 8.1b are dramatically dissimilar to those reported for the random graphs of Erdös and Rényi studied by Farkas et al. (2001, 2002). The classical Wigner semicircle distribution arises as the limiting distribution of eigenvalues of many random symmetric matrices as the size of the matrix approaches infinity (Sinai and Soshnikov 1998). The density distribution for the eigenvalues of the normalized Laplace operator for a random scale-free graph also follows the semicircle law (Chung et al. 2003). City spectra reveal the profound structural dissimilarity between urban networks and networks of other types studied before.

8.3.3 First-passage Times to Ghettos

The phenomenon of clustering of minorities, especially that of newly arrived immigrants, is well documented (Wirth 1928) (the reference appears in Vaughan 2005). Clustering is considering to be beneficial for mutual support and for the sustenance of cultural and religious activities. At the same time, clustering and the subsequent physical segregation of minority groups would cause their economic marginalization. The spatial analysis of the immigrant quarters (Vaughan 2005) and the study of London's changes over 100 years (Vaughan et al. 2005) shows that they were significantly more segregated from the neighboring areas, in particular, the number of street turning away from the quarters to the city centers were found to be less than in the other inner-city areas being usually socially barricaded by railways, canals and industries. It has been suggested (Hillier 2004) that space structure and its impact on movement are critical to the link between the built environment and its social functioning. Spatial structures creating a local situation in which there is no relation between movements inside the spatial pattern and outside it and the lack of natural space occupancy become associated with the social misuse of the structurally abandoned spaces.

We have analyzed the first-passage times to individual canals in the spatial graph of the canal network in Venice. The distribution of numbers of canals over the range of the first-passage time values is represented by a histogram shown in Fig. 8.2left. The height of each bar in the histogram is proportional to the number of canals in the canal network of Venice for which the first-passage times fall into the disjoint intervals (known as bins). Not surprisingly, the Grand Canal, the giant Giudecca Canal and the Venetian lagoon are the most connected. In contrast, the Venetian Ghetto (see Fig. 8.2right) – jumped out as by far the most isolated, despite being apparently well connected to the rest of the city – on average, it took 300 random steps to reach, far more than the average of 100 steps for other places in Venice.

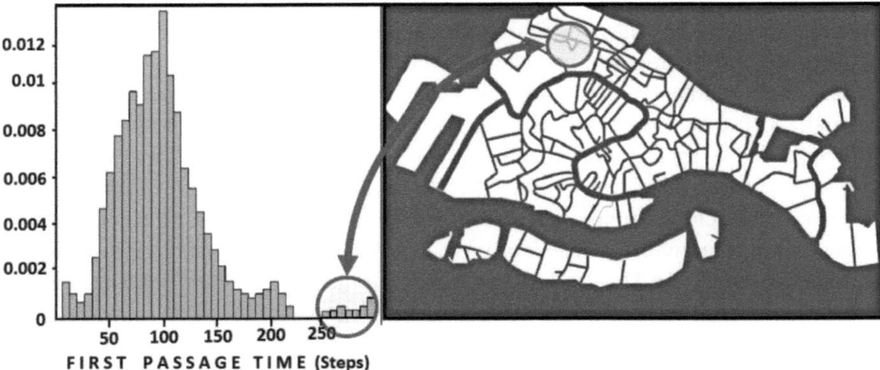

Fig. 8.2 The Venetian Ghetto jumped out as by far the most isolated, despite being apparently well connected to the rest of the city

The Ghetto was created in March 1516 to separate Jews from the Christian majority of Venice. It persisted until 1797, when Napoleon conquered the city and demolished the Ghetto's gates. Now it is abandoned.

8.3.4   Random Walks Estimate Land Value in Manhattan

The notion of isolation acquires the statistical interpretation by means of random walks. The first-passage times in the city vary strongly from location to location. Those places characterized by the shortest first-passage times are easy to reach while very many random steps would be required in order to get into a statistically isolated site.

Being a global characteristic of a node in the graph, the first-passage time assigns absolute scores to all nodes based on the probability of paths they provide for random walkers. The first-passage time can therefore be considered as a natural statistical centrality measure of the node within the graph (Blanchard and Volchenkov 2009b).

A visual pattern displayed on Fig. 8.3 represents the pattern of structural isolation (quantified by the first-passage times) in Manhattan (darker color corresponds to longer first-passage times). It is interesting to note that the spatial distribution of isolation in the urban pattern of Manhattan (Fig. 8.3) shows a qualitative agreement with the map of the tax assessment value of the land in Manhattan reported by B. Rankin (2006) in the framework of the RADICAL CARTOGRAPHY project being practically a negative image of that.

Recently, we have discussed in (Blanchard and Volchenkov 2009b) that distributions of various social variables (such as the mean household income and prison expenditures in different zip code areas) may demonstrate the striking spatial patterns which can be analyzed by means of random walks. In the present work, we analyze the spatial distribution of the tax assessment rate (TAR) in Manhattan.

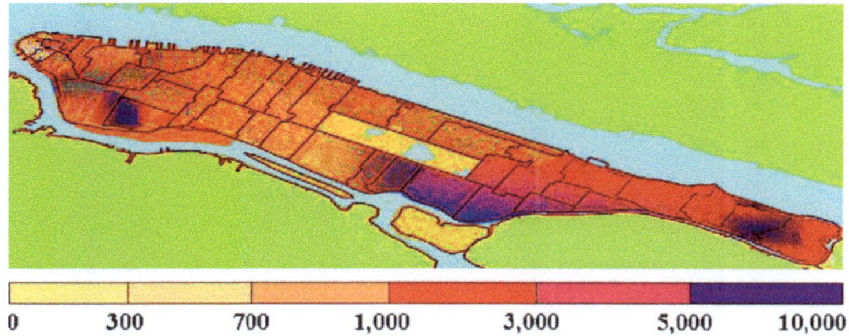

Fig. 8.3 Isolation map of Manhattan. Isolation is measured by first-passage times to the places. Darker color corresponds to longer first-passage times

The assessment tax relies upon a special enhancement made up of the land or site value and differs from the market value estimating a relative wealth of the place within the city commonly refereed to as the 'unearned' increment of land use (Bolton 1922). The rate of appreciation in value of land is affected by a variety of conditions, for example it may depend upon other property in the same locality, will be due to a legitimate demand for a site, and for occupancy and height of a building upon it.

The current tax assessment system enacted in 1981 in the city of New York classifies all real estate parcels into four classes subjected to the different tax rates set by the legislature: (i) primarily residential condominiums; (ii) other residential property; (iii) real estate of utility corporations and special franchise properties; (iv) all other properties, such as stores, warehouses, hotels, etc. However, the scarcity of physical space in the compact urban pattern on the island of Manhattan will naturally set some increase of value on all desirably located land as being a restricted commodity. Furthermore, regulatory constrains on housing supply exerted on housing prices by the state and the city in the form of 'zoning taxes' are responsible for converting the property tax system in a complicated mess of interlocking influences and for much of the high cost of housing in Manhattan (Glaeser and Gyourko 2003).

Being intrigued with the likeness of the tax assessment map and the map of isolation in Manhattan, we have mapped the TAR figures publicly available through the Office of the Surveyor at the Manhattan Business Center onto the data on first-passage times to the corresponding places. The resulting plot is shown in Fig. 8.4, in

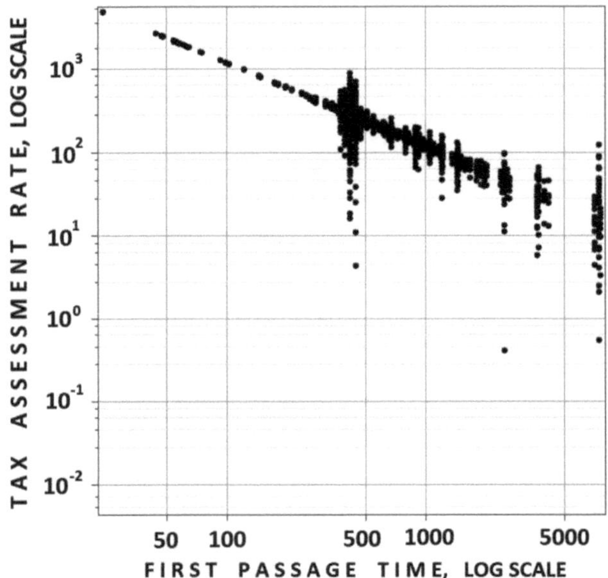

Fig. 8.4 Tax assessment rate (TAR) of places in Manhattan (*the vertical axes*, in $/fit^2) is shown in the logarithmic scale vs. the first-passage times (FPT) to them (*the horizontal axes*)

the logarithmic scale. The data presented in Fig. 8.4 positively relates the geographic accessibility of places in Manhattan with their 'unearned increments' estimated by means of the increasing burden of taxation. The inverse linear pattern dominating the data is best fitted by the simple hyperbolic relation between the tax assessment rate (TAR) and the value of first-passage time (FPT),

$$ \mathrm{TAR} \propto \frac{c}{\mathrm{FPT}}, \tag{8.31} $$

in which $c \simeq 120,000\ \$ \times \mathrm{Step}/\mathrm{fit}^2$ is a fitting constant.

8.4   Unraveling the Tangles of Language Evolution

Changes in languages go on constantly affecting words through various innovations and borrowings (Nichols and Warnow 2008). Although tree diagrams have become ubiquitous in representations of language taxonomies, they obviously fail to reveal full complexity of language affinity characterized by many phonetic, morphophonemic, lexical, and grammatical isoglosses; not least because of the fact that the simple relation of *ancestry* basic for a branching family tree structure cannot grasp complex social, cultural and political factors molding the extreme historical language contacts (Heggarty 2006). As a result, many evolutionary trees conflict with each other and with the traditionally accepted family arborescence (Nichols and Warnow 2008); the languages known as isolates cannot be reliably classified into any branch with other living languages (Gray and Atkinson 2003); the tree-reconstruction phylogenetic methods applied to the language families that do not develop by binary splitting lead to deceptive conclusions (Ben Hamed and Wang 2006).

Virtually all authors using the phylogenetic analysis on language data agree upon that a *network*, or a *web* rather than trees can provide a more appropriate representation for an essentially multidimensional phylogenetic signal (Heggarty 2008). Networks have already appeared in phylogenetic analysis (Forster and Toth 2003; Gray et al. 2007) either as a number of additional edges in the usual phylogenetic trees representing contacts and combined interactions between the individual languages and language groups, or as the considerable reticulation in a central part of the tree-like graphs representing a conflict between the different splits that are produced in the data analysis. However, the more comprehensive the graphical model is, the less clear are its visual apprehension and interpretation (Nichols and Warnow 2008).

In the what following, we show how the relationships between different languages in the language family can be represented geometrically, in terms of distances and angles, as in Euclidean geometry of everyday intuition. Our method is fully automated and based on the statistical analysis of orthographic realizations of the meanings of Swadesh vocabulary containing 200 words essentially resistant to changes. First, we have tested our method for the Indo-European language family by construction of language taxonomy for the fifty major languages spoken in Europe, on the Iranian plateau, and on the Indian subcontinent selected among about 450 languages and dialects of the whole family Gordon 2005. Second, we have investigated the Austronesian phylogeny considered again over 50 languages chosen among those 1,200 spoken by people in Indonesia, the Philippines, Madagascar, the central and southern Pacific island groups (except most of New Guinea), and parts of mainland Southeast Asia and the island of Taiwan.

8.4.1 Applying Phylogenetic Methods to Language Taxonomies

Applying phylogenetic methods to language taxonomies is a process containing a series of discrete stages (Nichols and Warnow 2008; Heggarty 2006), each one requires the application of techniques developed in different disciplines. In the first, *encoding* stage, the relations between languages is expressed in a numerical form suitable for further analysis. Various lexicostatistic techniques have been used in this stage so far (see Nichols and Warnow 2008, for a review). As a result each language is characterized by a vector (string), with components indicating the presence/absence of some features, traits, and other linguistic variables, readily converted into a matrix of lexical distances quantifying the perceptible affinity of languages in the group.

The numerical data containing the phylogenetic signal obtained in the first stage of the process lack a standard metric that makes a direct comparative analysis of the linguistic data impossible. Therefore, in the second stage, the various agglomerative clustering techniques are implemented in order to get the simplified *representations* of the data set. For example, the unweighted pair group method with arithmetic mean (UPGMA) is used in glottochronology to produce a tree from the distance matrix (Michener and Sokal 1957). The neighbor joining (NJ) (Saitou and Masatoshi 1987) and their variations are widely used for tree-like representations of language phylogeny. Since the phylogenetic signal is virtually multidimensional, trees and networks come at the cost of losing information. Eventually, in the *interpretation* stage, the linguistic meanings of the identified components have to be assessed. The last step is by no means trivial since, in the course of analysis, the initial linguistic features encoded in the data set appear to be strongly entangled due to the multiple transformations of coordinate systems and the phylogenetic signal may become unclear due to dramatic dimensionality reduction in the data set.

The phylogenetic methodology described above has been thoroughly criticized by linguists in each of its stages (Heggarty 2006). It is obvious that the direct use of techniques initially developed in genetics and palaeontology to language taxonomies is inappropriate, as the nature of interactions between the different languages in a language family and, say, between the genes in a genome is strikingly different.

Below, we present a fully automated method for building genetic language taxonomies that in many respects seems to be more relevant for the analysis of language data sets. Novelty of our approach is both in the *encoding* stage and in the *representation* one implying some novelty in the *interpretation* stage.

8.4.2   The Data Set We Have Used

The data set (Database) we have used in order to construct the language taxonomy is composed by 50 languages of the Indo-European group (IE) and 50 languages of the Austronesian group (AU). To minimize the effect of bias between orthographic and phonetic realizations of meanings, a short list of 200 words which are known to change at a slow rate are used, rather than a complete dictionary. The main source for the database for the IE group was the file prepared by Dyen et al. (1997). This database contains the Swadesh list of vocabulary with basic 200 meanings which seem maximally resistant to change, including borrowing (McMahon et al. 2005), for 96 languages. The words are given there without diacritics and adopted for using classic linguistic comparative methods to extract sets of 'cognates' – words that can be related by consistent *sound* changes. Some words are missing in Dyen et al. (1997) but for our choice of 50 languages we have filled most of the gaps and corrected some errors by finding the words from Swadesh lists and from dictionaries freely available on the web.

For the AU group, the huge database (Greenhill et al. 2008) has been used under the authors' permission that we acknowledge. The AU database is adopted to reconstruct systematic *sound correspondences* between the languages in order to uncover historically related 'cognate' forms and is under the permanent cleaning and development, with the assistance of linguistic experts correcting mistakes and improving the cognacy judgments. The lists in Greenhill et al. (2008) contain more than 200 meanings which do not completely coincide with those in the original Swadesh list. For our choice of fifty AU languages we have retained only those words which are included in the both data sets of Dyen et al. (1997) and of the original vocabulary Dyen et al. (1997) and Swadesh (1952). The resulting list has still many gaps due to missing words in the data set (Greenhill et al. 2008) and because of the incomplete overlap between the list of Greenhill et al. (2008) and the original Swadesh list Dyen et al. (1997) and Swadesh (1952). We have filled some of the gaps by finding the words from Swadesh's lists available on the web and by direct knowledge of the Malagasy language (by *M.S.*).

We used the English alphabet (26 characters plus *space*) in our work to make the language data suitable for numerical processing. Those languages written in the different alphabets (i.e., Greek, etc.) were already transliterated into English in Dyen et al. (1997). In Greenhill et al. (2008), many letter-diacritic combinations are used which we have replaced by the underlying letters reducing again the set of characters to the standard English alphabet. Interestingly, the abolition of all diacritics favoring a "simple" alphabet allowed us to obtain a reasonable result. The database modified by the Authors is available at Database. Readers are welcome to modify, correct and add words to the database.

8.4.3 The Relations Among Languages Encoded in the Matrix of Lexical Distances

Complex relations between languages may be expressed in a numerical form with respect to many different features (Nichols and Warnow 2008). In traditional glottochronology (Dyen et al. 1992), the percentage of significant words replaced while languages diverged from a common ancestor is counted. The concept of *cognates*, the words inherited from the ancestor language, as proved by regular sound correspondences, was introduced in the early work (d'Urville 1832) of D. d'Urville about the geographical division of the Pacific. The method used by modern glottochronology, developed by M. Swadesh in the 1950s, measures distances from the percentage of shared cognates (Swadesh 1952). Constructing ancestral forms of words requires trained and experienced linguists; it is very time consuming and cannot be automated. Statistical models used in language phylogeny (see for example Gray and Atkinson 2003; McMahon and McMahon 2005; Wang and Minett 2005; Warnow et al. 2006; Ellison and Kirby 2006) describe how a set of characters may randomly evolve within a family of languages provided the relevant substitution, replacement, or confusion probabilities are taken on. Usually, statistical models have been exploited within the tree-paradigm of the language data representations. Linguists have objected to a tacit assumption that real language data can be amenable to representation as opposition between two or more 'discrete states', all equally different from each other, related by means of some 'transition probabilities' (Heggarty 2006).

The standard Levenshtein (edit) distance accounting for the minimal number of insertions, deletions, or substitutions of single letters needed to transform one word into the other used previously in information theory (Levenshtein 1966) has also been implemented for the purpose of automatic clustering of languages (Nerbonne et al. 1999; McMahon and McMahon 2005; Kessler 2005) to compare the phonetic or phonological realizations of a particular vocabulary across the range of languages. The standard edit distance also gives deceptive results (Batagelj et al. 1992) if applied to the orthographic realizations of meanings in the different languages, since lengthy words provide more room for editing being therefore responsible for a decisive statistical impact distorting the results on language classification essentially. In order to compare two words having the same meaning albeit different lengths, the actual edit distance have to be normalized by the number of characters in these words. In Ellison and Kirby (2006), the original edit distance has been rescaled by the *average* length of the two words being compared. In our work, being guided by Petroni and Serva (2008) and Serva and Petroni (2008), while comparing two words, w_1 and w_2, we use the edit distance divided by the number of characters of the *longer of the two*,

$$D\left(w_1, w_2\right) = \frac{\|w_1, w_2\|_L}{\max\left(|w_1|, |w_2|\right)} \tag{8.32}$$

where $\|w_1, w_2\|_L$ is the standard Levenshtein distance between the words w_1 and w_2, and $|w|$ is the number of characters in the word w. For instance, according to (8.32) the normalized Levenshtein distance between the orthographic realizations of the meaning *milk* in English and in German (*Milch*) equals $2/5$. Such a normalization seems natural since the deleted symbols from the longer word and the empty spaces added to the shorter word then stand on an equal footing: the shorter word is supplied by a number of spaces to match the length of the longer one. The obvious advantage of (8.32) against the normalization used in Ellison and Kirby (2006) is that $D(w_1, w_2)$ takes values between 0 and 1 for any two words, w_1 and w_2, so that $D(w, w) = 0$, and $D(w_1, w_2) = 1$ when all characters in these words are different. Moreover, it is clear that the normalized edit distance defined in (8.32) is symmetric, i.e., $D(w_1, w_2) = D(w_2, w_1)$. The normalized edit distance between the orthographic realizations of two words (8.32) can be interpreted as the probability of mismatch between two characters picked from the words at random.

In order to obtain the lexical distances between the two languages, l_1 and l_2, we compute the average of the normalized Levenshtein distances (8.32) over Swadesh's vocabulary (Swadesh 1952) of 200 meanings – the smaller the result is, the more affine are the languages,

$$d(l_1, l_2) = \frac{1}{200} \cdot \sum_{\alpha \in \text{Swadesh list}} D\left(w_\alpha^{(l_1)}, w_\alpha^{(l_2)}\right), \tag{8.33}$$

where α is a meaning from Swadesh's vocabulary, and $w_\alpha^{(l)}$ is its orthographic realization in the language l. It is obvious that $d(l, l) = 0$, and $d(l_1, l_2) = 1$ if none of Swadesh's words belonging to the language l_1 has any common character with those words of the same meanings in the language l_2 that is already improbable even over the short list of 200 meanings. The lexical distance (8.33) between two languages, l_1 and l_2, can be interpreted as the average probability to distinguish them by a mismatch between two characters randomly chosen from the orthographic realizations of Swadesh's meanings. It is worth a mention that although the lexical distance defined by (8.33) can be calculated formally for any pair of languages, we have used it only for the evaluation of distances between the languages belonging to the same language family because of we like to construct the geometric representation of relations within the particular language families and not of relations between the different families, which is also possible in the framework of our method. As a result, for the two samples of 50 languages selected from the IE and AU language families, we obtained the two symmetric $50 \times 50-$matrices, $d(l_1, l_2) = d(l_2, l_1)$, with vanishing diagonal elements, $d(l, l) = 0$; each matrix therefore contains 1,225 independent entries. The encoding by lexical distances (8.33) is fully automated and therefore not time consuming at variance with the cognacy approach used in glottochronology. Comparing the edit distances between languages based on orthographic realizations might reflect different kinds of distances between languages (social, cultural, political) and not only genetic. The phylogenetic trees from the lexical distance matrices (8.33) were constructed in Serva and Petroni (2008) and Petroni and Serva (2008).

8.4.4 The Structural Component Analysis on Language Data

Component analysis is a standard tool in diverse fields from neuroscience to computer graphics. It helps to reduce a complex data set to a lower dimension suitable for visual apprehension and to reveal its simplified structures. Independent component analysis (ICA) (Hyvärinen et al. 2001) and Principal component analysis (PCA) (Jolliffe 2002) are widely used for separating a multivariate signal into additive subcomponents. The mutual statistical *independence* of the non-Gaussian source signals are supposed for the data subjected for the ICA analysis. The method finds the independent components by maximizing the statistical independence of the data instances being an efficient tool for separating independent signals mixed together like in the classical example of the "cocktail party problem", where a number of people are talking simultaneously in a room, and one is trying to follow one of the discussions. It is obviously inapplicable for reconstructing language phylogenies.

In the standard PCA analysis, the source signals are considered as simply *linearly* correlated, while all possible high-order dependencies are removed from the data set. In the course of the PCA method, the data instances are ordered according to their variance with respect to the mean by moving as much of the variance as possible into the first few dimensions. However, there is no reason to suggest that the directions of maximum variance recovered by the standard PCA method are good enough for identification of principal components in the linguistic data. Although both the statistical methods (Hyvärinen et al. 2001; Jolliffe 2002) are applied on a multitude of real world problems, their predictions largely fail not only on the essentially non-random, strongly correlated data sets, but even on multi-modal Gaussian data. It is clear that the standard techniques of component analysis have to be dramatically improved for any meaningful application on language data. Since all languages within a language family interact with each other and with the languages of other families in 'real time', it is obvious that any historical development in language cannot be described only in terms of 'pair-wise' interactions, but it reflects a genuine higher order influence among the different language groups. Generally speaking, the number of parameters describing all possible parallels we may observe between the linguistic data from the different languages would increase exponentially with the data sample size. The only hope to perform any useful data analysis in such a case relies upon a proper choice of features that re-expresses the data set to make all contributions from an asymptotically infinite number of parameters *convergent* to some non-parametric *kernel*.

It is important to mention that any symmetric matrix of lexical distances (8.33) uniquely determines a weighted undirected fully connected graph, in which nodes represent languages, and edges connecting them have weights equal to the relevant lexical distances between languages (8.33). Since the graph encoded by the matrix (8.33) is relatively small (of 50 nodes) and essentially not random, it is obviously out of the usual context of complex network theory (Dorogovtsev 2010). A suitable method for the structural analysis of networks (weighted graphs) by means of *random walks* (or Markov chains, in a more general context) has been formulated in

Blanchard and Volchenkov (2008b, 2009b) and Volchenkov (2010). Being a version of the *kernel PCA* method (Schölkopf et al. 1998), it generalizes PCA to the case where we are interested in principal components obtained by taking all higher-order correlations between data instances.

Before we explain how the most meaningful features of the lexical data encoded in the matrix (8.33) can be detected, let us note that there are infinitely many matrices that match all the structure of $d\left(l_i, l_j\right)$ and contain all the information about the relationships between languages estimated by means of the lexical distances (8.33). It is remarkable that all these matrices are related to each other by means of a linear transformation, which can be interpreted as a random walk (Blanchard and Volchenkov 2008b, 2009b) defined on the weighted undirected graph determined by the matrix of lexical distances $d\left(l_i, l_j\right)$. We have to emphasize that random walks appear in our approach in concern to neither any particular assumption regarding to evolutionary processes in language (as we do not concern ourselves with the problems of modeling contagion or the spread of information through a society), nor the Bayesian analysis used previously (Gray and Jordan 2000; Gray and Atkinson 2003; Gray et al. 2009) to construct the self-consistent tree-like representations in linguistic phylogenies, but as the *unique* linear transformation (in the class of stochastic matrices) consistent with all of the structure of the matrix of lexical distances calculated with respect to Swadesh's list of meanings.

A random walk associated to the matrix of lexical distances $d\left(l_i, l_j\right)$ calculated over the Swadesh vocabulary for a sample of N different languages (in our case, $N = 50$ for both language families) is defined by the transition probabilities

$$T\left(l_i, l_j\right) = \Delta^{-1} d\left(l_i, l_j\right) \tag{8.34}$$

where the diagonal matrix $\Delta = \mathrm{diag}\left(\delta_{l_1}, \delta_{d_2}, \dots \delta_{l_N}\right)$ contains the *cumulative* lexical distances $\delta_{l_i} = \sum_{j=1}^{N} d\left(l_i, l_j\right)$, for each language l_i. Diagonal elements of the matrix T are equal to zero, since $d\left(l_i, l_i\right) = 0$, for any language l_i. The matrix (8.34) is a stochastic matrix, $\sum_{j=1}^{N} T\left(l_i, l_j\right) = 1$, being nothing else, but the normalized matrix of lexical distances (8.33), in which a vector of probabilities $\mathbf{f}\left(l_i\right) \in [0, 1]^N$, a row of the matrix $T\left(l_i, l_j\right)$, is attributed to each language l_i, with respect to all other languages in the language family,

$$\mathbf{f}\left(l_i\right) = \left(\frac{d\left(l_i, l_1\right)}{\delta_{l_i}}, \frac{d\left(l_i, l_2\right)}{\delta_{l_i}}, \dots \frac{d\left(l_i, l_N\right)}{\delta_{l_i}}\right). \tag{8.35}$$

Each element of the vector (8.35) is a conditional probability describing the level of confidence that the language l_i can be identified successfully by comparing the orthographic representation of a randomly chosen Swadesh's meaning with that of the other language l_j, given the both languages belong to the same language family. It is worth a mention that since the sum of all elements in the probability vector, $\sum_{j=1}^{N} \left(f\left(l_i\right)\right)_j = 1$, for any language l_i, it is assumed that we can always confidently identify (with probability 1) a language by comparing its orthographic realizations with those from all other languages in the group.

Consequently, random walks defined by the transition matrix (8.34) describe the statistics of a sequential process of language classification. Namely, while the elements of the matrix $T\left(l_i, l_j\right)$ evaluate the successful identification of the language l_i provided the language l_j has been identified certainly, the elements of the squared matrix,

$$T^2\left(l_i, l_j\right) = \sum_{k=1}^{N} T\left(l_i, l_k\right) \cdot T\left(l_k, l_j\right), \tag{8.36}$$

ascertain the successful identification of the language l_i from l_j through an intermediate language, the elements of the matrix T^3 give the probabilities to identify the language through two intermediate steps, and so on. By the way, the whole host of complex and indirect relationships between orthographic representations of Swadesh's meanings encoded in our approach in the matrix of lexical distances (8.33) is uncovered by the powers T^n, $n \geq 1$ (Blanchard and Volchenkov 2008b, 2009b).

Under the successive actions of T, any probability distribution vector $\mathbf{f}$ converges to a *stationary* distribution,

$$\pi = \lim_{n \to \infty} \mathbf{f} T^n = \mathbf{f} T^\infty, \tag{8.37}$$

where

$$\pi = \left(\frac{\delta_{l_1}}{\delta}, \frac{\delta_{l_2}}{\delta}, \ldots \frac{\delta_{l_N}}{\delta}\right), \quad \delta \equiv \sum_{i=1}^{N} \delta_{l_i} \tag{8.38}$$

is the 'center of mass', which *does not* coincide with the simple centroid vectors (means) calculated with respect to either columns or rows of a data matrix, in the course of the standard PCA analysis.

Random walks ascribe the total probability of successful classification for any two languages in the language family,

$$P\left(l_i, l_j\right) = \lim_{n \to \infty} \sum_{k=0}^{n} T^k\left(l_i, l_j\right) = \frac{1}{1 - T}. \tag{8.39}$$

The operator $(1-T)^{-1}$ in the r.h.s. of the above equation diverges along the direction corresponding to the stationary distribution π (8.38) which belongs to the maximal eigenvalue 1 of the transition matrix (8.34), so that the last expression in (8.39) is formal. Nevertheless, we can use the Moore-Penrose generalized inverse matrix (Penrose 1955) instead of $(1 - T)^{-1}$. The use of generalized inverses is common in the study of finite Markov chains (Meyer 1975). Such a generalized inverse provides the unique best fit solution (with respect to least squares) to the system of linear equations described by the matrix $(1 - T)^{-1}$ that lacks a unique solution. Under the Moore-Penrose inverse, any probability distribution vector $\mathbf{f}\left(l_i\right)$ is naturally translated into a perspective projection

$$\phi_i = \mathcal{P}_\pi \left(\mathbf{f}\left(l_i \right) \right),$$

with the vector of stationary distribution π as the center of projection. We can use these projections in order to classify languages with respect to the center of mass π of the entire language family. The kernel function required for the kernel PCA component analysis is expressed as the dot product (see Schölkopf et al. 1998 and references therein)

$$J = \left(\phi_i, \phi_j \right) \tag{8.40}$$

and constitute a square symmetric Gram $N \times N$-matrix. Each diagonal element $\|\phi_i\|^2 \equiv J_{ii}$ is the *first-passage time* (Lovász 1993) of random walks to $\mathbf{f}\left(l_i \right)$ defined on the weighted undirected graph determined by the matrix of lexical distances (8.33). The off-diagonal entries J_{ij} quantify the interference of two random walks concluding at $\mathbf{f}\left(l_i \right)$ and $\mathbf{f}\left(l_j \right)$ respectively (Blanchard and Volchenkov 2008b, 2009b).

It is remarkable that the matrix J plays the essentially same role for the structural component analysis, as the covariance matrix does for the usual PCA analysis. Like the covariance values reflect the structure and redundancy in the linearly correlated data, the large diagonal values of J correspond to the notable heterogeneity of the data instances, while the large magnitudes of the off-diagonal terms correspond to high redundancy in the data sample. However, in contrast to the covariance matrix which best explains the variance in the data with respect to the mean, the matrix J traces out all higher order dependencies among data entities.

8.4.5   Principal Structural Components of the Lexical Distance Data

High-dimensional data, which require more than two or three dimensions to represent a complex nexus of relationships, are difficult to interpret. The standard goal of the component analysis is to minimize the redundancy in the data sample quantified by the off-diagonal elements J_{ij}. It is readily achieved by solving an eigenvalue problem for the real positive symmetric kernel matrix (8.40). Namely, there is a real orthogonal matrix Q, $Q^\top Q = 1$, (where $Q^\top$ stands for the transposed matrix Q) such that

$$\Lambda = Q^\top J Q \tag{8.41}$$

is a diagonal matrix. Each column vector q_k of the matrix Q is an eigenvector of the linear transformation that determines a direction where J acts as a simple rescaling, $J q_k = \lambda_k q_k$, with some real eigenvalue $\lambda_k \geq 0$ indicating the characteristic first-passage time associated to the virtually independent component q_k; each one represents an independent trait detected in the matrix of lexical distances $d\left(l_i, l_j \right)$ calculated over the Swadesh list of meanings.

The independent components $\{q_k\}$, $k = 1, \ldots, N$, define an orthonormal basis in $\mathbb{R}^N$ which specifies each language l_i by N numerical coordinates,

$l_i \rightarrow (q_{1,i}, q_{2,i}, \ldots q_{N,i})$, which are the signed distances from the point representing the language l_i to the axes associated to the virtually independent components. Languages that cast in the same mould in accordance with the N individual data features are revealed by geometric proximity in Euclidean space spanned by the eigenvectors $\{q_k\}$ that might be either exploited visually, or accounted analytically. The rank-ordering of data traits $\{q_k\}$, in accordance to their eigenvalues $0 = \lambda_1 < \lambda_2 \leq \ldots \leq \lambda_N$, provides us with the natural geometric framework for dimensionality reduction. The minimal eigenvalue $\lambda_1 = 0$ corresponds to the vector of stationary distribution $\pi = q_1^2$ containing no information about components.

At variance with the standard PCA analysis (Jolliffe 2002), where the *largest* eigenvalues of the covariance matrix are used in order to identify the principal components, as being characterized by the largest variance with respect to the mean, while building language taxonomy, we are interested in detecting the groups of the *most similar* languages, with respect to the selected group of features. The components of maximal similarity are identified with the eigenvectors belonging to the *smallest* non-trivial eigenvalues. In particular, we use the three consecutive components $(q_{2,i}, q_{3,i}, q_{4,i})$ as the three Cartesian coordinates of a language point $l_i(x, y, z)$ in order to build a three-dimensional geometric representation of language taxonomy. Points symbolizing different languages in space of the three major data traits are contiguous if the orthographic representations of Swadesh's meanings in these languages are similar. Although, we are doubtful of that such a statistical similarity detected automatically on a finite sample of lexicostatistical data can be directly related to the traditional isoglosses discussed by linguists, they would definitely help to formulate the plausible isogloss hypothesis for future testing (see Sect. 8.4.6).

8.4.6   *Geometric Representation of the Indo-European Family*

Many language groups in the IE family had originated after the decline and fragmentation of territorially-extreme polities and in the course of migrations when dialects diverged within each local area and eventually evolved into individual languages. In Fig. 8.5, we have shown the three-dimensional geometric representation of 50 languages of the IE language family in space of its three major data traits detected in the matrix of lexical distances calculated over the Swadesh list of meanings. Due to the striking central symmetry of the representation, it is natural to describe the positions of language points l_i with the use of spherical coordinates,

$$
\begin{aligned}
r_i &= \sqrt{q_{2,i}^2 + q_{3,i}^2 + q_{4,i}^2}, \\
\theta_i &= \arccos\left(\frac{q_{4,i}}{r_i}\right), \\
\phi_i &= \arctan\left(\frac{q_{3,i}}{q_{2,i}}\right),
\end{aligned}
\tag{8.42}
$$

rather than the Cartesian system.

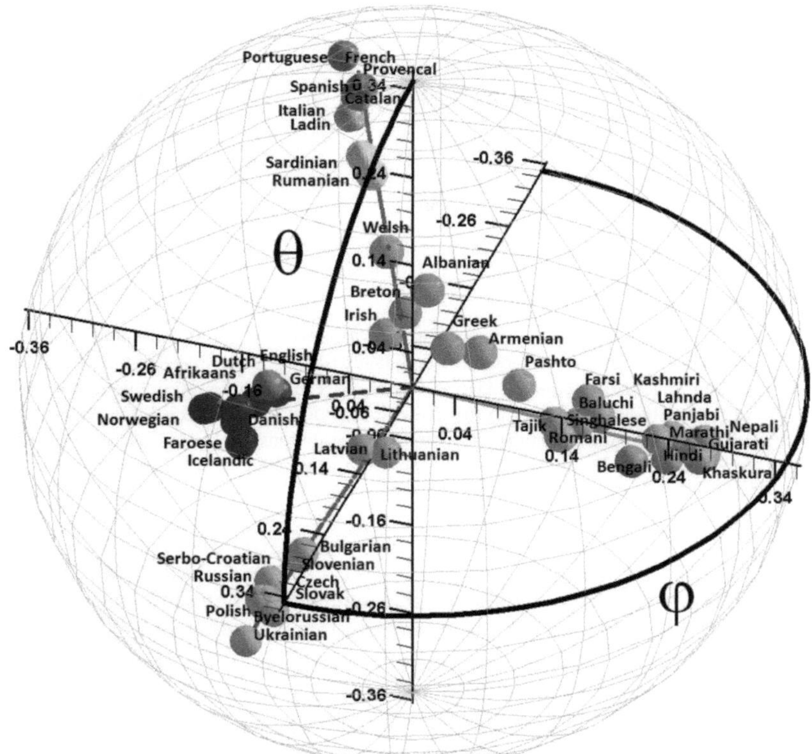

Fig. 8.5 The three-dimensional geometric representation of the IE language family in space of the major data traits (q_2, q_3, q_4) color coded. The origin of the graph indicates the 'center of mass' $q_1 = \pi$ of the matrix of lexical distances $d\left(l_i, l_j\right)$, not the Proto-IE language. Due to the central symmetry of representation, it is convenient to use the spherical coordinates to identify the positions of languages: the radius from the center of the graph, the inclination angle θ, and the azimuth angle φ

The principal components of the IE family reveal themselves in Fig. 8.5 by four well-separated spines representing the four biggest traditional IE language groups: Romance & Celtic, Germanic, Balto-Slavic, and Indo-Iranian. These groups are monophyletic and supported by the sharply localized distributions of the azimuth (φ) and inclination (zenith) angles (θ) over the languages shown in Fig. 8.6a and b respectively.

The Greek, Romance, Celtic, and Germanic languages form a class characterized by approximately the same azimuthal angle (Fig. 8.6a), thus belonging to one plane in the three-dimensional geometric representation shown in Fig. 8.5, while the Indo-Iranian, Balto-Slavic, Armenian, and Albanian languages form another class, with respect to the inclination (zenith) angle (Fig. 8.6b).

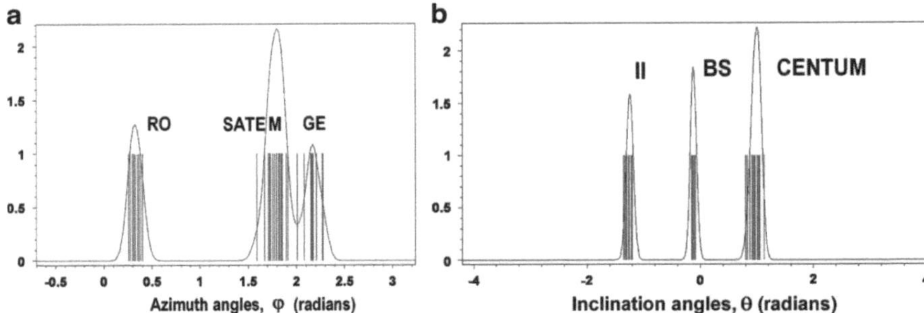

Fig. 8.6 (**a**) The kernel density estimates of the distributions of azimuthal angles in the three-dimensional geometric representation of 50 languages of the IE language family, together with the absolute data frequencies. Romance (RO), Germanic (GE), and the satem languages (SATEM) are easily differentiated with respect to the azimuthal angles. (**b**) The kernel density estimates of the distributions of inclination (zenith) angles in the three-dimensional geometric representation of 50 languages of the IE language family, together with the absolute data frequencies. Indo-Iranian (II), Balto-Slavic (BS), and the centum languages (CENTUM) are attested by the inclination (zenith) angles

It is remarkable that the division of IE languages with respect to the azimuthal and zenith angles evident from the geometric representation in Fig. 8.5 perfectly coincides with the well-known *centum-satem* isogloss of the IE language family (the terms are the reflexes of the IE numeral '100'), related to the evolution in the phonetically unstable palatovelar order (Gamkrelidze and Ivanov 1995). The palatovelars merge with the velars in centum languages sharing the azimuth angle, while in satem languages observed at the same zenith angle the palatovelars shift to affricates and spirants. Although the satem-centum distinction was historically the first original dialect division of the Indo-European languages (Renfrew 1987), it is not accorded much significance by modern linguists as being just one of many other isoglosses crisscrossing all IE languages (Baldi 2002). The basic phonetic distinction of the two language classes does not justify in itself the areal groupings of historical dialects, each characterized by some phonetic peculiarities indicating their independent developments. The appearance of the division similar to the centum-satem isogloss (based on phonetic changes only) may happen because of the systematic sound correspondences between the Swadesh words across the different languages of the same language family.

The projections of Albanian, Greek, and Armenian languages onto the axes of the principal components of the IE family are rather small, as they occupy the center of the diagram in Fig. 8.5. Being eloquently different from others, these languages can be resolved with the use of some minor components q_k, $k > 3$. Remarkably, the Greek and Armenian languages always remain proximate confirming Greeks' belief that their ancestors had come from Western Asia (Gamkrelidze and Ivanov 1990).

8.4.7   *In Search of Lost Time*

Geometric representations of language families can be conceived within the framework of various physical models that infer on the evolution of linguistic data traits. In traditional glottochronology (Swadesh 1952), the time at which languages diverged is estimated on the assumption that the core lexicon of a language changes at a constant average rate. This assumption based on an analogy with the use of carbon dating for measuring the age of organic materials was rejected by mainstream linguists considering a language as a social phenomenon driven by unforeseeable socio-historical events not stable over time (Heggarty 2006). Indeed, mechanisms underlying evolution of dialects of a proto-language evolving into individual languages are very complex and hardly formalizable.

In our method based on the statistical evaluation of differences in the orthographic realizations of Swadesh's vocabulary, a complex nexus of processes behind the emergence and differentiation of dialects within each language group is described by the single degree of freedom, along the radial direction (see (8.42)) from the origin of the graph shown in Fig. 8.5, while the azimuthal (φ) and zenith (θ) angles are specified by a language group.

It is worth a mention that the distributions of languages along the radial direction are remarkably heterogeneous indicating that the rate of changes in the orthographic realizations of Swadesh's vocabulary was anything but stable over time. Being ranked within the own language group and then plotted against their expected values under the normal probability distribution, the radial coordinates of languages in the geometrical representation Fig. 8.5 show very good agreement with univariate normality, as seen from the normal probability plots in Fig. 8.7a–d.

The hypothesis of normality of these distributions can be justified by taking on that for a long time the divergence of orthographic representations of the core vocabulary was a *gradual* change accumulation process into which many small, independent innovations had emerged and contributed additively to the outgrowth of new languages. Perhaps, the orthographic changes arose due to the fixation of phonetic innovations developed in the course of long-lasting interactions with non-IE languages in areas of their intensive historical contacts.

In physics, the univariate normal distribution is closely related to the time evolution of a mass-density function $\rho(r, t)$ under homogeneous diffusion in one dimension,

$$\rho(r, t) = \frac{1}{2\pi\sigma^2} \cdot \exp\left(-\frac{(r - \mu)^2}{2\sigma^2}\right),$$

in which the mean value μ is interpreted as the coordinate of a point where all mass was initially concentrated, and variance $\sigma^2 \propto t$ grows linearly with time. If the distributions of languages along the radial coordinate of the geometric representation do fit to univariate normality for all language groups, then in the long run the value of variance in these distributions grew with time at some approximately constant rate. We have to emphasize that the locations of languages might not be distributed normally if it were not true; we did not do any assumption

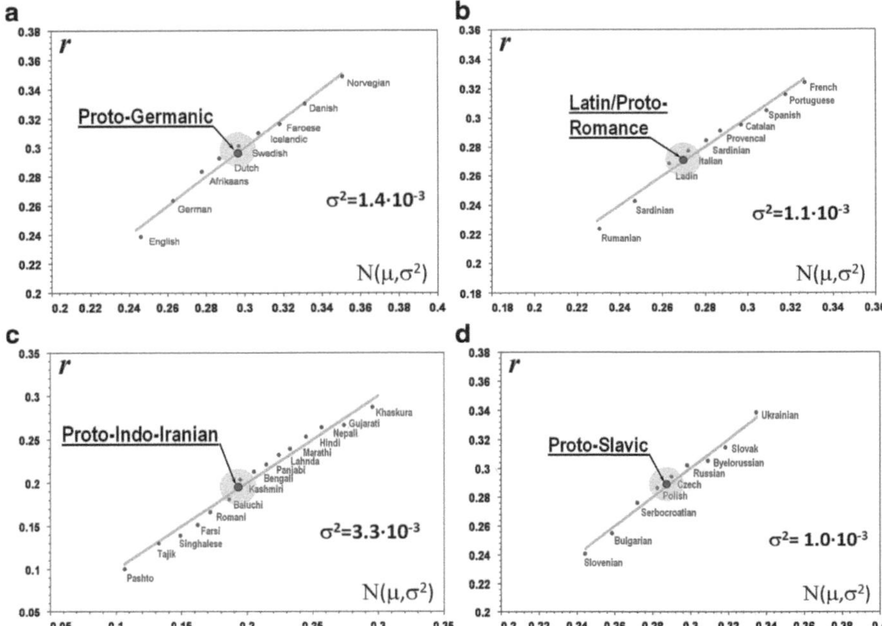

Fig. 8.7 The panels a–d show the normal probability plots fitting the distances r of language points from the 'center of mass' to univariate normality. The data points were ranked and then plotted against their expected values under normality, so that departures from linearity signify departures from normality. The values of variance are given for each language group. The expected locations of the proto-languages, together with the end points of the 95% confidence intervals, are displayed on the normal plots by circles

above. Again, the constant increment rates of variance of radial positions of languages in the geometrical representation Fig. 8.5 has nothing to do with the traditional glottochronological assumption about the steady borrowing rates of cognates (Embelton 1986). For clarity, we have used a simple code to produce a sequence of normally distributed integer numbers, with linearly growing variance: $[6, 7, 4, 3, 6, 4, 11, 7, 9, 4, 5, 1, 7, 2, 16]$; they obviously do not grow linearly. It is also important to mention that the values of variance σ^2 calculated for the languages over the individual language groups (see Fig. 8.7a–d) do not correspond to physical time but rather give a statistically consistent estimate of age for each language group. In order to assess the pace of variance changes with physical time and calibrate our dating method, we have to use the historically attested events.

Although historical compendiums report us on grace, growth, and glory succeeded by the decline and disintegration of polities in days of old, they do not tell us much about the simultaneous evolution in language. It is beyond doubt that massive population migrations and disintegrations of organized societies both destabilizing the social norms governing behavior, thoughts, and social relationships can be taken on as the chronological anchors for the onset of language differentiation. However, the idealized assumption of a punctual *split* of a proto-language into a number of successor languages shared implicitly by virtually all phylogenetic models is problematic for a linguist well aware of the long-lasting and devious process by which a real language diverges (Heggarty 2006). We do not aspire to put dates on such a fuzzy process but rather consider language as a natural appliance for dating of those migrations and fragmentation happened during poorly documented periods in history.

While calibrating the dating mechanism in our model, we have used the four anchor events (Fouracre 2007):

1. The last Celtic migration (to the Balkans and Asia Minor) (by 300 BC),
2. The division of the Roman Empire (by 500 AD),
3. The migration of German tribes to the Danube River (by 100 AD),
4. The establishment of the Avars Khaganate (by 590 AD) overspreading Slavic people who did the bulk of the fighting across Europe.

It is remarkable that a very slow variance pace of a millionth per year

$$\frac{t}{\sigma^2} = (1.367 \pm 0.002) \cdot 10^6 \tag{8.43}$$

is evaluated uniformly, with respect to all of the anchoring historical events mentioned above.

The time–variance ratio (8.43) deduced from the well attested events allows us to retrieve the probable dates for

1. The break-up of the Proto-Indo-Iranian continuum preceding 2400 BC, in a good agreement with the migration dates from the early Andronovo archaeological horizon (Bryant 2001);
2. The end of common Balto-Slavic history as early as by 1400 BC, in support of the recent glottochronological estimates (Novotná and Blažek 2007) well agreed with the archaeological dating of Trziniec-Komarov culture, localized from Silesia to Central Ukraine;
3. The separation of Indo-Arians from Indo-Iranians by 400 BC, probably as a result of Aryan migration across India to Ceylon, as early as in 483 BC (McLeod 2002);
4. The division of Persian polity into a number of Iranian tribes migrated and settled in vast areas of south-eastern Europe, the Iranian plateau, and Central Asia by 400 BC, shortly after the end of Greco-Persian wars (Green 1996).

8.4.8 Evidence for Proto-Indo-Europeans

The basic information about the Proto-Indo-Europeans arises out of the comparative linguistics of the IE languages. There were a number of proposals about early Indo-European origins in so far. For instance, the *Kurgan* scenario postulating that the people of an archaeological "Kurgan culture" (early fourth millennium BC) in the

Pontic steppe were the most likely speakers of the proto- IE language is widely accepted (Gimbutas 1982). The *Anatolian* hypothesis suggests a significantly older age of the IE proto-language as spoken in Neolithic Anatolia and associates the distribution of historical IE languages with the expansion of agriculture during the Neolithic revolution in the eighth and sixth millennia BC (Renfrew 1987).

It is a subtle problem to trace back the diverging pathways of language evolution to a convergence in the IE proto-language since symmetry of the modern languages assessed by the statistical analysis of orthographic realizations of the core vocabulary mismatches that in ancient time. The major IE language groups have to be reexamined in order to ascertain the locations of the individual proto-languages as if they were extant. In our approach, we associate the mean μ of the normal distribution of languages belonging to the same language group along the radial coordinate r with the expected location of the group proto-language. Although we do not know what the exact values of means were, the sample means calculated over the several extant languages from each language group give us the appropriate estimators. There is a whole interval around each observed sample mean within which, the true mean of the whole group actually can take the value.

In order to target the locations of the five proto-languages (the Proto-Germanic, Latin, Proto-Celtic, Proto-Slavic, and Proto-Indo-Iranian) with the 95% confidence level, we have supposed that variances of the radial coordinate calculated over the studied samples of languages are the appropriate estimators for the true variance values of the entire groups. The expected locations of the proto-languages, together with the end points of the 95% confidence intervals, are displayed on the normal plots, in Fig. 8.7a–d. Let us note that we did not include the Baltic languages into the Slavic group when computing the Proto-Slavic center point because these two groups exhibit different statistics, so that such an inclusion would dramatically reduce the confidence level for the expected locations of the proto-languages. Although the statistical behavior of the proto-languages in the geometric representation of the IE family is not known, we assume that it can be formally described by the 'diffusion scenario', as for the historical IE languages. Namely, we assume that the locations of the five proto-languages from a statistically determined central point fit to multivariate normality. Such a null hypothesis is subjected to further statistical testing, in which the chi-square distribution is used to test for goodness of fit of the observed distribution of the locations of the proto-languages to a theoretical one. The chi-square distribution with k degrees of freedom describes the distribution of a random variable $Q = \sum_{i=1}^{k} X_i^2$ where X_i are k independent, normally distributed random variables with mean 0 and variance 1.

In Fig. 8.8, we have used a simple graphical test to check three-variate normality by extending the notion of the normal probability plot. The locations of proto-languages have been tested by comparing the goodness of fit of the scaled distances from the proto-languages to the central point (the mean over the sample of the five proto-languages) to their expected values under the chi-square distribution with three degrees of freedom. In the graphical test shown in Fig. 8.8, departures from three-variant normality are indicated by departures from linearity. Supposing that the underlying population of parent languages fits to multivariate normality,

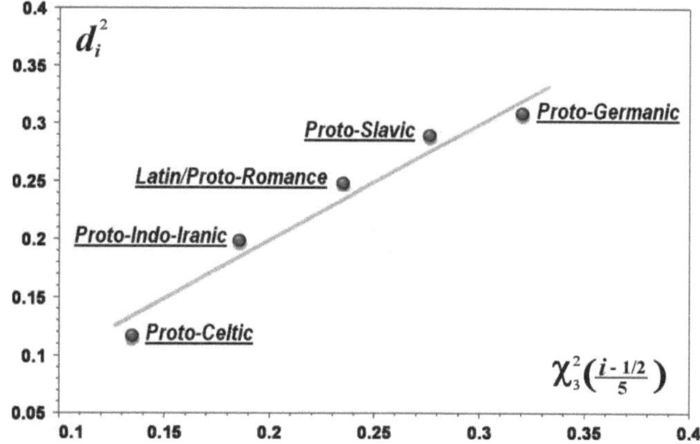

Fig. 8.8 The graphical test to check three-variate normality of the distribution of the distances d_i of the five proto-languages from a statistically determined central point is presented by extending the notion of the normal probability plot. The chi-square distribution is used to test for goodness of fit of the observed distribution: the departures from three-variant normality are indicated by departures from linearity

we conclude that the determinant of the sample variance-covariance matrix has to grow linearly with time. The use of the previously determined time–variance ratio (8.43) then dates the initial break-up of the Proto-Indo-Europeans back to 7000 BC pointing at the early Neolithic date, to say nothing about geography, in agreement with the Anatolian hypothesis of the early Indo-European origin (Renfrew 1987; Gamkrelidze and Ivanov 1990, 1995; Gray and Atkinson 2003; Renfrew 2003; Serva and Petroni 2008).

The linguistic community estimates of dating for the proto IE language lie between 4500 and 2500 BC, a later date than the Anatolian theory predicts. These estimations are primarily based on the reconstructed vocabulary (see Mallory 1991 and references therein) suggesting a culture spanning the Early Bronze Age, with knowledge of the wheel, metalworking and the domestication of the horse and thus favoring the Kurgan hypothesis. It is worth a mention that none of these words are found in the Swadesh list encompassing the basic vocabulary related to agriculture that emerged perhaps with the spread of farming, during the Neolithic era. Furthermore, the detailed analysis of the terms uncovered a great incongruity between the terms found in the reconstructed proto-IE language and the cultural level met with in the Kurgans lack of agriculture (Krell 1998). Let us note that our dating (2400 BC) for the migration from the Andronovo archaeological horizon (see Sect. 8.4.7) and the early break-up of the proto-Indo-Iranian continuum estimated by means of the variance (see Fig. 8.7c) is compatible with the Kurgan time frame. However, despite the Indo-Iranian group of languages being apparently the oldest among all other groups of the IE family, we cannot support the general claim of the Kurgan hypothesis, at least on the base of Swadesh's lexicon.

8.4.9 In Search of Polynesian Origins

The colonization of the Pacific Islands is still the recalcitrant problem in the history of human migrations, despite many explanatory models based on linguistic, genetic, and archaeological evidences have been proposed in so far. The origins, relationships, and migration chronology of Austronesian settlers have constituted the sustainable interest and continuing controversy for decades. The components probe for a sample of 50 AU languages immediately uncovers the both Formosan (F) and Malayo-Polynesian (MP) branches of the entire language family (see Fig. 8.9).

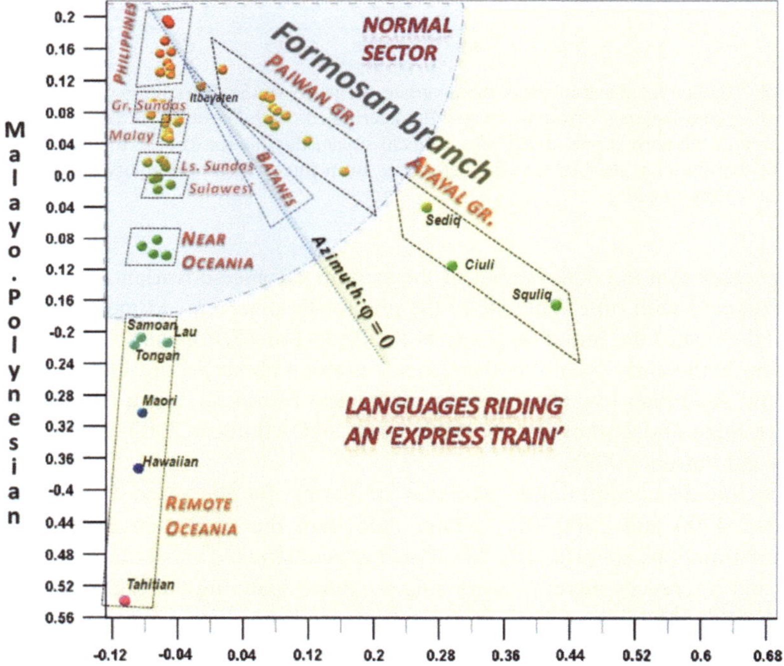

Fig. 8.9 The geometric representation of the 50 AU languages in space of the major data traits (q_2, q_3) shows the remarkable geographic patterning. It is convenient to use the polar coordinates: the radius from the center of the graph, $r_i = \sqrt{q_{2,i}^2 + q_{3,i}^2}$, and the azimuth angle $\varphi = \arctan\left(\frac{q_{3,i}}{q_{2,i}}\right)$, to identify the positions of languages. For languages in the 'normal sector', the distribution of radial coordinates conforms to univariate normality. At variance with them, languages located at the distant margins of the AU family apparently follow the 'express train' evolution model (see Sect. 8.4.11) The 'normal sector' consists of the following languages: from Philippines, *Bontoc, Kankanay, Ilokano, Hanunoo, Cebuano, Tagalog, Pangasinan, Mansaka, Maranao*; from Great Sunda and Malay, *Malagasy, Maanyan, Ngaiu dayak, Toba batak, Bali, Malay, Iban, Sasak, Sunda, Javanese*; from Lesser Sunda and Sulawesi, *Sika, Kambera, Wolio, Baree, Buginese, Manggarai, Sangir, Makassar*; from Near Oceania, *Manam, Motu, Nggela, Mota*; of Paiwan group (Taiwan) *Pazeh, Thao, Puyuma, Paiwan, Bunun, Amis, Rukai, Siraya, Kavalan*

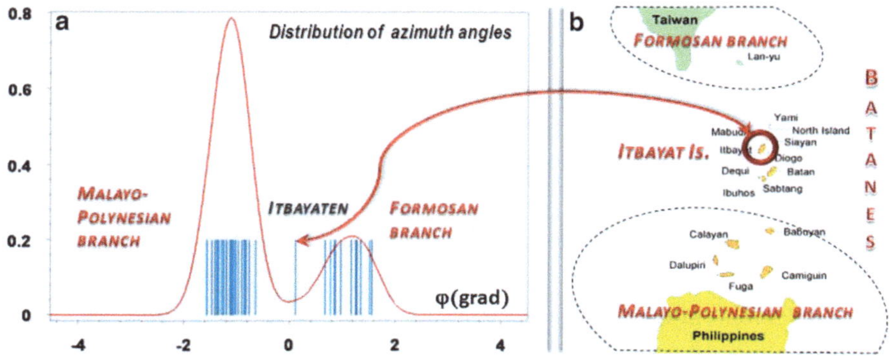

Fig. 8.10 (**a**) The distribution of azimuth angles in the geometric representation of the 50 AU languages shown in Fig. 8.9. (**b**) The Itbayaten language is pretty close to the azimuth, $\varphi = 0$, bridging over the language family branches lexically and geographically

The distribution of azimuth angles shown in Fig. 8.10. A identifies them as two monophyletic jets of languages that cast along either axis spanning the entire family plane. The clear geographic patterning is perhaps the most remarkable aspect of the geometric representation. It is also worth mentioning that the language groupings as recovered by the component analysis of lexical data reflect profound historical relationships between the different groups of AU population. For instance, the Malagasy language spoken in Madagascar casts in the same mould as the Maanyan language spoken by the Dayak tribe dwelling in forests of Southern Borneo and the Batak Toba language of North Sumatra spoken mostly west of Lake Toba.

Despite Malagasy sharing much of its basic vocabulary with the Maanyan language (Dahl 1951), many manifestations of Malagasy culture cannot be linked up with the culture of Dayak people: the Malagasy migration to East Africa presupposes highly developed construction and navigation skills with the use of outrigger canoes typical of many Indonesian tribes which the Dayak people however do not have, also some of the Malagasy cultivations and crop species (such as wet rice) cannot be found among forest inhabitants. In contrast, some funeral rites (such as the second burial, *famadihana*) typical of the leading entities of the Madagascar highlands are essentially similar to those of Dayak people. A possible explanation is that population of the Dayak origin was brought to Madagascar as slaves by Malay seafarers and unlikely realized the spectacular trip across the Indian Ocean (Petroni and Serva 2008). As the Dayak speakers formed the majority in the initial settler group, in agreement with the genetic parental lineages found in Madagascar (Hurles et al. 2005), their language could have constituted the core element of what later became Malagasy, while the language of the Malay dominators was almost suppressed, albeit its contribution is still recovered by the exploration of the leading traits on language data.

The AU language family forks at the northernmost tip of the Philippines, the Batanes Islands located about 190 km south of Taiwan (see Fig. 8.10b). On the

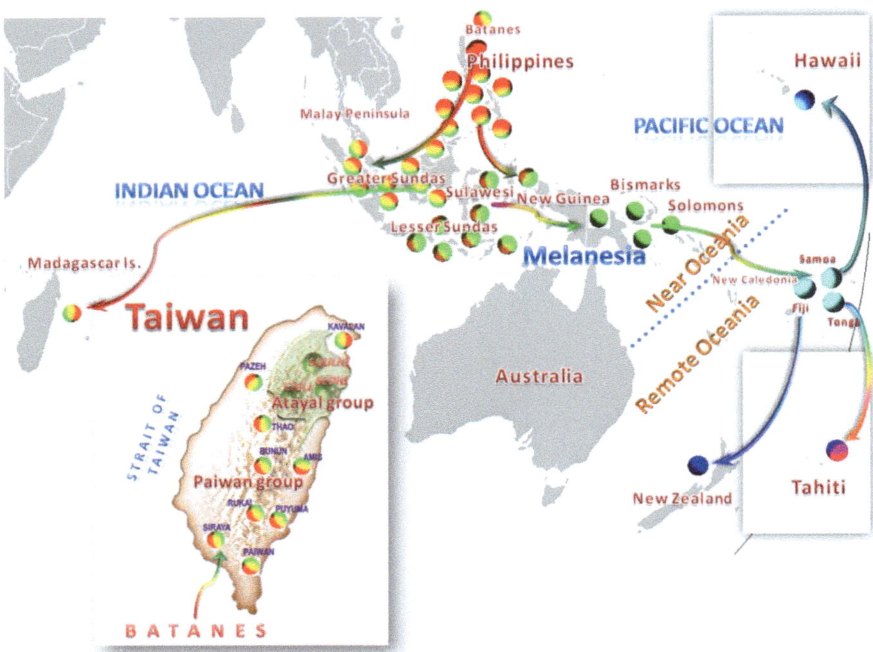

Fig. 8.11 The geometric representation of the 50 AU languages (Fig. 8.9) projected onto the geographic map uncovers the possible route of Austronesian migrations

distribution of azimuth angles shown in Fig. 8.10a, the Itbayaten language representing them in the studied sample is pretty close to the azimuth, $\varphi = 0$, bridging over the separating language family branches (Fig. 8.10b). By the way, the MP-offset descends from the northern Philippines (the northern Luzon Island) and springs forth eastward through the Malay Archipelago across Melanesia culminating in Polynesia (Fig. 8.11); in accordance with the famous 'express train' model of migrations peopled the Pacific (Diamond 1988). In its turn, the F-branch embarks on the southwest coast of Taiwan and finds its way to the northern Syueshan Mountains inhabited by Atayal people that compose many ethnic groups with different languages, diverse customs, and multiple identities. Evidently, both the offshoots derived their ancestry in Southeast Asia as strengthened by multiple archaeological records (Diamond 1988), but then evolved mostly independently from each other, on evidence of the Y-chromosome haplotype spread over Taiwanese and Polynesian populations (Su et al. 2000). The Bayesian methods for the language phylogeny trees (Gray and Jordan 2000) also evinced the earliest separation of these two branches of the AU language family. However, in the recent pulse-pause scenario (Gray et al. 2009), the Taiwanese origin of the entire AU family was suggested because of the "considerable diversity of Formosan languages". It is important to note that diversity itself is by no means a reliable estimate provided symmetry is downplayed (e.g., in spite of the greatest diversity, the Indo-Iranian language group is not an origin of the entire IE language family).

The distribution of languages spoken within Maritime Southeast Asia, Melanesia, Western Polynesia and of the Paiwan language group in Taiwan over the distances from the center of the diagram representing the AU language family in Fig. 8.9 conforms to univariate normality (see Fig. 8.12) suggesting that an interaction sphere had existed encompassing the whole region, from the Philippines and Southern Indonesia through the Solomon Islands to Western Polynesia, where ideas and cultural traits were shared and spread as attested by trade (Bellwood and Koon 1989; Kirch 1997) and translocation of farm animals (Matisoo-Smith and Robins 2004; Larson et al. 2007) among shoreline communities.

Although the lack of documented historical events makes the use of the developed dating method difficult, we may suggest that variance evaluated over Swadesh's vocabulary forges ahead approximately at the same pace uniformly for all human societies involved in trading and exchange forming a singular cultural continuum. Then, the time–age ratio (8.43) deduced from the previous chronological

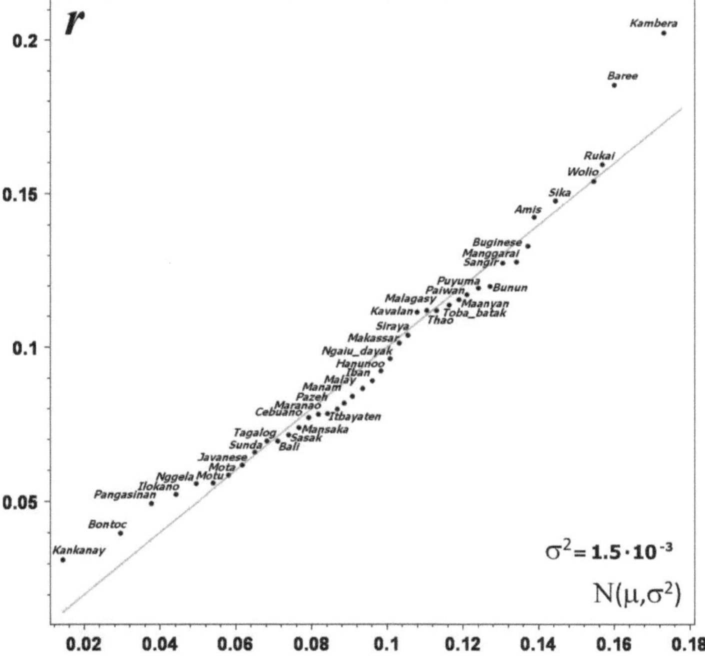

Fig. 8.12 The normal probability plot fitting the distances r of language points from the 'center of mass'of the geometrical representation of the AU language family to univariate normality. The data points for languages belonging to the 'normal sector' shown in Fig. 8.9 were ranked and then plotted against their expected values under normality, so that departures from linearity signify departures from normality. The value of variance over all languages belonging to the 'normal sector' is $\sigma^2 = 1.5 \cdot 10^{-3}$

estimates for the IE family returns 550 AD if applied to the Austronesians as the likely break-up date of their cultural continuum, pretty well before 600–1200 AD while descendants from Melanesia settled in the distant apices of the Polynesian triangle as evidenced by archaeological records (Kirch 2000; Anderson and Sinoto 2002; Hurles et al. 2003).

8.4.10 Geometric Representation of Malagasy Dialects

The Indonesian language most closely related to Malagasy is probably Maanyan of South-east Kalimantan with a 45% of shared basic vocabulary, but close languages can be found in Sulawesi, Malaysia and Sumatra. For this reason, the history of Madagascar peopling and settlement is subject to dispute and alternative interpretations among scholars. More mystery is added by the fact that the *Maanyan*, which seems to be the closest language to Malagasy, is spoken by a population which lives along the rivers of Kalimantan and which does not possess the necessary skill for long maritime navigation. A possible explanation is that they arrived as slaves of Malay sailors and, in this case, the dialects should show both a *Malay* and a *Maanyan* contribution (Petroni and Serva 2008).

In Fig. 8.13, we have shown the three-dimensional geometric representation of 23 dialects of the Malagasy language and the *Maanyan* language closely related to Malagasy, in space of the three major data traits ($\{q_2, q_3, q_4\}$, see the Appendix B for details) detected in the matrix of lexical distances calculated over the Swadesh list of meanings.

The clear geographic patterning is perhaps the most remarkable aspect of the geometric representation. The structural components reveal themselves in Fig. 8.13 by two well-separated spines representing both the Northern (red) and the Southern (blue) dialect groups of the entire language, which fork from the central part of the Island (the dialects spoken in the central part are green colored). It is remarkable that all Malagasy dialects belong to a single plain orthogonal to the data trait of the *Maanyan* language (q_2). The plain of Malagasy dialects is attested by the sharp distribution of the language points in Cartesian coordinates along the data trait q_2. It is important to mention that although the language point of *Antandroy* (*Ambovombe*) is located on the same plain as the rest of Malagasy dialects, it is situated far away from them and obviously belongs to neither of the dialect branches.

Due to the striking central symmetry of the representation, it is natural to describe the positions of language points l_i with the use of spherical coordinates (8.42) rather than the Cartesian system. The distribution of language points over the azimuth angles φ supports the main conclusion following from the UPGMA method on a triple division of the main group of Malagasy dialects (if *Antandroy* is excluded), as the three groups – Northern (red), Southern (blue), and Central (green) – are clearly evident from the representation shown in Fig. 8.14. However, the classification of dialects over those groups obtained by the SCA method differs from that one of the UPGMA method. Namely, the *Vangaindrano*, *Farafangana*, and *Ambatontrazaka*

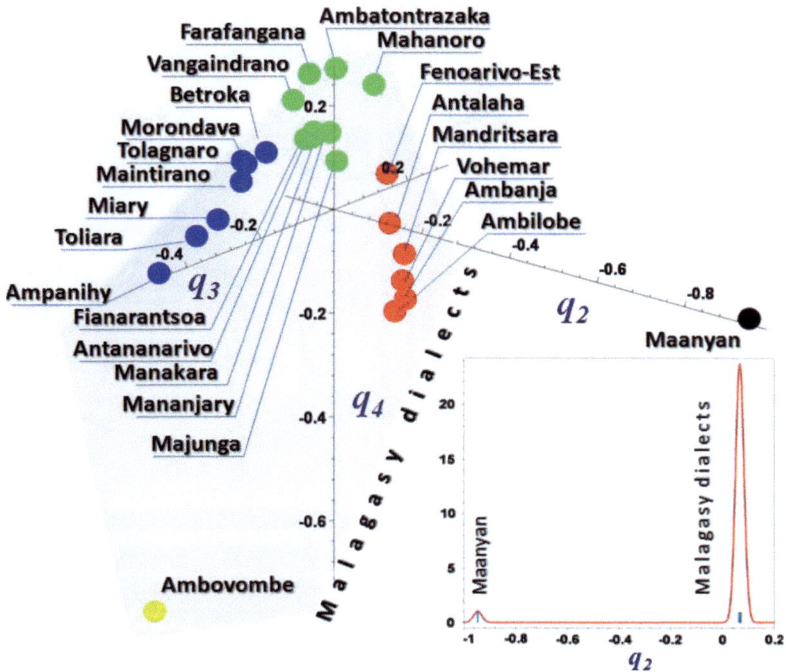

Fig. 8.13 The three-dimensional geometric representation of the Malagasy dialects and the *Maanyan* language in space of the major data traits (q_2, q_3, q_4) shows the remarkable geographic patterning: the Northern (*red*) and the Southern (*blue*) dialect groups of the entire language, which fork from the central part of the Island (the dialects spoken in the central part are green colored). In the outline, the kernel density estimate of the distribution of the q_2 coordinates, together with the absolute data frequencies, indicate that all Malagasy dialects belong to a single plain orthogonal to the data trait of the *Maanyan* language (q_2)

dialects spoken in the Central part of the Island are rather classified into the Southern dialects (blue) (as their azimuthal coordinates fit better the general trend of the Southern group) than to the Central group. The *Mahanoro* dialect is rather attested to the Northern group (red), as being best fitted to the Northern group azimuth angle. The rest five dialects of the Central group (green colored) are characterized by the azimuth angles close to a bisector ($\varphi = 0$).

It seems natural to suggest that namely the Central group of dialects was the origin of the diverging pathways of language evolution on Madagascar. As *Fianarantsoa* and *Antananarivo* are spoken in the depths of the Island, we can hypothesize that the landing place of the Austronesian colonists was probably close to *Manakara, Manajary* on the East coast, or to *Majunga*, on the West coast. This result was strongly confirmed by the analysis performed accordingly to the method of Wichmann et al. (2010).

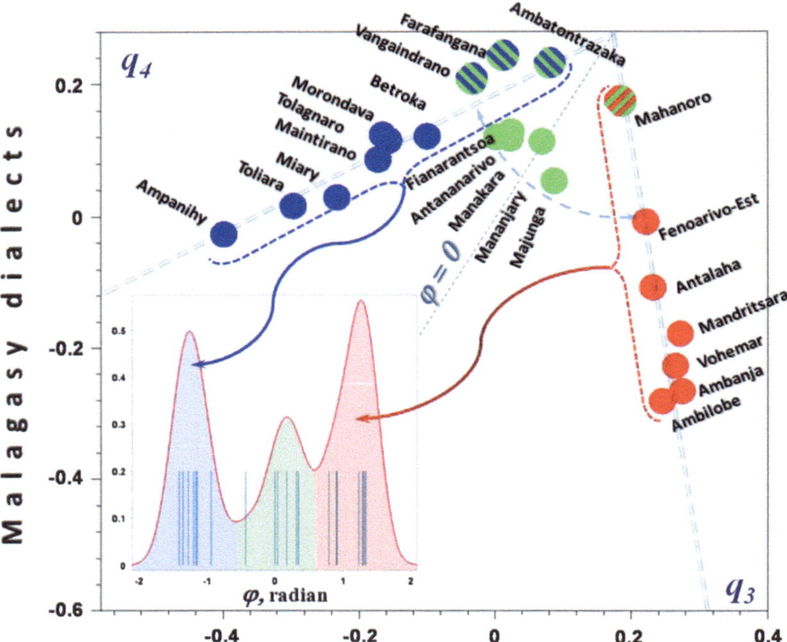

Fig. 8.14 The plain of Malagasy dialects (q_3, q_4); *Antandroy* (*Ambovombe*) is excluded. The kernel density estimate of the distribution over azimuth angles, together with the absolute data frequencies, allow to classify the rest of Malagasy dialects into the three groups: Northern (*red*), Southern (*blue*), and Central (*green*)

The distribution of Malagasy dialects along the radial direction is remarkably heterogeneous indicating that the rate of changes in the orthographic realizations of Swadesh's vocabulary was anything but constant violating the main assumption of the UPGMA method. Being ranked within the own dialect group and then plotted against their expected values under the normal probability distribution, the radial coordinates of Malagasy dialects in the geometrical representation of Fig. 8.13 show very good agreement with univariate normality with the value of variance $\sigma^2 = 0.99 \times 10^{-3}$, as seen from the normal probability plot Fig. 8.15 fitting the distances r of dialect points from the center of mass $(0,0)$ to univariate normality. The normality of distribution of dialect points over the radial coordinates can be justified by taking on that for a long time the divergence of representations of the core vocabulary was a gradual change accumulation process into which many small, independent innovations had emerged and contributed additively to the outgrowth of new dialects.

The univariate normal distribution (see Fig. 8.15) implies a homogeneous diffusion time evolution in one dimension, under which variance $\sigma^2 \propto t$ grows linearly with time. The locations of dialect points could not be distributed normally if in the long run the value of variance σ^2 did not grow with time at an approximately

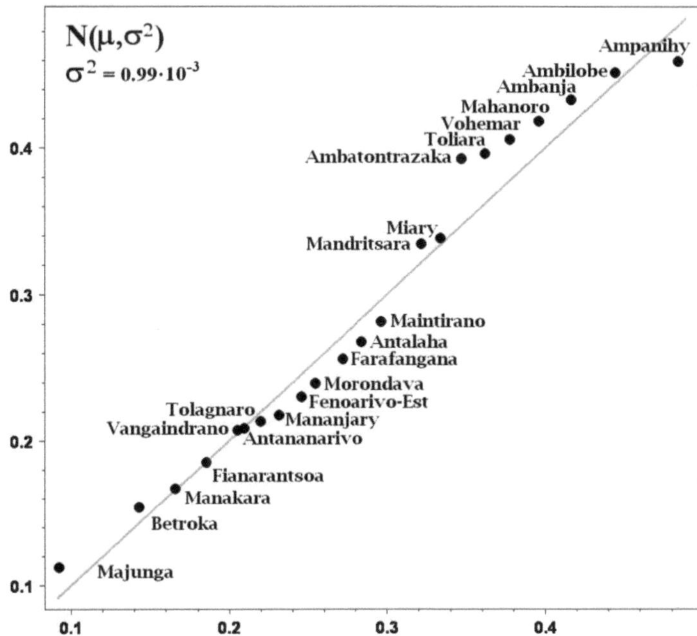

Fig. 8.15 The normal probability plot fitting the distances r of dialect points from the center of mass $(0, 0)$ to univariate normality. The data points were ranked and then plotted against their expected values under normality, so that departures from linearity signify departures from normality. The value of variance for the Malagasy dialects (excluding *Antandroy*) is $\sigma^2 = 0.99 \times 10^{-3}$

constant rate. Let us note that the constant increment rates of variance of radial positions of languages in the geometrical representation (Fig. 8.14) has nothing to do with the traditional glottochronological assumption about the steady borrowing rates of cognates standard for the UPGMA method. Although the lack of documented historical events makes the direct calibration of the method difficult, we may suggest (following Blanchard et al. 2010) that variance evaluated over Swadesh vocabulary forges ahead approximately at the same pace uniformly for all human societies involved in trading and exchange forming a singular cultural continuum. The time–age ratio $t/\sigma^2 = (1.367 \pm 0.002) \cdot 10^6$. deduced from the previous chronological estimates for the Indo-European language family returns $t = 1,353$ years if applied to the Malagasy dialects suggesting that landing in Madagascar was around 650 AD.

We also consider Malagasy together with other languages of the Greater Barito East group. In Fig. 8.16, we have shown the angular sections of the three-dimensional geometric representations of 23 Malagasy dialects, together with the *Malay* language (left) and the *Maanyan* language (right). The geometric patterns of relations (shown in the plane of azimuth (φ) and zenith (θ) angles, see (8.42)) between the Malagasy dialects and the two of the Austronesian languages are strikingly similar indicating that the roles of them for the entire language evolution process

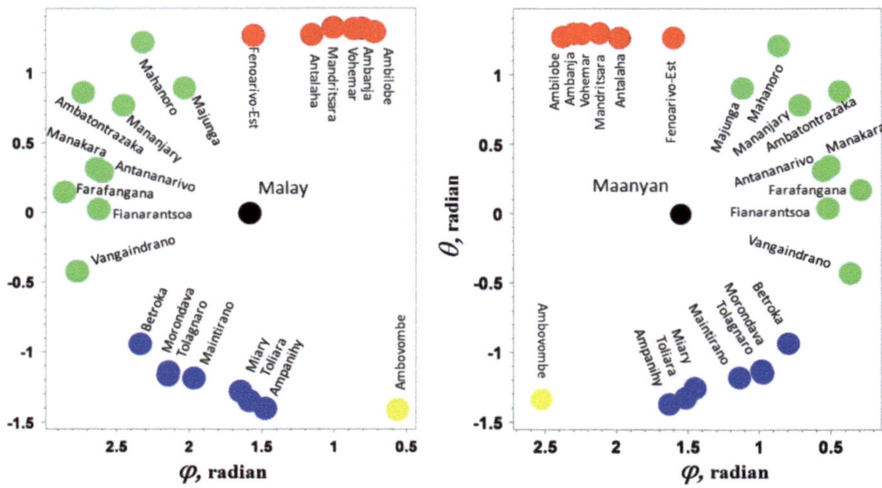

Fig. 8.16 The angular sections of the three-dimensional geometric representations of 23 Malagasy dialects, together with the *Malay* language (*left*) and the *Maanyan* language (*right*) are mirror symmetric with respect to the azimuth angle φ

on Madagascar was alike. Interestingly, the left and right figures shown in Fig. 8.16 are mirror symmetric with respect to the azimuth angle φ.

8.4.11 Austronesian Languages Riding an Express Train

The distributions of languages spoken in the islands of East Polynesia and of the Atayal language groups in Taiwan over the radial coordinate from the center of the geometric representation shown in Fig. 8.9 break from normality, so that the general 'diffusive scenario' of language evolution used previously for either of the chronological estimates is obviously inapplicable to them. For all purposes, the evolution of these extreme language subgroups cannot be viewed as driven by independent, petty events. Although the languages spoken in Remote Oceania clearly fit the general trait of the entire MP-branch, they seem to evolve without extensive contacts with Melanesian populations, perhaps because of a rapid movement of the ancestors of the Polynesians from South-East Asia as suggested by the 'express train' model (Diamond 1988) consistent with the multiple evidences on comparatively reduced genetic variations among human groups in Remote Oceania (Lum et al. 2002; Kayser et al. 2006; Friedländer et al. 2008).

In order to obtain reasonable chronological estimates, an alternative mechanism on evolutionary dynamics of the extreme language subgroups in space of traits of the AU language family should be reckoned with. The simplest 'adiabatic' model entails that no words had been transferred to or from the languages riding the express train to Polynesia, so that the lexical distance among words of the most distanced languages tends to increase primarily due to random permutations, deletions or substitutions of phonemes in the words of their ancestor language. Under such circumstances the radial coordinate of a remote language riding an 'express train' in the geometric representation (see Fig. 8.9) effectively quantifies the duration of its relative isolation from the Austronesian cultural continuum. Both of the early colonization of a secluded island by Melanesian seafarers and of the ahead of time migration of the indigenous people of Taiwan to highlands can be discerned by the excessively large values of the radial coordinates r of their languages. In Fig. 8.17, we have presented the log-linear plot, in which the radial coordinates of remote languages were ranked and then plotted against their expected values under the exponential distribution (shown by the dash-dotted line in Fig. 8.17).

The radial coordinates of the languages at the distant margins of the AU family diagram shown in Fig. 8.17 may be deduced as evolving in accordance with the simple differential equation

$$\dot{r} = ar \tag{8.44}$$

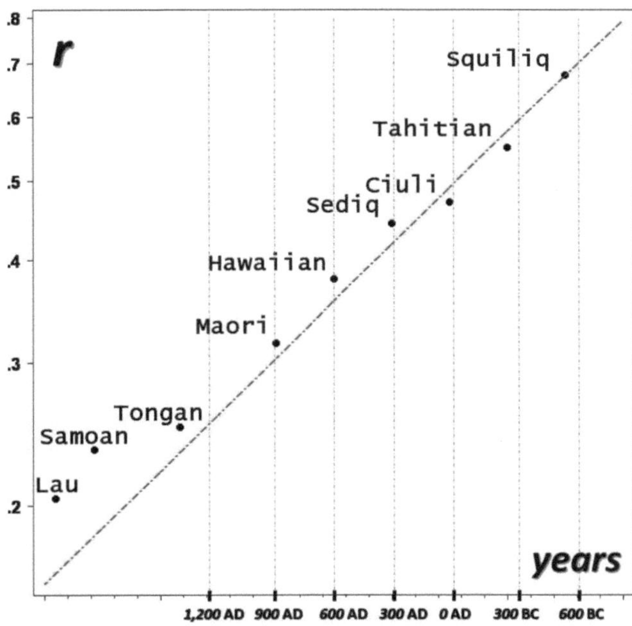

Fig. 8.17 The log-linear plot fitting the distances r to remote languages riding an 'express train' in the geometric representation (see Fig. 8.9) to an exponential distribution. The radial coordinates of the languages were ranked and then plotted against their expected values under the exponential distribution. As usual, the departures from linearity signify departures from the tested distribution (given by the dash-dotted line)

where $\dot{r}$ means the derivative of r with respect to isolation time, and $a > 0$ is some constant quantifying the rate of radial motion of a language riding the express train in space of the major traits of the AU family. The suggested model of language taxonomy evolution is conceived by that while the contact borrowings are improbable the orthographic realizations of Swadesh's meanings would accumulate emergent variations in spellings, so that the radial coordinate of a remote language can formally grow unboundedly with isolation time.

A simple equation mathematically similar to (8.44) has been proposed by M. Swadesh (Swadesh 1952) in order to describe the change of cognates in time, in the framework of the glottochronological approach. In our previous work (Petroni and Serva 2008), another similar equation has been suggested for the purpose of modeling the time evolution of normalized edit distances between languages. We have to emphasize that the statistical model (8.44) can be hardly related to them both, as the radial coordinate r in the geometrical representations of language families described above does not have a direct relation to neither the percentage of cognates, nor the edit distance.

Then the relative dates estimating the duration of relative isolation of the distant languages from the extensive contacts with other Austronesian languages can be derived basing on the assumption (8.44) as

$$t_1 - t_2 = \frac{1}{a} \cdot \ln \frac{r_1}{r_2} \tag{8.45}$$

where $r_2 > r_1$ are the radial coordinates of the languages from the center of the sample diagram shown in Fig. 8.9.

Tahiti located in the archipelago of Society Islands is the farmost point in the geometric representation of the Austronesian family and the foremost Austronesian settlement in the Remote Oceania attested as early as 300 BC (Kirch 2000), the date we placed the incipience of the Tahitian society. According to many archaeological reconstructions (Kirch 2000; Anderson and Sinoto 2002; Hurles et al. 2003), descendants from West Polynesia had spread through East Polynesian archipelagos and settled in Hawaii by 600 AD and in New Zealand by 1000 AD testifying the earliest outset dates for the related languages. It is worth mentioning that all stride times between the offsets of these three Polynesian languages hold consistently the same rate

$$a = (4.27 \pm 0.01) \cdot 10^{-4} \tag{8.46}$$

affirming the validity of the 'adiabatic' conjecture described above and allowing us to assign the estimated dates to the marks of the horizontal axis of the timing diagram presented in Fig. 8.17. The language divergence among Atayal people distributed throughout an area of rich topographical complexity is neatly organized by the myths of origin place, consanguine clans, and geographical barriers that have lead to the formation of a unique concept of ethnicity remarkable for such a geographically small region as Taiwan. The complexity of the Atayal ethnic system and the difficulty of defining the ethnic borders hindered the classification of the

Atayal regional groups and their dialects which has been continuously modified throughout the last century.

In our work, we follow the traditional classification Utsurikawa 1935 of the Atayal group into three branches based on their places of origin: Sediq (Sedek), Ciuli (Tseole) Atayal, and Squiliq (Sekilek) Atayal. In account with the standard lexicostatistic arguments (Li 1983), the Sediq dialect subgroup could have split off from the rest of the Atayal groups about 1,600 years ago, as both the branches share up to a half of the cognates in the 200 words of basic vocabulary. This estimated date is very tentative in nature and calls for a thorough crosschecking. The Atayal people had been recognized as they had started to disperse to the northern part of Taiwan around 1750 AD (Li 2001). Being formed as the isolated dialect subgroups in island interiors, they showed the greatest diversity in race, culture, and social relations and sometimes considered each other as enemies and prime head hunting targets.

Given the same rate of random phonetic changes as derived for the Polynesian languages, the 'adiabatic' model of language evolution returns the stride times of 1,000 years between the Sediq dialect subgroup and Squiliq Atayal and of 860 years between the Ciuli and Squiliq Atayal languages. Consistently, Sediq is estimated to have branched off from the other Atayal languages 140 years before the main Atayal group split into two. The Squiliq subgroup had been attested during the latest migration of Atayal people, as late as 1820 AD (Li 2001). Perhaps, a comprehensive study of the Atayal dialects by their symmetry can shed light on the origins of the Atayal ethnic system and its history.

We have presented the new paradigm for the language phylogeny based on the analysis of geometric representations of the major traits on language data. The proposed method is fully automated.

On the encoding stage, we evaluated the lexical distances between languages by means of the mean normalized edit distances between the orthographic realizations of Swadesh's meanings. Then, we considered an infinite sequential process of language classification described by random walks on the matrix of lexical distances. As a result, the relationships between languages belonging to one and the same language family are translated into distances and angles, in multidimensional Euclidean space. The derived geometric representations of language taxonomy are used in order to test the various statistical hypotheses about the evolution of languages.

Our method allows for making accurate inferences on the most significant events of human history by tracking changes in language families through time. Computational simplicity of the proposed method based primarily on linear algebra is its crucial advantage over previous approaches to the computational linguistic phylogeny that makes it an invaluable tool for the automatic analysis of both the languages and the large document data sets that helps to infer on relations between them in the context of human history.

8.5 Markov Chain Analysis of Musical Dice Games

The Internet based economy calls for robust recommendation engines for appreciating and predicting the musical taste of customers, since any improvement in the accuracy of predictions might have an immense economic value. From one hand, studies of Markov chains aggregating pitches in musical pieces might provide a neat way to efficient algorithms for identifying musical features important for a listener. From another hand, the analysis of weighted directed graphs correspondent to the time-irreversible random walks defined on a finite set of states (pitches) belonging to a cyclic group, under the assumption of octave equivalency, (the cyclic group is $\mathbb{Z}/12\,\mathbb{Z}$, as the perceived fundamental frequency of a sound $f = 440\,\mathrm{Hz} \times 2^{n/12}, \quad n \in \mathbb{Z}$) is a daunting task for the contemporary theory of networks being therefore of a special theoretical interest.

Interactions between humans via speech and music constitute the unifying theme of research in modern communication technologies. As with music, speech and written language also have the sets of rules (crucial for establishing efficient communication) that determine which particular combinations of sounds and letters may or may not be produced. However, while communications by the spoken and written forms of human languages have been paid much attention from the very onset of information theory (Shannon 1948, 1951), not very much is known about the relevant information aspects of music (Wolfe 2002). Although we use the acoustic channel in both music and speech, the acoustical and structural features we implement to encode and perceive the signals in music and speech are dramatically different, as "speech is communication of world view as the intellection of reality while music is communication of world view as the feeling of reality" (Seeger 1971).

A Markov chain model allows to appraise tonal music as a generalized communication process, in which a composer sends a message transmitted by a performer to a listener. The applications of Markov chains in music have a long history dating back to 1757 (see the next subsection). In modern times, the first computer program that used Markov chains to compose a string quartet (*Illiac Suite*) was developed in 1957 (Hiller and Isaacson 1959). The further developments of computer musical data formats also called for a formalization of musical events either in terms of *frequency, duration, and intensity* (Xenakis 1971), or in terms of *pitch, duration, amplitude, instrument* (Jones 1981) – in the both cases, the musical events were naturally treated as the states of hierarchical Markov chains. Although much work using Markov chains for compositional purposes have been done in so far, including a real–time interactive control of the Markov chains (Zicarelli 1987), less researches have been focused on the detailed analysis of musical compositions by means of Markov chains. The reason for such a discrepancy is quite simple: Markov chains encoding musical compositions might be not ergodic thus being difficult to analyze.

Although stochastic techniques have the certain advantage over other compositional approaches polishing computer–generated music that might sound rather artificial otherwise, there has been some criticism of using Markov chains to create music. In Levitt (1993) and Moorer (1993), it has been pointed that the reliance on random note selection tends to obscure the practices of music compositions and thus can not be considered as consistent with a composition of high quality music. Perhaps, partially in response to such a criticism, in Marom (1997) and Franz (1998), Markov chains have been used as tools for the jazz improvisation analysis. Indeed, a melodic pattern generated randomly in accordance to a given aggregated transition matrix would hardly resemble the original composition, as being just a particular random realization over an ensemble of statistically identical musical pieces which we call the *Musical dice game* (MDG) throughout the section.

Here, we report some results on the Markov chain analysis of MDG encoded by the transition matrices between pitches in the MIDI representations of the 804 musical compositions attributed to 29 composers: J.S. Bach (371), L.V. Beethoven (58), A.Berg (7), J. Brahms (8), D. Buxtehude (3), F. Chopin (26), C. Debussy (26), G. Fauré (5), C. Franck (7), G.F. Händel (45), F. Liszt (4), F. Mendelssohn Bartholdi (19), C. Monteverdi (13), W.A. Mozart (51), J. Pachelbel (2), S. Rachmaninoff (4), C. Saint-Saëns (2), E. Satie (3), A. Schönberg (2), F. Schubert (55), R. Schumann (30), A. Scriabin (7), D. Shostakovitch (12), J. Strauss (2), I. Stravinsky (5), P. Tchaikovsky (5), J. Titelouze (20), A. Vivaldi (4), R. Wagner (8). The MIDI representations of many musical pieces are freely available on the Web (Mutopia).

8.5.1   *Musical Dice Game as a Markov Chain*

A system for using dice to compose music randomly, without having to know neither the techniques of composition, nor the rules of harmony, named *Musikalisches Würfelspiel* (*Musical dice game*) had become quite popular throughout Western Europe in the eighteenth century (Noguchi 1996). Depending upon the results of dice throws, the certain pre-composed bars of music were patched together resulting in different, but similar, musical pieces. "*The Ever Ready Composer of Polonaises and Minuets*" was devised by Ph. Kirnberger, as early as in 1757. The famous chance music machine attributed to W.A. Mozart (K 516f) consisted of numerous two-bar fragments of music named after the different letters of the Latin alphabet and destined to be combined together either at random, or following an anagram of your beloved had been known since 1787.

Contrary to the alphabets used in human languages, the sets of pitches underlying the different musical compositions can be very distinct and may not overlap (even under chromatic transposition). The cardinality R of the set of pitches used to compose a piece is one of its important characteristics, as the melody obviously centers around a few given notes when this cardinality is small, but calling for a wide range, from low to high pitches, otherwise. The cardinality R changes from piece to piece demonstrating a tendency of slow growth, with the length of arrangement N. In Fig. 8.18, we have sketched how the number of different pitches R used to compose a piece depends upon its size N. The data collected over 610

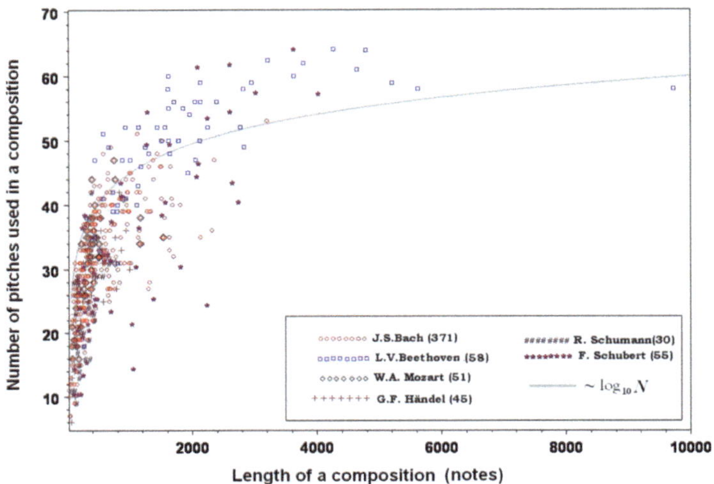

Fig. 8.18 The number of different pitches used in a composition grows approximately logarithmically with its size. The data have been collected over 371 compositions of J.S. Bach, 58 various compositions of L.V. Beethoven, 55 compositions of F. Schubert, 51 compositions of W.A. Mozart, 45 compositions of G.F. Händel, and 30 compositions of R. Schumann

pieces created by the six classical music composers show that the growth can be well approximated by a logarithmic curve indicating that

$$R \simeq \log N \tag{8.47}$$

can be used as the simplest parameter assessing complexity of a musical composition.

In the MDG, we consider a note as an elementary event providing a natural discretization of musical phenomena that facilitate their performance and analysis. Namely, given the entire keyboard $\mathcal{K}$ of 128 notes (standard for the MIDI representations of music) corresponding to a pitch range of 10.5 octaves, each divided into 12 semitones, we regard a note as a discrete *random variable X* that maps the musical event to a value of a R-set of pitches $\mathcal{P} = \{x_1, \ldots, x_R\} \subseteq \mathcal{K}$. In the musical dice game, a piece is generated by patching notes X_t taking values from the set of pitches $\mathcal{P}$ that sound good together into a temporal sequence $\{X_t\}_{t \geq 1}$. Herewith, two consecutive notes, in which the second pitch is a *harmonic* of the first one are considered to be pleasing to the ear, and therefore can be patched to the sequence. Harmony is based on consonance, a concept whose definition changes permanently in musical history. Two or more notes may sound consonant for various reasons such as luck of perceptual roughness, spectral similarity of the sequence to a harmonic series, familiarity of the sound combination in contemporary musical contexts, and eventually for a personal taste, as there are consonant and dissonant harmonies, both of which are pleasing to the ears of some and not others. A detailed statistical analysis of subtle harmony conveyed by melodic lines in tonal music certainly calls for the complicated stochastic models, in which successive notes in the sequence $\{X_t\}_{t \geq 1}$ are not chosen independently, but their probabilities depend on preceding notes. In the general case, a set of n-note probabilities

$$\Pr\left[X_{t+1} = x \mid X_t = y, X_{t-1} = z, \ldots, X_{t-n} = \upsilon\right]$$

might be required to insure the resemblance of the musical dice games to the original compositions. However, it is rather difficult to decide *a priori* upon the enough memory depth n in the stochastic models required to compare reliably the pieces of tonal and atonal music created by different composers, with various purposes, in different epochs, for diverse musical instruments subjected to the dissimilar tuning techniques. Under such circumstances, it is mandatory to identify some meaningful blocks of musical information and to detect the hierarchical tonality (basic for perception of harmony in Western music Dahlhaus 2007) in a simplified statistical model, as the first step of statistical analysis. For this purpose, in the present work, we neglect possible statistical influences extending over than the only preceding note and limit our analysis to the simplest time – homogeneous model called *Markov chain* (Markov 1906),

$$\Pr\left[X_{t+1} = x \mid X_t = y, X_{t-1} = z, \ldots\right] = \Pr\left[X_{t+1} = x \mid X_t = y\right] \\ = T_{yx}, \tag{8.48}$$

where the elements of the stochastic transition matrix T_{yx},

$$\sum_{x \in \mathcal{P}} T_{yx} = 1,$$

weights the chance of a pitch x going directly to another pitch y independently of time.

It is worth mentioning that the model (8.48) obviously does not impose a severe limitation on melodic variability, since there are many possible combinations of notes considered consonant, as sharing some harmonics and making a pleasant sound together. The relations between notes in (8.48) are rather described in terms of probabilities and expected numbers of random steps than by physical time. Thus the actual length N of a composition is formally put $N \to \infty$, or as long as you keep rolling the dice. Markov chains are widely used in algorithmic music composition, as being a standard tool, in music mix and production software (Reaktor 2005; Reason 2007; Ableton 2009).

8.5.2 Encoding of a Discrete Model of Music (MIDI) into a Transition Matrix

While analyzing the statistical structure of musical pieces, we used the MIDI representations providing a computer readable *discrete time model* of music by a sequence of the 'note' events, `note_on` and `note_off`:

In the MIDI representation, each note event (like that one shown in Table 8.1) is characterized by the four variables: 'time', 'channel', 'note', and 'velocity'. Motivated by the logarithmic pitch perception in humans, music theorists represent pitches using a numerical scale based on the logarithm of fundamental frequency,

$$\text{note} = 69 + 12 \times \log_2 \left(\frac{f}{440\,\text{Hz}} \right). \tag{8.49}$$

The resulting *linear* pitch space in which octaves have size 12, semitones have size 1, and the number 69 is assigned to the note A4. The linear distance in the pitch space corresponds to the musical distance as measured in psychological experiments and allows a MIDI file to have a specific value of discreteness 'ticks/quarter' indicating the number of 'ticks' that make up a quarter note. The value of 'time' then gives the number of 'ticks' between two consequent note events. In the example given in Table 8.1, the event of C4 starts after 192 'ticks' have passed. The 'channel' indicate one of 16 channels (0–15) this event may belong to.

Notes are not encoded by their names like C or A. Instead, the harmonic scale is mapped onto numbers from 0 to 127 with chromatic steps. For instance, the identification number 60 corresponds to the C4, in musical notation. Then, note number 61 is C4#, 62 is D4 etc. (see Table 8.2 for some octaves and their MIDI note ID numbers)

Finally, the 'velocity' (0–127) describes the strength with which the note is played. As MIDI files contain all musically relevant data, it is possible to determine the probabilities of getting from one note to another for all notes in a musical composition by analyzing its MIDI file with a computer program. To get transition matrices (8.48) for tonal sequences, we need only 'time', 'channel', and 'note' to be considered.

Table 8.1 MIDI events for the note C4

Event type	Time	Channel	Note	Velocity
note_on	192	0	60	127
note_off	192	0	60	64

Table 8.2 MIDI note ID numbers corresponding to musical notation

Octave	C	C#	D	D#	E	F	F#	G	G#	A	A#	B
3	48	49	50	51	52	53	54	55	56	57	58	59
4	60	61	62	63	64	65	66	67	68	69	70	71
5	72	73	74	75	76	77	78	79	80	81	82	83

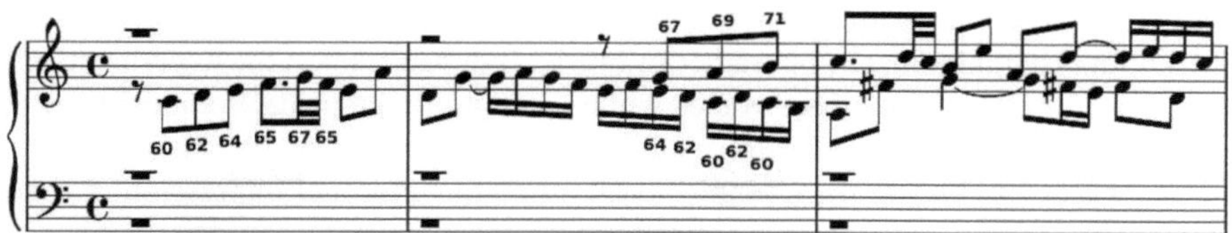

Fig. 8.19 The first three bars from the fugue of BWV846. Also shown are MIDI note numbers

Table 8.3 MIDI and MIDI::Score data from the beginning of fugue I BWV846

"MIDI"				"MIDI::Score format"				
Event type	Time	Ch	Note	Event type	Time	Dur	Ch	Note
note_on	192	0	60	note	192	192	0	60
note_off	192	0	60					
note_on	0	0	62	note	384	192	0	62
note_off	192	0	62					
note_on	0	0	64	note	576	192	0	64
note_off	192	0	64					

The MIDI files of 804 musical compositions were processed by a program written in `Perl`; the MIDI parsing was done using the module `Perl::MIDI` (Perl MIDI), which allowed the conversion of the MIDI data into a more convenient form called `MIDI::Score` where each two consequent `note_on` and `note_off` events are combined to a single `note` event. Each `note` event contains an absolute `time`, the starting time of the event, and a `duration` which gives the duration of the event in ticks.

To give an example of the process of getting to a transition matrix from a musical score, we consider the first bars of the fugue from `BWV846` of J.S. Bach shown in Fig. 8.19. The numbers below the first notes in Fig. 8.19 indicate the corresponding MIDI ID note numbers. In Table 8.3, we show the representation of the first three notes in MIDI and in `MIDI::Score` format. Here, the value of velocity is omitted.

For the first notes shown in Fig. 8.19, the definition of a transition is easy as there is only one voice. In particular, from Table 8.3, we can conclude that there would be the consequent transitions $60 \to 62$ and $62 \to 64$. However, like most musical pieces, this Fugue then contains several voices that play simultaneously, so that an additional convention is required to define a transition from note to note.

In the middle of the second bar shown in Fig. 8.19, a second voice is starting. Some note events starting from there are given in Table 8.4 in `MIDI::Score` form. From Table 8.4, it is clear that for a MIDI representation it is not necessary to put the upper voice into a different channel than that of the lower voice. In the example shown in Table 8.4, the notes 67 and 64 both start at time 2,496. As note 64 has a duration of 96 ticks, it is obvious that note 62 at time 2,592 belongs to the same voice as note 64. However, for the notes 69 and 60 starting at 2,688, it is unclear to which voice each note belongs to, and how they might be encoded into a transition matrix. It is important to note that such an ambiguity is not a problem of MIDI representation itself, but rather of music. It depends upon the experience of a listener how she distinguishes voices while listening to a musical composition that contains several simultaneous voices. Even if the musical score explicitly separates those voices by placing them atop of each other, our personal impression of them might not coincide with that one notated, rather arising from live audio mixing of all simultaneous voices during the performance. Thus, to get transition matrices from MIDI files, we have to answer the following important question: "Which transitions between which note events have to be accounted?"

In our approach, we sort note events ascending by time and channel. By surfing over the list of events, a transition between two subsequent occurrences is accounted when the moment of time of the second event is greater than that of the first one. When several events occur simultaneously, we give a priority to the event belonging to a small number channel. Let us emphasize that under the used method not all possible transitions between note events contribute into the transition matrix.

For example, let us consider the notes shown in Fig. 8.20; their list of events is given in the adjacent table. The resulting transitions accounted in the matrix would

Table 8.4 MIDI::Score data from the middle of the second bar of fugue I BWV846 where the second voice starts playing. The note names and the voices of the events are also shown

Time	Dur	Ch	Note	Name	Voice
2496	192	0	67	G4	upper
2496	96	0	64	E4	lower
2592	96	0	62	D4	lower
2688	192	0	69	A4	upper
2688	96	0	60	C4	lower
2784	96	0	62	D4	lower
2880	192	0	71	B4	upper
2880	96	0	60	C4	lower

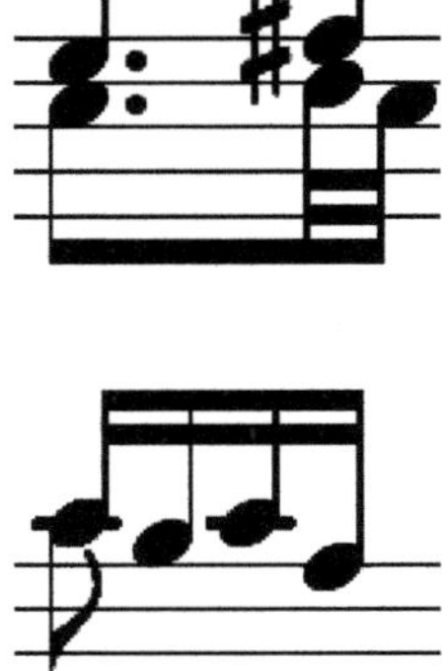

Time	Dur	Ch	Note	Name	
13056	288	0	76	(E5)	*
13056	288	0	72	(C5)	
13056	192	1	60	(C4)	
13152	96	1	59	(B3)	*
13248	96	1	60	(C4)	*
13344	96	0	78	(F#5)	*
13344	48	0	74	(D5)	
13344	96	1	57	(A3)	
13392	48	0	72	(C5)	*

Fig. 8.20 Example from our fugue, Mutopia 0000

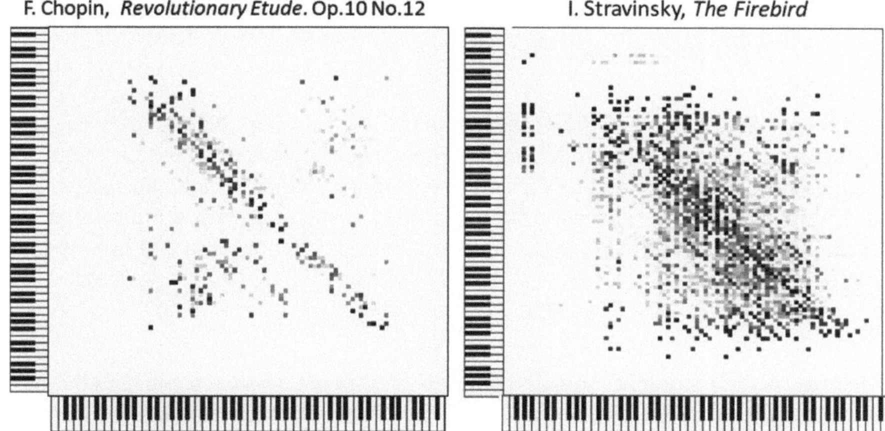

Fig. 8.21 Transition matrices for the MDG based on the F.Chopin "Revolutionary Etude" (Op.10, No 12) (*left*) and the I. Stravinsky "The Fire-bird" suite (*right*)

be those between events marked with '*': $76 \to 59$, $59 \to 60$, $60 \to 78$, $78 \to 72$. Note events with small channel values are favored over those with higher values. For simultaneous note events occurring in the same channel, only the first one is considered that mostly means the topmost voice, in musical notation.

We believe that the encoding method we use is quite efficient for unveiling the individual melodic lines and identifying a creative character of a composer from musical compositions because of the appearance of the resulting transition matrices. Those matrices generated with respect to the chosen encoding method look differently, from piece to piece and from composer to composer (see the examples shown in Fig. 8.21). However, if we were treated each voice in a musical composition separately (the transitions of the upper voice and those of the lower voice might be accounted independently while computing the probabilistic vector forming a row of the transition matrix), the transition matrices were clearly dominated by a region along the main diagonal, similarly for all compositions. The last but not least remark upon the second algorithm is that although the statistics of transitions in those matrices is much more rich, they formally account for a number of odd transitions between the simultaneous voices definitely inconsistent with the main melody. It is important to mention that no matter which encoding method is used the resulting transition matrices appear to be essentially not symmetric: if $T_{xy} > 0$, for some x, y, it might be that $T_{yx} = 0$. A musical composition can be represented by a weighted directed graph, in which vertices are associated with pitches and directed edges connecting them are weighted accordingly to the probabilities of the immediate transitions between those pitches. Markov's chains determining random walks on such graphs are not ergodic: it may be impossible to go from every note to every other note following the score of the musical piece.

8.5.3 Musical Dice Game as a Generalized Communication Process

A musical piece generated as an output of the musical dice game (8.48) is a stochastic process which can be encoded by a sequence of independent and identically-distributed random variables representing notes which can take values of different pitches. To measure the uncertainty associated with a pitch in such a stochastic process, we can use the Shannon entropy (Schürmann and Grassberger 1996),

$$H = -\sum_{x \in \mathcal{P}} \pi_x \log_R \pi_x \tag{8.50}$$

where π_x is the probability to find the note $x \in \mathcal{P}$ in the musical score, and the base of the logarithm is $R = |\mathcal{P}|$. Since the entropy of a musical piece defined by (8.50) is affected by the number of used pitches R, the parameter of information redundancy, $\mathcal{R} = 1 - H/\max H$, $\max H = \log R$, where $\max H$ is the theoretical maximum entropy, might be used for comparing different musical compositions. Accordingly to information theory (Cover and Thomas 1991), redundancy quantifies predictability of a pitch in the piece, as being a natural counterpart of entropy.

As we have mentioned above, a Markov chain encoding the musical dice game might be not ergodic, and therefore the probability to find a pitch in the musical score cannot be found simply as the entry in the left eigenvector (which now might be non-unique) of the transition matrix $\mathbf{T}$ belonging to the maximal eigenvalue $\mu = 1$. In order to find the probability of observing the note in the musical score, we can use the method of group generalized inverse (Meyer 1975, 1982) that might be applied for analyzing every Markov chain regardless of its structure. As the Laplace operator corresponding to the Markov chain (8.48), $\mathbf{L} = \mathbf{1} - \mathbf{T}$, where $\mathbf{1}$ is a unit matrix, is always a member of a multiplicative matrix group, it always possesses a group inverse $\mathbf{L}^{\sharp}$, a special case of the Drazin generalized inverse (Drazin 1958; Ben-Israel and Greville 2003; Meyer 1975) satisfying the Erdélyi conditions (Erdelyi 1967):

$$\mathbf{L}\mathbf{L}^{\sharp}\mathbf{L} = \mathbf{L}, \quad \mathbf{L}^{\sharp}\mathbf{L}\mathbf{L}^{\sharp} = \mathbf{L}^{\sharp}, \quad \left[\mathbf{L}, \mathbf{L}^{\sharp}\right] = 0, \tag{8.51}$$

where $[\mathbf{A}, \mathbf{B}] = \mathbf{A}\mathbf{B} - \mathbf{B}\mathbf{A}$ denotes the commutator of the two matrices. The methods for computing the group generalized inverse for matrices of rank($\mathbf{L}$) $= N - 1$ by considering the eigenprojection of the matrix $\mathbf{L}$ corresponding to the smallest eigenvalue $\lambda_1 = 0$ have been developed in Campbell et al. (1976), Hartwig (1976) and Agaev and Chebotarev (2002),

$$\mathbf{L}^{\sharp} = (\mathbf{L} + \mathbf{Z})^{-1} - \mathbf{Z}, \quad \mathbf{Z} = \prod_{\lambda_i \neq 0} \left(1 - \frac{1}{\lambda_i}\mathbf{L}\right) \tag{8.52}$$

where the product in the idempotent matrix $\mathbf{Z}$ is taken over all nonzero eigenvalues of $\mathbf{L}$. The role of group inverses (8.51) in the analysis of Markov chains has been discussed in details in Meyer (1975, 1982) and Campbell and Meyer (1979). Here, we only mention that a generalized inverse provides the *unique* best fit (with respect to least squares) approximation for a solution (among infinitely many) to the system of linear equations described by the matrix $\mathbf{L}^{-1}$. Then the correspondent best fit approximation to the stationary distribution can be calculated as

$$\pi_{x_i} = \left(1 - \mathbf{L}\mathbf{L}^{\sharp}\right)_{x_i x_j} ; \tag{8.53}$$

the rows of (8.53) are all equal to the corresponding components of the vector π. It is important to mention that in the framework of the method of group generalized inverses, we account for all possible sequences of pitches that would arise if we start the musical dice game infinitely many times from the randomly chosen notes. Thus, the inverse value $\pi_{x_i}^{-1}$ calculated in accordance to (8.53), i.e., the recurrence time to the state x_i, would formally be finite even if the state is transient (i.e., for which there is a non-zero probability that we will never return to the state); in the same way, the recurrence time would not equal zero for absorbing states (which are impossible to leave).

Determining the entropy of texts written in a natural language is an important problem of language processing. The entropy of current written and spoken languages (English, Spanish) has been estimated experimentally as ranged from 0.5 to 1.3 bit per character (Shannon 1951; Lin 1973). An approximately even balance (50:50) of entropy and redundancy is supposed as necessary to achieve effective communication in transmitting a message, as it makes easier for humans to perceive information (Lin 1973).

For all MDG we studied, the magnitudes of entropy fluctuate in a range between 0.7 and 1.1 bit per note well fitting with the entropy range of usual languages. It is important to note that pieces involving more pitches appear to be characterized with lower magnitudes of entropy but higher values of redundancy (predictability). In Fig. 8.22, we have presented the statistics of entropy and redundancy vs. the number of pitches through their five-number summaries, for 371 chorales of J.S. Bach (left) and 7 compositions of A. Berg (right). A central line of each box in the box plot (Fig. 8.22) shows the median (not the mean), the value separating the higher half of the data sample from the lower half, that is found by arranging all the observations from lowest value to highest value and picking the middle one. Other lines of the box plot indicate the quartile values which divide the sorted data set into four equal parts, so that each part represents one fourth of the sample. A lower line in each box shows the first quartile, and an upper line shows the third quartile. Two lines extending from the central box of maximal length 3/2 the interquartile range but not extending past the range of the data. The outliers are those points lying outside the extent of the previous elements.

The entropy and redundancy statistics shown in Fig. 8.22 suggests that musical compositions generated as an output of the MDG might contain some repeated patterns, or motives in which certain combinations of notes are more likely to occur than others. In particular, the dramatic increase of redundancy as the range of pitches expands up to 7.5 octaves implies that musical compositions involving many

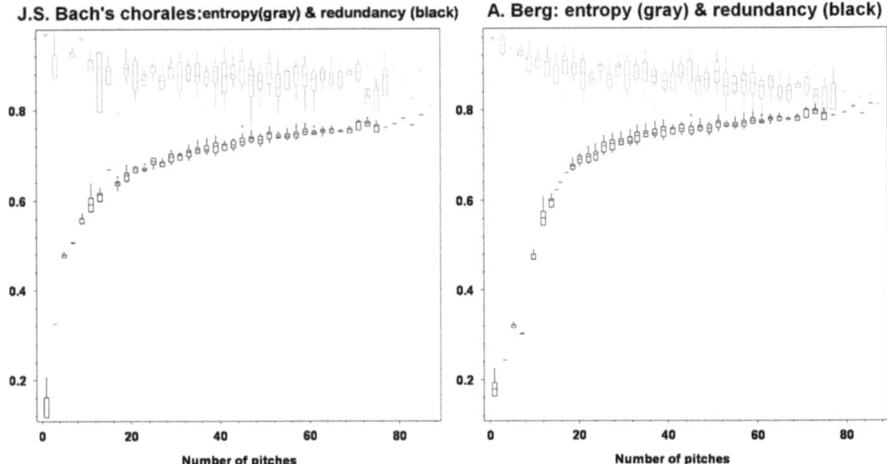

Fig. 8.22 The box plots show the statistic of the magnitudes of entropy and redundancy vs. the number of pitches used in a composition, for the MDG generated over the 371 compositions of J.S. Bach (*left*) and the 7 compositions of A. Berg (*right*). In a box plot, a central line of each box shows the median; a lower line shows the first quartile; an upper line shows the third quartile; two lines extending from the central box of maximal length 3/2 the interquartile range but not extending past the range of the data; eventually, the outliers are those points lying outside the extent of the previous elements

pitches might convey mostly conventional, predictable blocks of information to a listener. However, in contrast to human languages where entropy and redundancy are approximately equally balanced (Lin 1973), in the MDG pieces based on classical music entropy clearly dominates over redundancy. While decoding a musical message requires the listener to invest nearly as much efforts as in everyday decoding of speech, the successful understanding of the composition would call for an experienced listener ready to invest his or her full attention to a communication process that would span across cultures and epochs.

Another possible information measure that can be applied to the analysis of MDG is the past-future mutual information (*complexity*) introduced in studies of the symbolic sequences generated by dynamical systems (Shaw 1984) (see also Cover and Thomas 1991). It estimates the information content of the patches of notes and can be formally derived as the limiting excess of the block entropy

$$H(S^m) = -\sum_{S^m} P(S^m) \log_R P(S^m),$$

in which $P(S^m)$ is the probability to find a patch S^m of m notes, over the m times Shannon entropy H, as the size of the block $m \to \infty$,

$$C = \lim_{m \to \infty} (H(S^m) - mH). \tag{8.54}$$

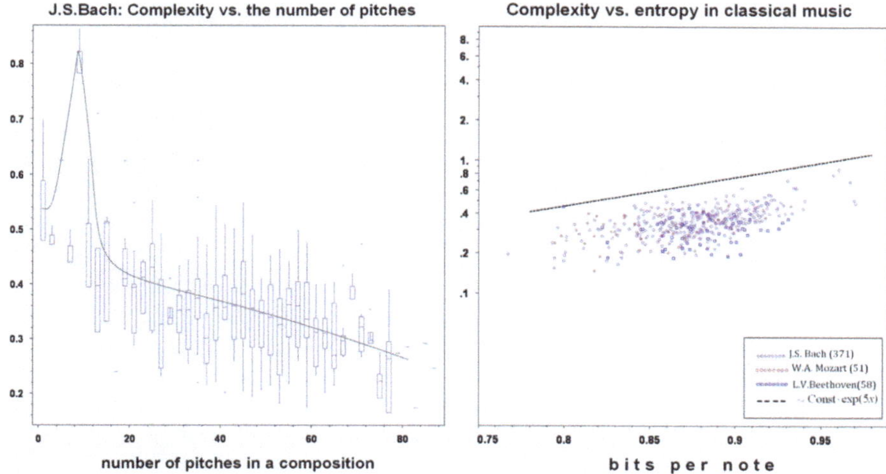

Fig. 8.23 The box plot (*left*) represents complexity (measured by the past-future mutual information) vs. the number of pitches used in a composition, for the MDG inspired by the 371 pieces of
J.S. Bach. The trend (shown by a solid line) stands for the cubic splines interpolating between the
mean values of complexity over the data ranges. The scatter plot of complexity vs. the magnitude
of entropy in 480 pieces generated by the MDG (given in the log-linear scale) suggests that a
strong positive correlation exists between the value of entropy and the logarithm of complexity.
The reference line indicates an exponential growth, in the log-linear scale

Following Li (1991), we use the fact that the transition probability between states
in a Markov chain determined by the matrix (8.48) is independent of m, so that
complexity (8.54) can be computed simply as

$$C = -\sum_{x \in \mathcal{P}} \pi_x \, \log_R \frac{\pi_x}{\prod_{y \in \mathcal{P}} T_{xy}^{T_{xy}}}. \tag{8.55}$$

In Fig. 8.23 (left), we have presented the statistics of complexity values for the
random pieces generated by the MDG based on Bach's compositions. The main
trend (shown in Fig. 8.23 (left) by a solid line) represents the cubic splines
interpolating between the mean (not the median) complexity values and indicates
that patches consisting of 8 notes are characterized by the maximum past-future
mutual information. Then complexity decreases rapidly with the number of pitches
used in a composition suggesting that the musical pieces might contain a few
types of melodic prototypes translated over the entire diapason of pitches by
chromatic transposition. Finally, in Fig. 8.23 (right), we have sketched a scatter
plot showing the pace of complexity with entropy in the MDG generated over the
480 compositions of classical music that implies that a strong positive correlation
exists between the value of entropy and the logarithm of complexity, in those
compositions.

8.5.4　First Passage Times to Notes Resolve Tonality of Musical Dice Games

Statistics of entropy, redundancy, and complexity in the MDG over classical musical compositions suggests that tonal music generated by the musical dice game (8.48) constitutes the well structured data that contain conventional patterns of information. Obviously, some notes might be more "important" than others, with respect to such a structure.

In music theory (Thomson 1999), the hierarchical pitch relationships are introduced based on a *tonic* key, a pitch which is the lowest degree of a scale and that all other notes in a musical composition gravitate toward. A successful tonal piece of music gives a listener a feeling that a particular (tonic) chord is the most stable and final. The regular method to establish a tonic through a cadence, a succession of several chords which ends a musical section giving a feeling of closure, may be difficult to apply without listening to the piece.

While in a MDG, the intuitive vision of musicians describing the tonic triad as the "center of gravity" to which other chords are to lead acquires a quantitative expression. Namely, every pitch in a musical piece is characterized with respect to the entire structure of the Markov chain by its level of accessibility estimated by the first passage time to it (Blanchard and Volchenkov 2009b; Volchenkov 2010) that is the expected length of the shortest path of a random walk toward the pitch from any other pitch randomly chosen over the musical score. Analyzing the first passage times in scores of tonal musical compositions, we have found that they can help in resolving tonality of a piece, as they precisely render the hierarchical relationships between pitches. It is interesting to note that from the physical point of view the first passage time can be naturally interpreted as a potential, as being equal to the diagonal elements of the generalized inverse of the Laplace operator. For example, the first passage time to a node precisely equals to the electric potential of the node, in an electric resistance network (Doyle and Snell 1984; Volchenkov 2010). Thus, in the framework of musical dice games, the role of a note in a tonal scale can be understood as its potential.

The majority of tonal music assumes that notes spaced over several octaves are perceived the same way as if they were played in one octave (Burns 1999).

Using this assumption of octave equivalency, we can chromatically transpose each musical piece into a single octave getting the 12×12 transition matrices, uniformly for all musical pieces, independently of the actual number of pitches used in the composition. Given a stochastic matrix $\mathbf{T}$ describing transitions between notes within a single octave $\mathcal{O}$, the first passage time to the note $i \in \mathcal{O}$ is computed (Volchenkov 2010) as the ratio of the diagonal elements,

$$\mathcal{F}_i = \left(\mathbf{L}^{\#}\right)_{ii} / \left(1 - \mathbf{L}\mathbf{L}^{\#}\right)_{ii}, \tag{8.56}$$

where $\mathbf{L}$ is the Laplace operator corresponding to the transition matrix $\mathbf{T}$, and $\mathbf{L}^{\#}$ is its group generalized inverse. Let us note that in the case of ergodic Markov chains

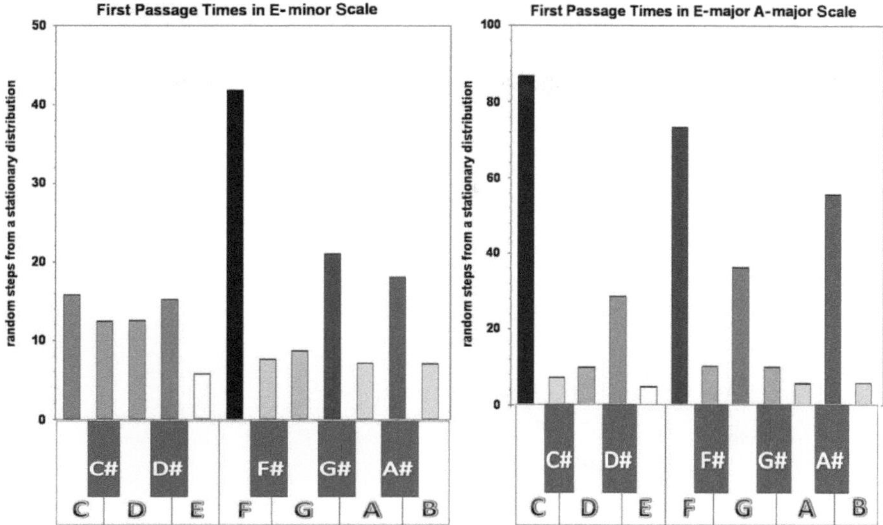

Fig. 8.24 The histograms show the first passage times to the notes for the MDG over a part of Duet I of J.S. Bach (BWV 802) written in E minor (*left*) and over a part of the Cello Sonata No.3, Op.69 of L.V. Beethoven written in E major, A major (*right*) mapped into a single octave. Bars are shaded with the intensity of gray scale 0–100%, in proportion to the magnitude of the first passage time. Therefore, the basic pitches of a tonal scale are rendered with light gray color, as being characterized by short first passage times, and the tonic key by the smallest magnitude of all

the result (8.56) coincides with the classical one about the first passage times of random walks defined on undirected graphs (Lovász 1993).

In Fig. 8.24, we have shown the two examples of the arrangements of first passage times to notes in one octave, for the E minor scale (left) and E major, A major scales (right). The basic pitches for the E minor scale are E, F#, G, A, B, C, and D. The E major scale is based on E, F#, G#, A, B, C#, and D#. Finally, the A major scale consists of A, B, C#, D, E, F#, and G#. The values of first passage times are strictly ordered in accordance to their role in the tone scale of the musical composition. Herewith, the tonic key is characterized by the shortest first passage time (usually ranged from 5 to 7 random steps), and the values of first passage times to other notes collected in ascending order reveal the entire hierarchy of their relationships in the musical scale.

It is intuitive that the time of recurrence to a note estimated by π_i^{-1} is related positively to the first passage time to it, $\mathcal{F}_i$: the faster a random walk over the score hits the pitch for the first time, the more often it might be expected to occur again. The time of recurrence equals the first passage time in a salient recurring succession of notes (a motif) the pattern of three short notes followed by one long that opens the Fifth Symphony of L.V. Beethoven and reappears throughout the work is a classic example.

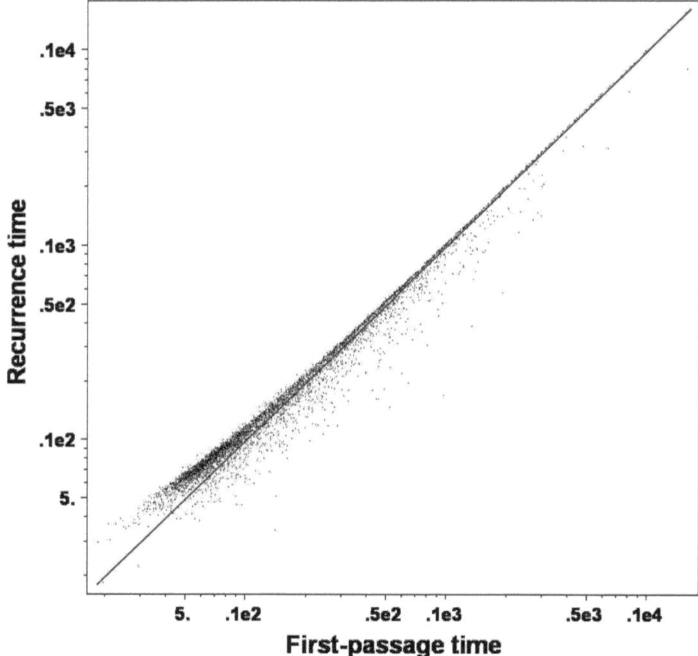

Fig. 8.25 The log-log scatter plot contains 12 × 804 points representing the recurrence time vs. the first passage time to the 12 notes of one octave, over the MDG based on 804 compositions of 29 composers. The straight line is given for a reference indicating the horizon of intelligibility (when equality of recurrence times and first passage times is achieved); departures from linearity signify departures from intelligibility

The log-log scatter plot shown in Fig. 8.25 represents the relation between the recurrence time and the first passage time to the 12 notes of one octave in all MDG over musical compositions we studied. The straight line indicates equality of recurrence times and first passage times. The data provide convincing evidence for the systematic departure of recurrence times from first passage times for those pitches characterized by the relatively short first passage times (recurrence times to them are typically longer than first passage times). The excess of recurrence times over first passage times quantifies musical development encompassing distinct musical figures that are subsequently altered and sequenced throughout a piece of music. It is not a surprise that such a musical development is essentially visible in the range of the relatively short first passage times, as they play an important role in the tonal scale structure of a piece guaranteeing its unity. The important conclusion of this section is that the frequency analysis of note occurrences alone is not complete enough to reliably resolve the tonality of a musical composition.

8.5.5   First Passage Times to Notes Feature a Composer

By analyzing the typical magnitudes of first passage times to notes in one octave, we can discover an individual creative style of a composer and track out the stylistic influences between different composers.

The box plots shown in Fig. 8.26 depict the data on first passage times to notes in the arrangements generated by the MDG over a number of compositions written by J.S. Bach, A. Berg, F. Chopin, and C. Franck through their five-number summaries: 3/2 the interquartile ranges, the lower quartile, the third quartile, and the median. In tonal music, the magnitudes of first passage times to the notes are completely determined by their roles in the hierarchy of tone scales. Therefore, a low median in the box plot (Fig. 8.26) indicates that the note was often chosen as a tonic key in many compositions.

Correlation and covariance matrices calculated for the medians of the first passage times in a single octave provide the basis for the classification of composers, with respect to their tonality preferences. For our analysis, we have selected only those musical compositions, in which all 12 pitches of the octave were used. The tone scale symmetrical correlation matrix has been calculated for 23 composers, with the elements equal to the Pearson correlation coefficients between the medians of the first passage times. For exploratory visualization of the tone scale correlation matrix, we arranged the "similar" composers contiguously. Following Friendly (2002), while ordering the composers, we considered the eigenvectors (principal components) of the correlation matrix associated with its three largest eigenvalues. Since the cosines of angles between the principal components approximate the correlations between the tonal preferences, we used an ordering based on the angular positions of the three major eigenvectors to place the most similar composers contiguously, as it is shown in Fig. 8.27. The correlogram presented on Fig. 8.27 allows for identifying the three groups of composers exhibiting similar preferences in the use of tone scales, as correlations are positive and strong within each tone group while being weak or even negative between the different groups. The smaller subgroups might be seen within the first largest group (from J. Strauss to G. Fauré), in the left upper corner of the matrix on Fig. 8.27. Most of the composers that appeared in the largest group are traditionally attributed to the Classical Period of music. The strongest positive correlations we observed in the choice of a tonic key (about 97%) is between the compositions of J. Strauss and A. Vivaldi who led the way to a more individualistic assertion of imaginative music. The tonality statistics in the masterpieces of R. Wagner appears also quite similar to them. Other subgroups are formed by G.F. Händel and D. Shostakovitch, J.S. Bach and R. Schumann. The Classical Period boasted by L.V. Beethoven and W.A. Mozart who led the way further to the Romantic period in classical music. F. Mendelssohn Bartholdi was deeply influenced by the music of J.S. Bach, L.V. Beethoven, and W.A. Mozart, as often reflected by his biographers (Brown 2003) – not surprisingly, he found his place next to them. Furthermore, the piano concerts of C. Saint-Saëns were known to be strongly influenced by those of W.A. Mozart,

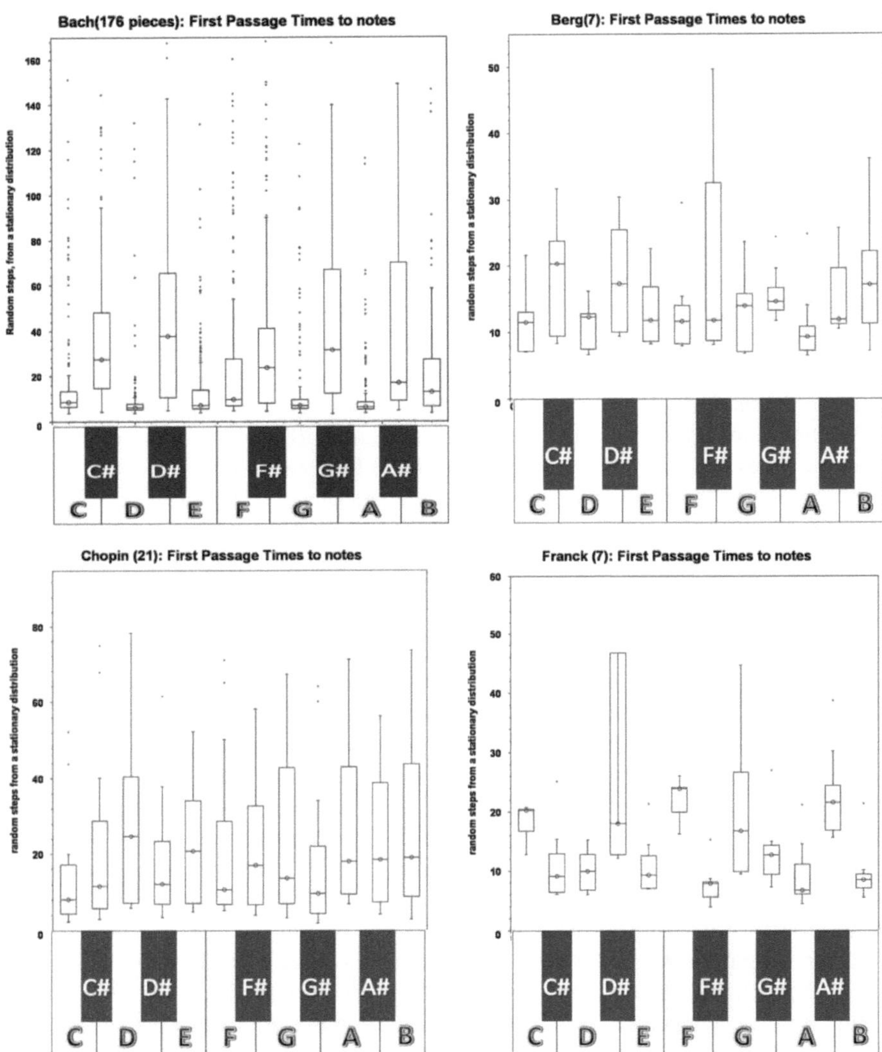

Fig. 8.26 Statistics of first passage times of in the musical pieces generated by the MDG over the original compositions of J.S. Bach, A. Berg, F. Chopin, and C. Franck are represented through their five-number summaries in the box plots

and, in turn, appear to have influenced those of S. Rachmaninoff that receives full exposure in the correlogram (Fig. 8.27). Moreover, we also get the evidence of affinity between I. Stravinsky and A. Berg, F. Schubert, F. Chopin, and G. Faure, as well as of the strong correlation between the tonality styles of A. Scriabin and F. Liszt. The last group, in the lower right corner of the matrix are occupied by the

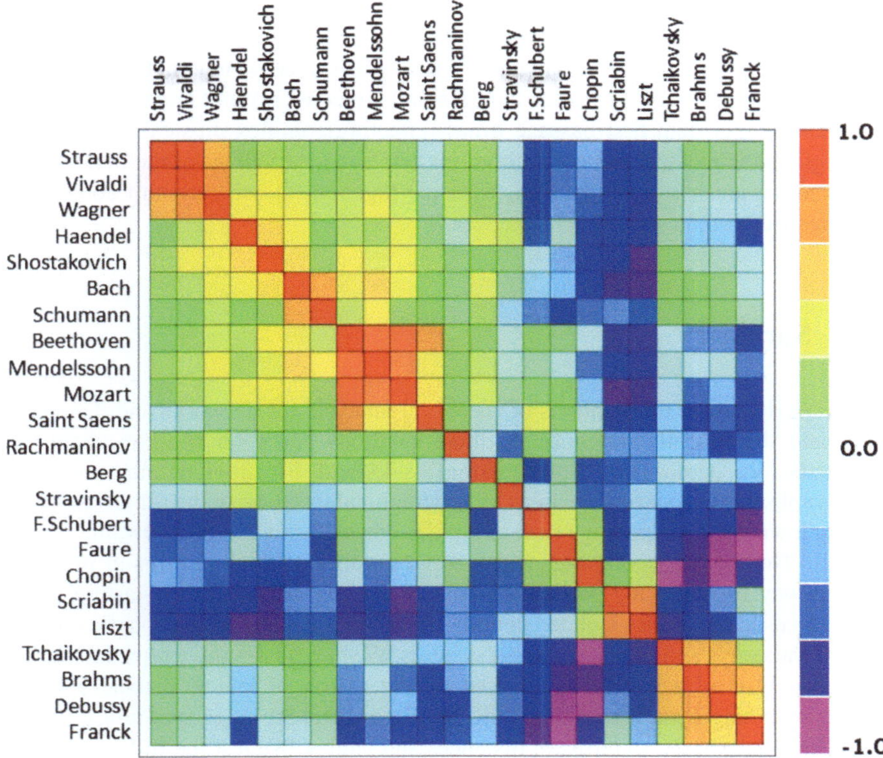

Fig. 8.27 The correlogram displays the correlation matrix for the medians of the first passage times to notes of one octave, for 23 composers. In the shaded rows, each cell is shaded from violet to red depending on the sign of the correlation, and with the intensity of color scaled 0–100%, in proportion to the magnitude of the correlation

Middle and Late Romantic era composers: P. Tchaikovsky, J. Brahms, C. Debussy, and C. Franck. Interestingly, the names of composers that are contiguous in the correlogram (Fig. 8.27) are often found together in musical concerts and on records performed by commercial musicians.

We have studied the MDG encoded by the transition matrices between pitches over the 804 musical compositions. Contrary to human languages where the alphabet is independent of a message, musical compositions might involve different sets of pitches; the number of pitches used to compose a piece grows approximately logarithmically with its size. Entropy dominates over redundancy in the MDG based on the compositions of classical music. Statistics of complexity in the note sequences generated by the MDG suggests that the maximum of past-future mutual information is achieved on the blocks consisting of just a few notes (8 notes, for the MDG generated over Bach's chorales) which might serve as a base for the melody prototypes. Pieces in classical music might contain a few melodic prototypes translated over the diapason of pitches by chromatic transposition. The hierarchical relations between pitches in tonal music might be rendered by means of first passage times to them. The frequency analysis of note occurrences is not enough to reliably resolve the tonality of a musical composition, since recurrence times to notes are typically longer than first passage times reflecting complex musical development throughout the musical piece. Correlations between the medians of the first passage times to the notes provide the basis for a phylogenetic classification of composers, with respect to their tonality preferences.

8.6 Summary

We have discussed the methods of structural analysis for networks and databases. We paid the essential attention to those methods based on Markov chains. In particular, we demonstrated that random walks and diffusions defined on spatial city graphs might spot hidden areas of geographical isolation in the urban landscape going downhill. First passage time to a place correlates with assessed value of land in that. The method accounting the average number of random turns at junctions on the way to reach any particular place in the city from various starting points could be used to identify isolated neighborhoods in big cities with a complex web of roads, walkways and public transport systems. We have also shown how the Markov chain analysis of a network generated by the matrix of lexical distances allows for representing complex relationships between different languages in a language family geometrically, in terms of distances and angles. We test the fully automated method for construction of language taxonomy on a sample of fifty languages of the Indo-European language group and applied to a sample of fifty languages of the Austronesian language group. The Anatolian and Kurgan hypotheses of the Indo-European origin and the 'express train' model of the Polynesian origin are thoroughly discussed. Finally, we have presented our study of entropy, redundancy, complexity, and first passage times to notes for the musical dice games generated by the transition matrices between pitches encoded for 804 musical pieces of 29 composers. The successful understanding of tonal music calls for an experienced listener, as entropy dominates over redundancy in musical messages. First passage times to notes resolve tonality and feature a composer.

Chapter 9
When Feedbacks Matter: Epidemics, Synchronization, and Self-regulation in Complex Networks

Graph theory has been developed in regard to its applications in travel, biology, energetics, and many other fields. The general optimization mindset dominating these researches has addressed to graph theory the questions which were often related to finding a shortest path between nodes, as being of the minimum time delay for information transmission and of the minimum cost for connection maintenance. Not surprisingly, the very definition of distance between two vertices in a graph is given as the *geodesic distance*, i.e., the shortest path connecting them. With respect to the graph metric, a complex network of weighted edges is rather considered as a *minimum weight spanning tree* of the underlying graph, i.e., a subset of paths that has no cycles but still connects to every vertices at the lowest total cost. However, in many problems of practical interest found in biology, sociology, and economics, the existence of many paths of different lengths as well as a nexus of cycles traversing the nodes in complex interaction graphs do matter.

In the present chapter, we discuss the problems of epidemic spreading, synchronization, and self-regulation in complex networks. The dynamical behavior in these models is driven substantially by the *feedback circuits* which combine some of their output with input changing the performance of the entire network in a strongly nonlinear way.

Historically, the earliest class of relevant mathematical models accounting for the spread of disease was carried out in 1766 by Daniel Bernoulli who created a mathematical model to defend the practice of inoculating against smallpox (Hethcote 2000). Nowadays, stochastic models defined on networks which depend on the chance variations are widely used to estimate the risks of exposure, disease and other illness dynamics.

P. Blanchard and D. Volchenkov, *Random Walks and Diffusions on Graphs and Databases*, Springer Series in Synergetics 10, DOI 10.1007/978-3-642-19592-1_9,
© Springer-Verlag Berlin Heidelberg 2011

9.1   Susceptible-Infected-Susceptible Models in Epidemics

One of the key models used in the epidemiological studies is the *susceptible-infected-susceptible* (SIS) model (Murray 1993), in which individuals represented by nodes exist in either "healthy" or "infected" discrete states, and each link represents a connection along which the infection can spread. At each time step, each healthy individual is infected with rate v if it is connected at least to one infected individual. At the same time, infected individuals are cured with rate δ regaining susceptibility. One defines an effective spreading rate as $\lambda = v/\delta$.

The first step for the understanding of epidemic spreading on scale free networks has been made in Pastor-Satorras and Vespignani (2001) where the SIS model has been studied on the scale-free graphs generated in accordance to the preferential attachment approach (Barabasi and Albert 1999). An important conclusion reported in Pastor-Satorras and Vespignani (2001) on the epidemic spreads observed over the preferential attachment scale-free graphs for $\gamma \leq 3$ states the absence of a critical epidemic threshold, $\lambda_c = 0$.

It implies that the preferential attachment scale-free networks are disposed to the spreading and the persistence of infections at whatever spreading rate the epidemic agents possess if $\gamma \leq 3$. For $\gamma > 4$, epidemics on the preferential attachment scale-free networks have the same properties as on random graphs of Erdös – Reiny, i.e., there exists $\lambda_c > 0$ such that the infection spreads and becomes persistent if $\lambda \geq \lambda_c$ and dies out fast when $\lambda < \lambda_c$ (Pastor-Satorras and Vespignani 2001).

In this section, we consider two models for the partner choice preference. In the first model, we assume that the society is unstructured in a sense that an individual is chosen as a partner depending only on its connectivity degree k. Namely this type of models has been at a focus of studies (Pastor-Satorras and Vespignani 2001). However, from the sociological point of view, another model taking a possible social structure into account seems much more natural. The coupling strength between vertices $v \in C_k$ and $w \in C_s$ belonging to the different connectivity classes is supposed to depend on the difference $|k - s|$, and fades out if $|k - s| \gg 1$. In particular, we demonstrate that the epidemic spreadings in these two types of scale-free networks are dramatically different, even if their indexes γ are equal. Among the striking distinctions, we can point out that an average fraction of infected individuals in the unstructured societies grows up with γ for any immunization administrative policy, but in the structured communities it decreases with γ that is the good news, indeed.

9.1.1   Dynamical Equation of the Epidemic Spreading in Scale Free Networks

Let us consider a *scale free* network (SF) of N nodes spanned with the graph $\mathbb{G}(N, \gamma)$ with $\gamma > 1$. In general, one can partition the set of vertices V ($|V| = N$)

into $N - 1$ different classes C_k comprising of vertices having the same degree k,

$$C_k = \{v \in V : \deg(v) = k\}, \quad k \in [1 \ldots N - 1]. \tag{9.1}$$

A *configuration* of a scale-free graph $\mathbb{G}(N, \gamma)$ is the string

$$\xi = (n_1, \ldots n_k \ldots n_{N-1})$$

where $n_k = \text{card}\,(C_k)$ are the random variables asymptotically distributed accordingly to a power law,

$$p_\gamma(k) \sim_{N \to \infty} (\gamma - 1)k^{-\gamma}. \tag{9.2}$$

The structural properties of networks (9.2) depend upon the certain social strategy chosen by individuals establishing a pair formation process generating edges of the graph $\mathbb{G}(N, \gamma)$. It can be defined by the matrix $\widehat{\sigma}_{sk}$ which elements are the probabilities that the vertex $v \in C_s$ chooses some other vertex $w \in C_k$ as a partner. For instance, in the popular preference attachment model of the scale free random graphs proposed by Barabási and Albert (1999), the elements of $\widehat{\sigma}$ depend only on one variable k,

$$\sigma_{sk} = \frac{k}{\langle k \rangle}, \quad \text{for any } s, \tag{9.3}$$

where $\langle k \rangle$ is the average number of connections between vertices. For $1 < \gamma < 2$, the average connectivity $\langle k \rangle$ diverges, and therefore such graphs do not exist. However, for other generating algorithms with alternative $\widehat{\sigma}$ the scale free networks exist even for $1 < \gamma < 2$.

We suppose that $N \to \infty$ and treat the connectivity k as a continuous variable taking values in between 1 and ∞. The edge generating rule is given by an arbitrary positive integrable function (generally speaking, of two variables) σ satisfying the normalization condition

$$1 = \int_1^\infty \!\!\! \int_1^\infty \sigma(k, s) p_\gamma(s) dk ds. \tag{9.4}$$

In the problem of epidemic spreading in SF networks, we are interested in the fraction of infected individuals at time t, $0 < F(t) < 1$ which is given by

$$F_\gamma(t) = \int_1^\infty p_\gamma(k)\phi(t, k) dk \tag{9.5}$$

where $\phi(t, k)$ is the probability function that an arbitrary node $v \in C_k$ is infected, we suppose that the initial probability function $\phi(0, k)$ is known.

In accordance to the SIS model (Murray 1993), at each time step the total fraction of infected individuals in a scale free community is changed by the quantity

$$\Delta F = \nu F_h - \delta F_i \tag{9.6}$$

where $0 < \nu < 1$ is the infection rate, $0 < \delta < 1$ is the rate of which the infected nodes are cured, F_h is the fraction of healthy nodes connected at least to one infected node, F_i is the fraction of infected nodes. In general, the probability to be linked to an infected node depends upon the social strategy of individuals σ,

$$\Theta_\gamma(t,k) = \int_1^\infty \sigma(k,s) p_\gamma(s) \phi(t,s) ds. \tag{9.7}$$

Since the balance equation (9.6) is satisfied for any $p_\gamma(k)$, one can write down the evolution equation for the probability functions $\phi(t,k)$ in the following form:

$$\partial_t \phi(t,k) = -\delta \cdot \mu(k)\phi(t,k) + (1 - \phi(t,k))\left[1 - \left(1 - \nu\Theta_\gamma(t,k)\right)^k\right]. \tag{9.8}$$

The infecting term considers the probability that a node with k links is healthy $(1 - \phi(t,k))$ and gets infection, proportional to the rate $\nu > 0$, via an infected connected node chosen with the probability $\sigma(k,s)$. The recovering term describes the probability that an infected node chosen with the probability $\mu(k)$ is cured proportional to the rate $\delta > 0$. One can think of $\mu(k)$ as a distribution of funds destined for a recovering of individuals from the class C_k provided the total scope is taken as 1.

The stationary solution of the (9.8) $(\partial_t \phi_{st}(k,t) = 0)$ is given by the formula

$$\phi_{st}(k) = \frac{1 - (1 - \nu\Theta_\gamma(k))^k}{1 - (1 - \nu\Theta_\gamma(k))^k + \delta\,\mu(k)}, \tag{9.9}$$

in which the stationary probability function $\Theta_\gamma(k)$ satisfies the self-consistency equation

$$\Theta_\gamma(k) = \int_1^\infty \frac{\sigma(k,s) p_\gamma(s)\left(1 - (1 - \nu\Theta_\gamma(k))^k\right)}{1 - (1 - \nu\Theta_\gamma(k))^k + \delta\mu(k)}\, ds. \tag{9.10}$$

The above equation has a countable number of solutions, but only one of them belongs to the unit interval. For large connectivities $k \gg 1$, the solution (9.9) behaves like

$$\phi_{st} \simeq_{k\gg1} \begin{cases} 1, & \text{if } \delta \cdot \mu(k) = o(1), \\ 0, & \text{if } \delta \cdot \mu(k) = O(1). \end{cases} \tag{9.11}$$

9.1.2　Simplified Equation for Low Infection Rates

In almost all papers devoted to the problem of epidemic spreading in the SF networks, a simplified version of the (9.8) is considered. Namely, it is supposed that the probability that a chosen node will be infected via connection with other infected nodes is very small, $\nu\Theta_\gamma(k,t) \ll 1$. In this case, the r.h.s. of the (9.8) is expanded into the power series in $\nu\Theta_\gamma$ and then only the linear term is retained. As a result, one arrives at the following simplified equation:

$$\partial_t \phi(t,k) = -\delta \cdot \mu(k)\phi(t,k) + \nu\,(1 - \phi(t,k))\,k\Theta_\gamma(k,t). \tag{9.12}$$

Let us note that the solutions of the exact (9.8) differ from those of (9.12), nevertheless, the critical epidemic threshold λ_c predicted by these equations is the same.

9.1.3   Stationary Solution of the Epidemic Equation for Low Infection Rates

The stationary solutions, $\partial_t f_\gamma(\lambda,k) = 0$, of the (9.12) is given by the function

$$f_\gamma(\lambda,k) = \frac{\lambda k \Theta_\gamma(k)}{\lambda k \Theta_\gamma(k) + \mu(k)}. \tag{9.13}$$

The asymptotic behavior of $f_\gamma(\lambda,k)$ as $k \gg 1$ depends essentially upon the large scale behavior of $\mu(k)$ and $\sigma(k,s)$,

$$f_\gamma(\lambda,k) \simeq_{k\gg 1} \begin{cases} 1, & \mu(k)/\lambda k\Theta_\gamma(k) = o(1), \\ 0, & \mu(k)/\lambda k\Theta_\gamma(k) = O(1). \end{cases} \tag{9.14}$$

The stationary probability that any given link points to an infected node, $0 < \Theta_\gamma(k) \leq 1$, satisfies the self-consistency equation

$$\Theta_\gamma(k) = \int_1^\infty \frac{\lambda s \Theta_\gamma(s)\sigma(k,s)p_\gamma(s)}{\lambda s \Theta_\gamma(s) + \mu(s)}\, ds. \tag{9.15}$$

Trivial solution $\Theta_\gamma(k) = 0$ is always satisfying the above equation and giving a zero stationary prevalence, $f_\gamma = 0$. A non-zero stationary prevalence $f_\gamma(k) \neq 0$ is obtained when the (9.15) has a nontrivial solution in the interval $0 < \Theta_\gamma(k) \leq 1$ that takes place if

$$\delta_\Theta \left[\int_1^\infty \frac{\lambda s \Theta_\gamma(s,\lambda)\sigma(k,s)p_\gamma(s)}{\lambda s \Theta_\gamma(s,\lambda) + \mu(s)}\, ds \right]\Bigg|_{\Theta=0} \geq 1. \tag{9.16}$$

The above inequality defines the critical epidemic threshold such that $f_\gamma(k) > 0$ as $\lambda > \lambda_c(k)$,

$$\lambda_c(k) \cdot \left[\int_1^\infty \frac{s\sigma(k,s)p_\gamma(s)}{\mu(s)}\, ds \right] = 1. \tag{9.17}$$

The (9.17) shows that if $\sigma(k,s)$ depends on two variables, the critical epidemic threshold depends on k, i.e., in a structured society, the different classes of vertices C_k can possess different critical epidemic thresholds. Otherwise, if the probability to be chosen as a partner depends merely on the connectivity of a node, the same critical epidemic threshold λ_c stays for all vertices of the network.

As an example of the epidemic spreading in such a homogeneous SF network modeling the unstructured society, let us consider a generalized power law model with the social and immunization preference functions given by

$$\sigma(k) = \frac{\gamma - \beta - 1}{\gamma - 1}k^\beta, \quad \mu(k) = \frac{\gamma - \alpha - 1}{\gamma - 1}k^\alpha, \tag{9.18}$$

$$\alpha, \beta < \gamma - 1, \quad \gamma \geq 1.$$

Then the integral in the (9.17) defining the critical epidemic threshold converges if $\gamma + \alpha - \beta > 2$ and results in

$$\lambda_c = \frac{(\gamma - \alpha - 1)(\gamma + \alpha - \beta - 2)}{(\gamma - \beta - a)(\gamma - 1)}. \tag{9.19}$$

In the Fig. 9.1a, we have presented the critical domains in which the critical epidemic threshold exists for different values of γ. Let us note that one of the phase (α, β)-diagrams presented in Fig. 9.1a (for $\gamma = 3$) passes through the point $(\beta = 1, \alpha = 0)$ relevant to the Barabási-Albert scale-free network exactly as predicted in Pastor-Satorras and Vespignani (2001).

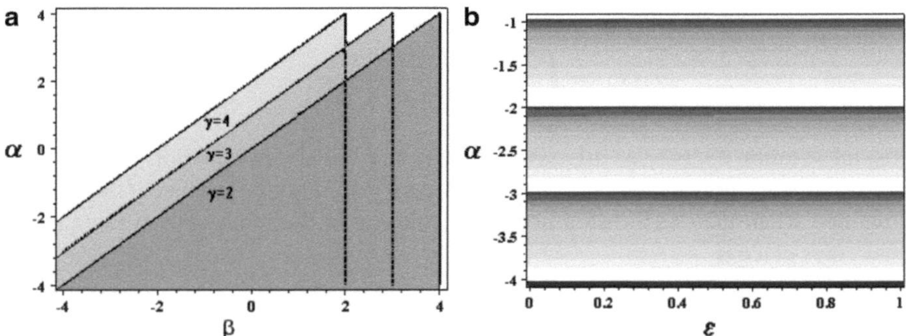

Fig. 9.1 The critical domains in which the critical epidemic threshold exists for (**a**) the power law model (9.18), for $\gamma = 3$, the phase diagram passes through the point $(\beta = 1, \alpha = 0)$ which corresponds to the Barabási-Albert scale free network exactly as predicted in Pastor-Satorras and Vespignani (2001); and (**b**) the hierarchical society with the social strategy of individuals given by the function (9.20). The critical epidemic threshold is infinite as α takes negative integer values and have zeros somewhere in between negative integers. The parameter ε determines the width of the band in critical domains

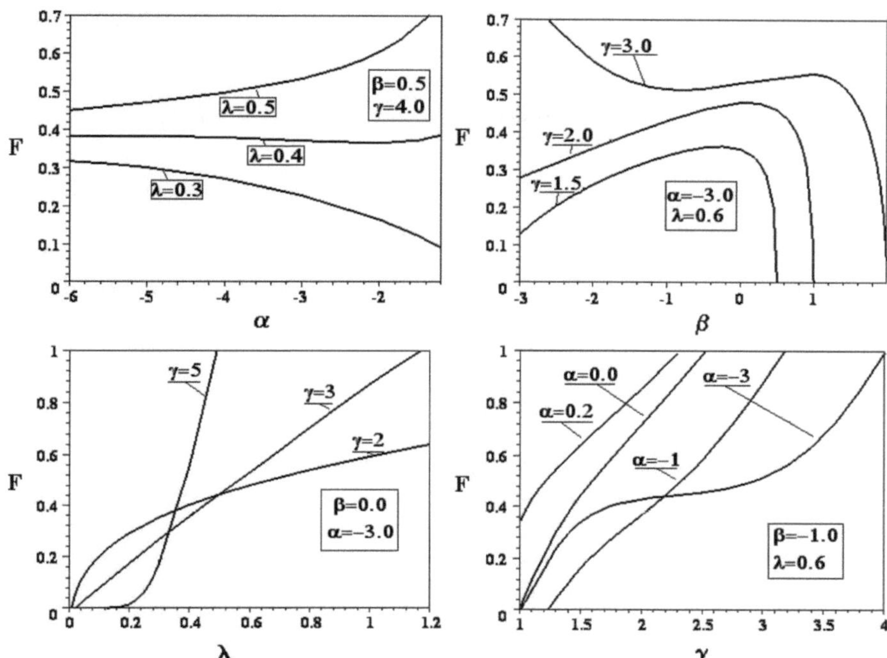

Fig. 9.2 The stationary fraction of infected individuals $F_\gamma > 0$ in the "power law" model (9.18) depending on the affinity parameters α and β characterizing the immunization policy and the social strategy chosen by individuals, effective spreading rate λ and the index of the degree statistics γ. The figures reveal an absence of a plain immunization strategy which can be efficient simultaneously for all types of scale free networks with power law preferences (9.18). The stationary fraction F_γ increases with the effective spreading rate of viruses λ, and the index of degree statistics γ. The model predicts that the healing of the certain classes of individuals does not help much in eradication of epidemics except for the small virus spreading rates $\lambda < 0.2$

In Fig. 9.2, we have sketched out some patterns of complicated behavior of the stationary fraction of infected individuals $F_\gamma > 0$ depending on the affinity parameters α and β characterizing the immunization policy and the social strategy chosen by individuals, effective spreading rate λ and the index of the degree statistics γ. These figures reveal the absence of a plain immunization program which can be effective simultaneously for all types of scale free networks and for all effective spreading rates of viruses. One can see that, in general, the stationary fraction of infected individuals for the model (9.18) increases with λ and γ. The model predicts that the healing of the certain classes of individuals does not help much in eradication of epidemics except for the small virus spreading rates $\lambda < 0.2$. An effective immunization program in this case would assume a decentralization of the network providing a course of remedial treatment to everybody.

In a structured society, the coupling between vertices of different classes C_k and C_s depends on the distance $|k - s|$ between them and fades out for $|k - s| \gg 1$.

A possible social strategy of individuals can be modeled by the function of two variables

$$\sigma(k,s) = \frac{(\gamma - \varepsilon - 1)\Gamma(\gamma)}{(\gamma - 1)\,\Gamma(\gamma - \varepsilon)\Gamma(\varepsilon)} \frac{\theta(k - s)}{(k - s)^{1-\varepsilon}}, \quad 0 < \varepsilon < 1, \tag{9.20}$$

satisfying the normalization condition (9.4). Here $\theta(x)$ is the step function (we need to include it to make the normalization integral (9.4) to converge at ∞). It is also required that $\gamma > 1 + \varepsilon$. We use the power law model (9.18) for the immunization preference function $\mu(k)$ with $\alpha < \gamma - 1$. Then the critical epidemic threshold given by the (9.17) is

$$\lambda_c(k) = \frac{(\gamma - \alpha - 1)}{(\gamma - 1)(\gamma - \varepsilon - 1)} \cdot \frac{\Gamma(\varepsilon)\Gamma(\gamma - \varepsilon)\Gamma(\gamma + \alpha - 1)}{\Gamma(\gamma)\Gamma(\gamma + \alpha - \varepsilon - 1)} k^{\gamma + \alpha - \varepsilon - 1} \tag{9.21}$$

and is different for the different classes of vertices C_k exhibiting the power law behavior with k. The domains where the non-trivial critical epidemic threshold exists are determined by the poles of $\Gamma(x)$ and have a band-like structure. They are displayed on the Fig. 9.1b. The critical epidemic threshold is infinite as α takes negative integer values and have zeros somewhere in between negative integers. The parameter ε determines the width of the band in critical domains.

The behavior of the stationary fraction of infected population at different values of λ, γ, α, and ε is displayed on the Fig. 9.3. In general, it is rather different from the behavior observed for the unstructured community described by the power law model (9.18). Similarly to the model (9.18), the fraction F_γ is increasing with the effective spreading rate λ, but decreases with γ and α. An efficient immunization program would be to vaccinate hubs and consolidate them to enlarge γ.

9.1.4   Dynamical Solution of the Evolution Equation for Low Infection Rates

Given the initial probability distribution of infected nodes $\phi(0, k)$, the dynamical solution $\phi(t, k)$ of the (9.12) can be obtained in the following form

$$\begin{aligned}
\phi(t,k) = vk \int_0^t \Theta(k, \tau) \exp\left[\frac{\tau - t}{T(k, \tau)}\right] d\tau \\
+ \phi(k, 0) \exp\left[-\frac{t}{T(k, t)}\right]
\end{aligned} \tag{9.22}$$

where the inverse relaxation time is

$$\frac{1}{T(k,t)} = \frac{1}{T(k)} + \frac{vk}{t} \int_0^t \widetilde{\Theta}(k, \tau) d\tau, \tag{9.23}$$

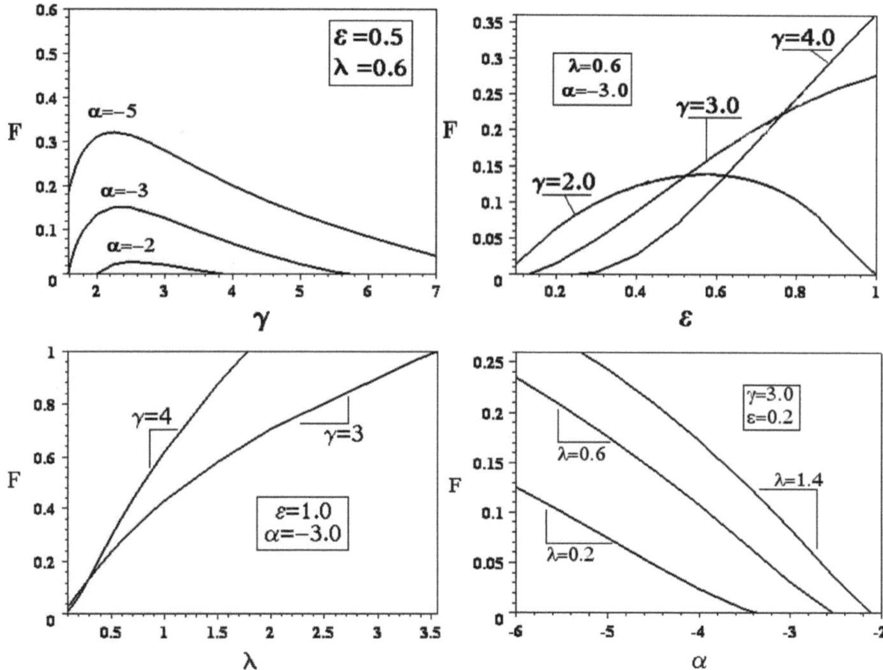

Fig. 9.3 The behavior of the stationary fraction of infected population at different values of λ, γ, α, and ε for the model (9.20) in which individuals chose their partners from the classes of the similar communication ability. Similarly to the model (9.18), the fraction F_γ is increasing with the effective spreading rate λ, but decreases with γ and α

$$\frac{1}{T(k)} = \delta\mu(k) + \nu k\Theta(k),$$

and $\Theta(k,t)$ is presented as the sum of the stationary probability $\Theta(k)$ satisfying the self-consistency (9.15) and the time-dependent part $\widetilde{\Theta}(k,t)$:

$$\Theta(k,t) = \Theta(k) + \widetilde{\Theta}(k,t).$$

Neglecting this time dependent part $\widetilde{\Theta}(k,t)$, we obtains:

$$\phi(k,t) = f(\lambda,k) + \Delta(k)\exp\left[-\frac{t}{T(k)}\right] \tag{9.24}$$

where

$$\Delta(k) = \phi(k,0) - f(\lambda,k)$$

is the departure of the initial probability distribution of infected population from the stationary solution.

The solution (9.24) is trivial as $\lambda < \lambda_c$. If $T(k) \gg 1$, the contribution coming from the initial distribution drives $\phi(k, t)$ out from the stationary solution even for very large time t. For the low infection rates $|\nu\Theta| \ll 1$, and if $\delta \ll 1$, the initial distribution $\phi(k, 0)$ features the epidemic spreading over almost all vertices except may be a few hubs accumulating a considerable fraction of connections.

9.2   Epidemic Spreading in Evolutionary Scale Free Networks

In the present section, we give an example of a scale free network, which has no critical epidemic thresholds at any γ. A flexible algorithm generating SF networks based on the principle of evolutionary selection of a common large-scale structure of biological networks has been proposed in Volchenkov et al. (2002). We briefly reproduce this algorithm:

Let us consider three random variables x, y, and z that are the real numbers distributed in accordance to the probability distribution functions f, g, and v within the unit interval $[0, 1]$. We assume that x represents the current performance of a biological network (say, the protein-protein interaction map), while y and z are the thresholds for outgoing and incoming edges respectively. The network is supposed to be stable until $x < y$ and $x < z$, and is condemned otherwise. Fluctuations of thresholds reflect the changes of an environment.

The random process begins on the set of N vertices with no edges at time 0, at a chosen vertex i. Given two fixed numbers $\eta \in [0, 1]$ and $v \in [0, 1]$, the variable x is chosen with respect to pdf f, y is chosen with pdf g, and z is chosen with pdf v, we draw e_{ij} edge outgoing from i vertex and entering j vertex if $x < y$ and $x < z$, and continue the process to time $t = 1$. Otherwise, if $x \geq y$ ($x \geq z$), the process moves to other vertices having no outgoing (incoming) links yet.

At time $t \geq 1$, one of the three events happens:

1. with probability η, the random variable x is chosen with pdf f but the thresholds y and z keep their values they had at time $t - 1$.
2. with probability $1 - \eta$, the random variable x is chosen with pdf f, and the thresholds y and z are chosen with pdf g and v respectively.
3. with probability v, the random variable x is chosen with pdf f, and the threshold z is chosen with pdf v but the threshold y keeps the value it had at time $t - 1$.

If $x \geq y$, the process stops at i vertex and then starts at some other vertex having no outgoing edges yet. If $x \geq z$, the accepting vertex j is blocked and does not admit any more incoming link (provided it has any). If $x < y$ and $x < z$, the process continues at the same vertex i and goes to time $t + 1$.

It has been shown in Volchenkov et al. (2002) that the above model exhibits a multi-variant behavior depending on the probability distribution functions f, g, and v chosen and values of relative frequencies η and v. In particular, if $v = 0$, both thresholds y and z have synchronized dynamics, and sliding the value of η form 0 to 1, one can tune the statistics of out-degrees and in-degrees simultaneously out from the pure exponential decay (for $\eta = 0$) to the power laws (at $\eta = 1$) provided f, g, and v belong to the class of power law functions. For instance, by choosing the probability distribution functions in the following forms

$$f(u) = (1 + \alpha)u^\alpha, \qquad\qquad \alpha > -1,$$
$$v(u) \equiv g(u) = (1 + \beta)(1 - u)^\beta, \ \beta > -1, \tag{9.25}$$

one obtains that

$$p_{\eta=1}(k) \simeq_{k \gg 1} \frac{(1 + \beta)\ \Gamma(2 + \beta)\ (1 + \alpha)^{-1-\beta}}{k^{2+\beta}} \left(1 + 0\left(\frac{1}{k}\right)\right). \tag{9.26}$$

For different values of β, the exponent of the threshold distribution, one gets all possible power law decays of $p_{\eta=1}(k)$. Notice that the exponent $\gamma = 2 + \beta$ characterizing the decay of $p_{\eta=1}(k)$ is independent of the distribution $f(u)$ of the state variable x. In the uncorrelated case, $\eta = 0$, the degree distribution functions decays exponentially (for instance, $p_{\eta=0} = 2^{-k}$ for $f = g = v = 1$) (Volchenkov et al. 2002). For the intermediate values of η, the decay rate is mixed.

In Fig. 9.4a, we have plotted these asymptotic profiles $p(k)$ vs. k in the log-linear scale for the case of uniform densities $f = g = v = 1$, for the consequent frequency values $\eta = 0$, $\eta = 0.5$, $\eta = 0.7$, $\eta = 0.9$, $\eta = 1$ (bottom to top). In Fig. 9.4b, we have presented the distribution $p(k)$ vs. k in the log-log scale over $N = 10^5$ vertices for

$$f = g = v = 1, \quad \eta = 1, \quad v = 0.$$

Here, the circles stay for outgoing degrees, and diamonds are for incoming degrees. For $k \gg 1$, both profiles enjoy a power law decay with

$$\gamma_{\mathrm{in}} = \gamma_{\mathrm{out}} = 2.$$

Interesting in epidemic spreading properties in such a evolutionary network, we note that the preference choice function for the above model is

$$\sigma(k) = (1 + \beta)\left(1 - \frac{k}{N - 1}\right)^\beta, \quad \beta > -1. \tag{9.27}$$

Expanding the binomial in the above equation, one gets the leading term $\propto (k/N-1)^\beta$. Consequently, the integral determining the critical epidemic threshold diverges and $\lambda_c = 0$ for any γ.

The Figs. 9.5a and b illustrate the absence of the critical epidemic threshold for the model (9.27) (we have checked this fact for γ up to several tenth) and reveal the complexity of the epidemic spreading in evolutionary SF network as a dynamical system. We have presented the results of numerical simulations for the epidemic

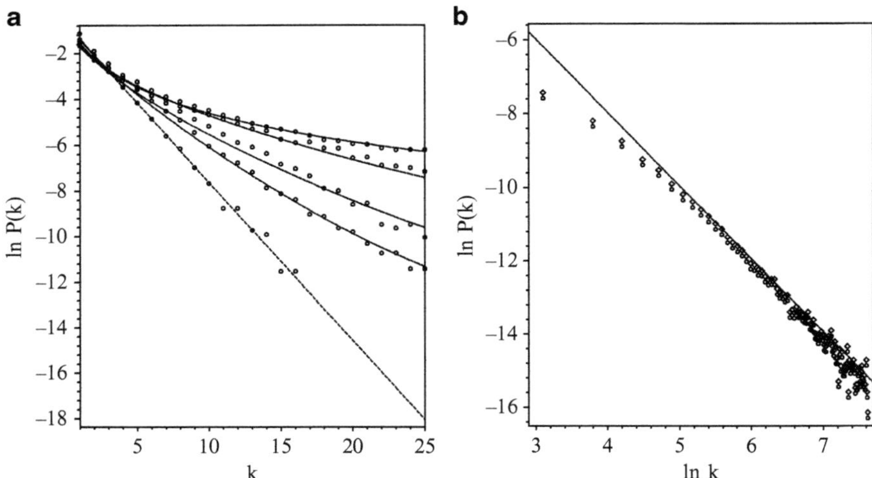

Fig. 9.4 Statistics of the evolutionary scale free random graphs. (**a**) The probability degree distributions $p(k)$ in the log-linear scale for different values of the parameter η, $\eta = 0$, $\eta = 0.5$, $\eta = 0.7$, $\eta = 0.9$ and $\eta = 1.0$ (*bottom to top*). Straight line (*bottom*) corresponds to the pure exponential decay ($\sim 2^{-k}$) observed in the case $\eta = 0$; the top line ($\eta = 1.0$) corresponds to the power law decay $\sim k^{-2}$. (**b**) The probability degree distribution $p(k)$ vs. k in the log-log scale generated on $N = 10^5$ nodes for $f = g = v = 1$, $\eta = 1$, $\nu = 0$. Here, the circles stay for the outgoing degrees k_{out} and diamonds are for the incoming degrees k_{in}. Both profiles enjoy a power law decay with $\gamma = 2$

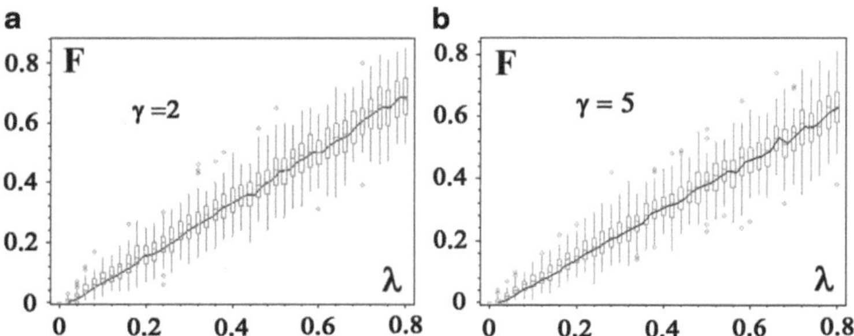

Fig. 9.5 The fraction of infected agents in the evolutionary scale free networks vs. the effective spreading rate λ for different values of γ: (**a**) $p(k) \propto k^{-2}$, (**b**) $p(k) \propto k^{-5}$. At the onset of the process, the initial state ("healthy" or "infected") has been assigned to each of $1,000$ nodes by the coin tossing procedure. Starting from such a random configuration of initially infected individuals, each infection spreading process was simulated in $1,000$ consequent iterations and then got started again with a different random initial configuration. The bold lines represent the mean infected fraction averaged over 500 spreading processes vs. the effective spreading rate λ. The error bars correspond to the standard deviations, herewith, a lower line of boxes is showing the first quartile of data, and an upper line is showing the third quartile

spreading process in the above network (for $\gamma = 2$ and $\gamma = 5$) starting from the randomly chosen configuration of initially infected individuals. At the onset of the process, the initial state ("healthy" of "infected") has been assigned to each of $N = 10^3$ nodes by the coin tossing procedure with a probability $1/2$. Each infection spreading process has taken 1,000 consequent iterations and then get started again with a different random initial configuration of infected individuals. The bold lines represent the mean infected fraction averaged over 500 spreading processes vs. the effective spreading rate λ. The error bars correspond to the standard deviations, herewith, a lower line of boxes is showing the first quartile of data, and an upper line is showing the third quartile.

In a conclusion, we have shown that the epidemic spreading in scale free networks is very sensitive to the statistics of degree distribution characterized by the index γ, the effective spreading rate of a virus, λ, the social strategy using by individuals to choose a partner, and the policy of administrating a cure to an infected node. Depending on the interplay of these four factors, the stationary fractions of infected population F_γ as well as the epidemic threshold properties can be essentially different.

We have considered two alternative models for the partner choice preference and demonstrated that the epidemic spreadings in these two types of SF networks are dramatically different even if their indexes γ are equal. Probably, it is impossible to obtain a plain immunization program which can be simultaneously effective for all types of SF networks.

Finally, we have given the example of the evolutionary SF network which is disposed to the spreading and the persistence of infections at whatever spreading rate $\lambda > 0$ for any value of the index γ. We have demonstrated that such a network is strongly influenced by the initial configuration of the infected nodes even for long time.

9.3   Transitions to Intermittency and Collective Behavior in Randomly Coupled Map Networks

In the present section, we discuss the transition to spatio-temporal intermittency in random network of coupled Chaté-Manneville maps following our work (Sequeira et al. 2002). The relevant parameters are the network connectivity, coupling strength, and the local parameter of the map. We show that spatiotemporal intermittency occurs for some intervals or windows of the values of these parameters. Within the intermittency windows, the system exhibits periodic and other nontrivial collective behaviors. The detailed behavior depends crucially upon the topology of the random graph spanning the network.

Partial differential equations describing continuous models and real physical systems can, in many cases, be discretized into a system of *coupled map lattices* (CML). Coupled map lattices are spatiotemporal dynamical systems comprised of an interacting array of discrete-time maps. Much attention to these systems has been drawn in virtue of studies of generic properties of spatiotemporal chaos (Kaneko 1984, 1985, 1986). A mean-field extension of CML is the globally coupled map lattice introduced by Kaneko (1990). Here, we consider another one which refers to the random networks of the chaotic coupled maps.

Although just a few studies devoted to the *randomly coupled map networks* (RCMN) have been reported in so far, it is beyond a dispute that such systems would be very rich in practical applications. To motivate the increasing interest in RCMN, let us note that many real-world networks have a rather complicated topology. Social networks (Wasserman and Faust 1994), food webs formed by the biological communities (Paine 1992; McCann et al. 1998), communication networks (Albert et al. 1999; Huberman and Adamic 1999) have plenty of shortcuts inconsistent with any regular structure. Previous studies of the chaotic coupled maps defined on the Sierpinski gasket (Cosenza and Kapral 1992) and the Cayley tree (Gade et al. 1995) show convincingly that the topology of networks affects the spatiotemporal behavior of these systems crucially.

Randomly coupled logistic maps

$$f(x) = a\,x(1 - x) \tag{9.28}$$

have been considered first in Chaté and Manneville (1992a). The emergence of synchronization in random networks of logistic maps with nonlocal couplings has been investigated in Gade (1996), and more recently, the dynamical clustering has been observed in the coupled maps connected symmetrically at random (Manrubia and Mikhailov 1999). Here, we study the collective behavior and phase transitions in the random networks of coupled maps different from those considered in Chaté and Manneville (1992a), Gade (1996) and Manrubia and Mikhailov (1999). As a local evolution law, we use the *Chaté-Manneville map* (CM) (Chate and Manneville 1988a) which is the "minimal" one demonstrating the spatio-temporal intermittency. Depending on the value of the variable, this piecewise linear map can either be in a "turbulent" (excited) state or in a "laminar" (inhibited) state. The network topology in our model is spanned by the random graph $\mathbb{G}(N, k)$ such that each of N nodes has precisely k outgoing edges (so called "k-out model", Janson et al. 2000). Our model is similar to the *susceptible-infected-susceptible models* (SIS) (Bailey 1975; Murray 1993) and can be applied to the study of epidemic spreading in the unstructured societies characterized by the random contacts between individuals.

In the present section, we show that the entire collective behavior observed in RCMN is a net result of the interplay between properties of the local map, the probabilistic topology of random graphs, and the strength of couplings. For some ranges of coupling and connectivity parameters, a valuable fraction of nodes triggers to a sustained turbulent state. We also observe a phase transition to the global periodic motions known as the *non-trivial collective behavior* (NTCB) within the sustained turbulent state. Bifurcation diagrams of RCMN present the same self-similarity as that of their local maps, even for rather large values of coupling. The same behavior has been reported for the CML with a regular diffusive coupling

(Lemaître and Chaté 1998). However, in contrast to the regular lattice, the detailed structure of bifurcation diagrams for RCMN depends very much on the random graph topology.

9.3.1 The Model of Random Networks of Coupled Maps

Let $\Omega \subset \mathbb{Z}$ be a finite lattice of $N \in \mathbb{N}$ sites. At each site $\omega \in \Omega$ we define a local phase space X_ω with an uncountable number of elements.

The global phase space $\mathfrak{M} = \Pi_{\omega \in \Omega} \mathrm{X}_\omega$ is a direct product of local phase spaces such that a point $\mathrm{x} \in \mathcal{M}$ can be represented as $\mathrm{x} = (x_\omega)$. A *coupled map lattice* is any mapping $\Phi : \mathfrak{M} \to \mathfrak{M}$ which preserves the product structure, $\Phi \mathrm{x} = (\Phi_\omega \mathrm{x})_{\omega \in \Omega}$, in which $\Phi_\omega : \mathfrak{M} \to \mathrm{X}_\omega$. The mapping $\Phi = G \circ F$ is a composition of the independent local mapping $(Fx)_\omega = f_\omega(x_\omega)$ where $f_\omega : \mathrm{X}_\omega \to \mathrm{X}_\omega$ and the interaction $(Gx)_\omega = g_\omega(x)$.

We study the following coupled map lattice supplied with some boundary conditions,

$$(\Phi x)_\omega = \left[(1 - \varepsilon)\mathbf{I} + \frac{\varepsilon}{k}\mathbf{M} \right] f(x_\omega), \tag{9.29}$$

where $\varepsilon \in [0, 1]$ is the coupling strength parameter, $0 < k < N - 1$ is the connectivity, $\mathbf{I}$ is a unit matrix, and $\mathbf{M}$ is a traceless random connectivity matrix ($\mathrm{M}_{jj} = 0$) determining the network topology. A local map possessing the minimal requirements for observing spatiotemporal intermittency is the Chaté-Manneville map (Chaté and Manneville 1992a),

$$f(x) = \begin{cases} \dfrac{r}{2}\left(1 - |1 - 2x|\right), & \text{if } x \in [0, 1] \\[2mm] x, & \text{if } x > 1, \end{cases} \tag{9.30}$$

with $r > 2$. This map is chaotic for $f(x)$ in $[0, 1]$. However, for $f(x) > 1$ the iteration is locked on a fixed point. The local state can thus be seen as a continuum of stable "laminar" fixed points ($x > 1$) adjacent to a chaotic repeller or "turbulent" state ($x \in [0, 1]$).

Several different models for $\mathbf{M}$ have been proposed in the literature. For instance, in Chaté and Manneville (1992a) the total number of edges E is considered fixed and the symmetry is not required for $\mathbf{M}$. Such a random network is related to a *uniform directed random graph* $\mathrm{G}(N, E)$. Alternatively, in Manrubia and Mikhailov (1999), it was supposed that $M_{ij} = M_{ji}$, and the matrix elements are Either 0 (when the connection between maps i and j is absent) or 1 (if otherwise), while loops are not allowed, $M_{ii} = 0$. The main advantage of this model is the independent presence of edges, but the drawback is that the number of edges is not fixed, but varies according to a binomial distribution with an expectation

$$\mathbb{E}(\#\,edges) = \binom{N}{2}\, p, \quad 0 \le p \le 1.$$

This model relies upon a symmetric binomial random graph, $\mathbb{G}(N, p)$. Finally, in Gade (1996) M_{ij} is equal to the number of times vertex i is connected to vertex j, and multiple edges and loops are allowed. Therefore, M_{ij} is not necessarily symmetric, and

$$\sum_i M_{ij} = k$$

for any j, i.e., each map is coupled to k maps chosen randomly. We suppose that each vertex j has always k outgoing edges, $\sum_i M_{ij} = k$, and do not allow for loops, $M_{ii} = 0$. Double connections, however, are possible. The number of incoming edges is a random Poisson distributed variable with a mean

$$\mathbb{E}(\#\,edges) = \frac{kN}{(N-1)}.$$

Such random graphs are known as the *k-out model*, we shall denote them as $\mathbb{G}(N, k)$.

Random graphs $\mathbb{G}(N, 1)$ have been extensively studied in Kolchin (1986) and Aldous and Pitman (1994), however many properties of $\mathbb{G}(N, k)$ for arbitrary $k > 1$ remain to be investigated. A nice property of $\mathbb{G}(N, k)$ is that they allow for an explicit computation of the graph entropy as $h(k) = \log k$ (Lind and Marcus 1995). In the limit $N \to \infty$, the graph $\mathbb{G}(N, k)$ is asymptotically equivalent to that considered in Gade (1996) since either possibility, that two vertices will be connected more than once or that one vertex will be coupled to itself, are negligible. If $\binom{N}{2} p \approx k$, the graph $\mathbb{G}(N, k)$ is also asymptotically equivalent to $\mathbb{G}(N, p)$. However, their probabilistic geometries are rather different. For examples, graphs in $\mathbb{G}(N, k)$ are typically sparse but connected.

9.3.2   *Spatiotemporal Intermittency and Collective Behavior*

Spatiotemporal intermittency in extended systems consists of a sustained regime where coherent and chaotic domains coexist and evolve in space and time. The transition to turbulence via spatiotemporal intermittency has been studied in coupled map lattices whose spatial supports are Euclidean (Chate and Manneville 1988a,b; Kaneko 1985; Stassinopoulos and Alstrøm 1992), and also in nonuniform lattices such as fractals (Cosenza and Kapral 1994) and hierarchical lattices (Cosenza and Tucci 2000).

In regular arrays, the turbulent state can propagate through the lattice in time for a large enough coupling, producing sustained regimes of spatiotemporal intermittency

(Chate and Manneville 1988a,b). Here, we investigate the phenomenon of transition to turbulence in random networks $G(N, k)$ using the local map f (9.30) in the coupled system described by (9.29). As observed for regular lattices, starting from random initial conditions and after some transient regime, our systems settle in a stationary statistical behavior. The transition to turbulence can be characterized through the average value of the instantaneous fraction of turbulent sites F_t, a quantity that serves as the order parameter (Chaté and Manneville 1992a). We have calculated $\langle F \rangle$ as a function of the coupling parameter ϵ for several random networks from a time average of the instantaneous turbulent fraction F_t, as

$$\langle F \rangle = \frac{1}{T} \sum_{t=1}^{T} F_t. \tag{9.31}$$

About 10^4 iterations were discarded before taking the time average in (9.31), and T was typically taken at the value 10^4.

We consider Chaté-Manneville maps coupled on a random network $G(N, k)$ for different parameter values. As initial conditions, we use random cell values uniformly distributed over the interval $[0, r/2]$. Some minimum number of initially excited cells is always required to reach the sustained turbulent state. The typical system size used in the calculations was $N = 10^4$. We have verified that increasing the averaging time T or the network size N do not have appreciable effects on the results.

Two models of random topological configuration have been studied. Model A proposes a random graph to be fixed while the maps are updating. It is, in fact, equivalent to a model of "frozen disorder" proposed in Chaté and Manneville (1992a). Model B possesses a random graph which is changed at each time step simultaneously with the updating of the maps.

We have calculated $\langle F \rangle$ vs. ϵ for random networks with different connection numbers k. The local parameter has been kept fixed at $r = 3$ in most of the calculations. Fig. 9.6 shows the mean turbulent fraction $\langle F \rangle$ versus ε for $G(10^4, 2)$. One can see that, as $\varepsilon > \varepsilon_c \approx 0.145$, the excitation occupies a significant fraction of vertices. The random graph $G(10^4, 2)$ consists of a set of small disjoint subgraphs of the length $(m \ll N)$ and the largest connected component which includes about $O(N^{2/3})$ vertices (Janson et al. 2000). The transition to spatiotemporal intermittency for $k = 2$ is characterized by the scaling relation

$$\langle F \rangle \propto (\varepsilon - \varepsilon_c)^\beta$$

near the critical value ϵ_c, where the critical exponent is $\beta = 0.55 \pm 0.03$ for $r = 3$.

A power law behavior of mean turbulent fraction near the onset of spatiotemporal intermittency also occurs for diffusively coupled CM maps in regular Euclidean lattices (i.e., nearest neighbor coupling) (Chate and Manneville 1988a,b). The value of the critical exponent β for the random network with $k = 2$ coincides with that

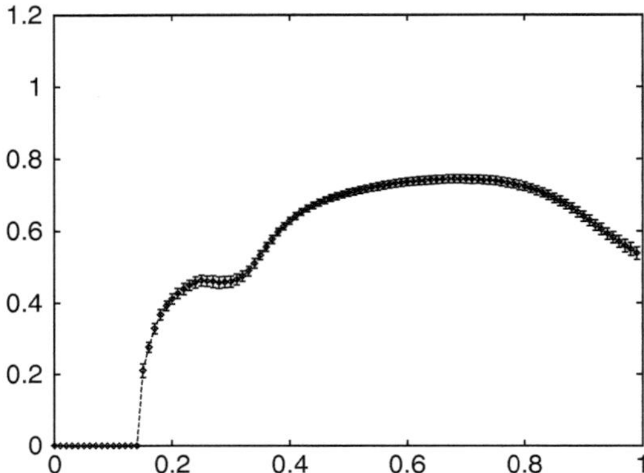

Fig. 9.6 The mean turbulent fraction $< F >$ vs. ε, for $r = 3$. Model A, $\mathbb{G}(10^4, 2)$. Onset of spatiotemporal intermittency

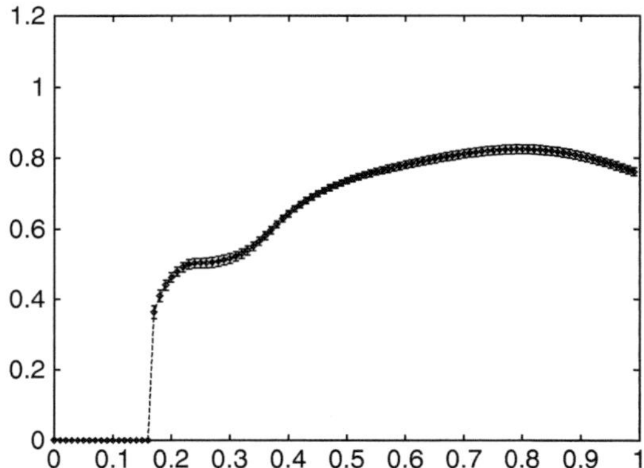

Fig. 9.7 Model A, $\mathbb{G}(10^4, 3)$. The mean fraction $< F >$ vs. ε, for $r = 3$. The critical coupling for the onset of intermittency is $\varepsilon_c \approx 0.161$

found for the two-dimensional lattice (Chate and Manneville 1988b; Houlrik et al. 1990).

For $k = 3$, a Hamilton cycle traversing all vertices in the network appears for the first time. There is no isolated vertex in the graph $\mathbb{G}(10^4, 3)$. Figure 9.7 shows that the onset of intermittency for the case $k = 3$ occurs more abruptly as k is increased.

Figures 9.8, 9.9 display the mean turbulent fraction $\langle F \rangle$ versus the coupling ε for both Model A and Model B in the RCMN induced by realizations of the random graph $\mathbb{G}(10^4, 4)$. Figures 9.8, 9.9 show that the onset of intermittency when $k = 4$ occurs as a discontinuous jump in the order parameter $\langle F \rangle$ at the critical value of the coupling. A discontinuous jump of $\langle F \rangle$ at the onset of spatiotemporal intermittency has also been observed for globally coupled Chaté-Manneville maps and interpreted as a first order phase transition in Cosenza and Parravano (1996).

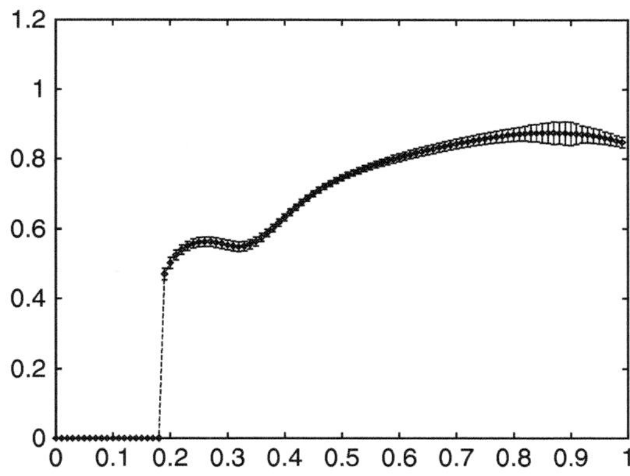

Fig. 9.8 The average turbulent fraction $< F >$ vs ε in the network $\mathbb{G}(10^4, 4)$, Model A, with $r = 3$. The onset of nontrivial collective behavior is observed as an emerging "bulb" around $\varepsilon = 0.85$

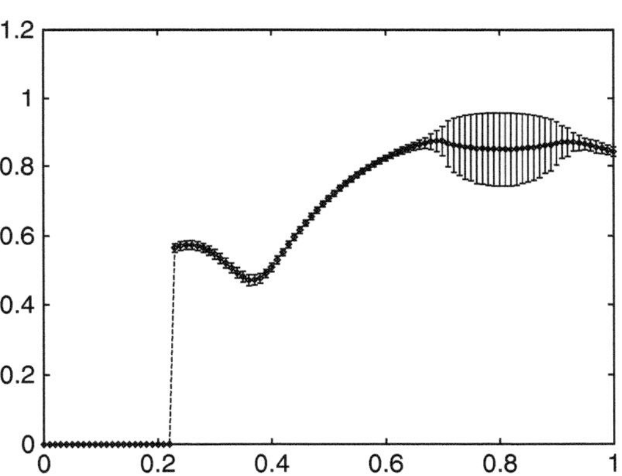

Fig. 9.9 The average turbulent fraction $< F >$ vs. ε with fixed $r = 3$ for $\mathbb{G}(10^4, 4)$, Model B. Nontrivial collective behavior occurs in the "bulb" region

The error bars shown on $\langle F \rangle$ in Figs. 9.8, 9.9 correspond to the standard deviation (the square root of the variance) of the time series of the instantaneous fraction F_t at each value of ε. With increasing system size N, some of those fluctuations do not fade out. Large, non-statistical fluctuations in the time series of the instantaneous turbulent fraction F_t persist with increasing connectivity k in the networks. For $\varepsilon > 0.5$ these fluctuations appear as large "bulbs" around $\langle F \rangle \approx 1$. This phenomenon is associated to the emergence of nontrivial collective behavior commonly observed in CML systems (Chaté and Manneville 1992b). In fact, the observed large amplitudes of the standard deviations reflect collective periodic states of the system.

In Figs. 9.10, 9.11 we show the bifurcation diagram of the instantaneous turbulent fraction F_t as a function of the coupling ε for RCMN induced by the random graphs $\mathbb{G}(10^4, 25)$ and $\mathbb{G}(10^4, 30)$, respectively. Figures 9.10, 9.11 reveal a bifurcating band structure for the range of coupling corresponding to the observed large fluctuations in $\langle F \rangle$, reminiscent of the pitchfork bifurcations of unimodal maps. This observation provides an argument in favor of similarity between the bifurcation diagrams of CMLs and that of their local maps reported for the diffusively coupled maps defined on regular lattices (Lemaître and Chaté 1998). However, one can see

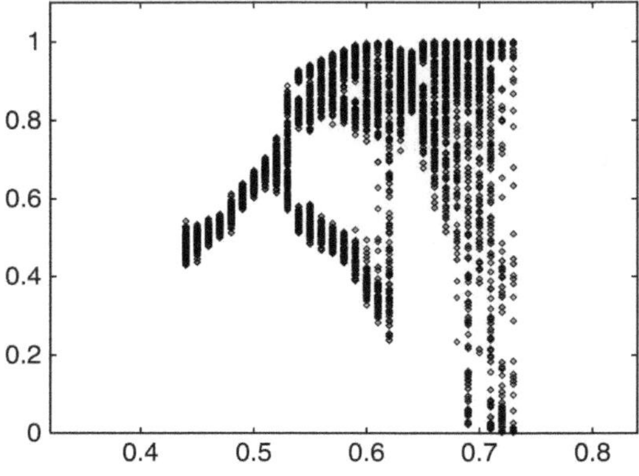

Fig. 9.10 Model A. $\mathbb{G}(10^4, 25)$. Bifurcation diagram of the instantaneous turbulent fraction F_t as a function of ε

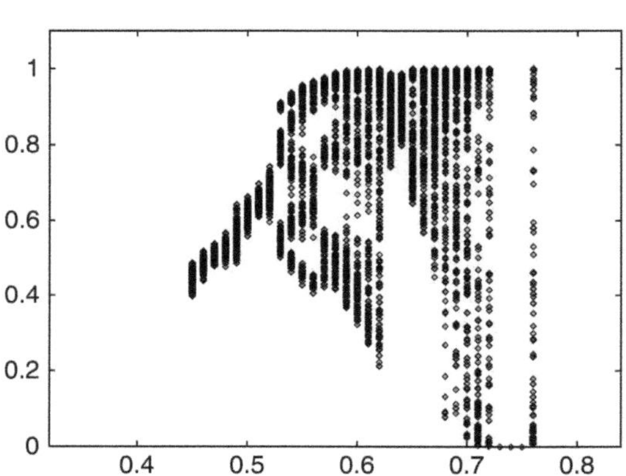

Fig. 9.11 Model A. $\mathbb{G}(10^4, 30)$. Bifurcation diagram of F_t vs. ε

that the detailed bifurcation picture in the case of RCMN depends very much on the particular random graph spanning the network.

From the bifurcation theory point of view, this fact means that the map $\Psi : F_t \to F_{t+1}$ has the sustained uniformly turbulent state $F_t = 1$ as a metastable fixed point. Close to this fixed point, Ψ is a polynomial unimodal map in F_t characterized by the negative Schwarzian derivative,

$$\mathcal{S}\Psi \equiv \frac{\Psi'''}{\Psi'} - \frac{3}{2}\left(\frac{\Psi''}{\Psi'}\right)^2 < 0. \tag{9.32}$$

In this case, Ψ displays an infinite sequence of pitchfork bifurcations when the attractor relevant to the unique chaotic state losses its stability. An example of such a map has been presented in Cosenza and Parravano (1996). These bifurcations are actually observable in some intervals of the parameter values constituting a bifurcation route to collective periodic behavior.

In the diagrams shown on Figs. 9.10, 9.11, the bifurcation branches which draw away from the stable point $F_t = 1$ terminate soon, while for the branches which

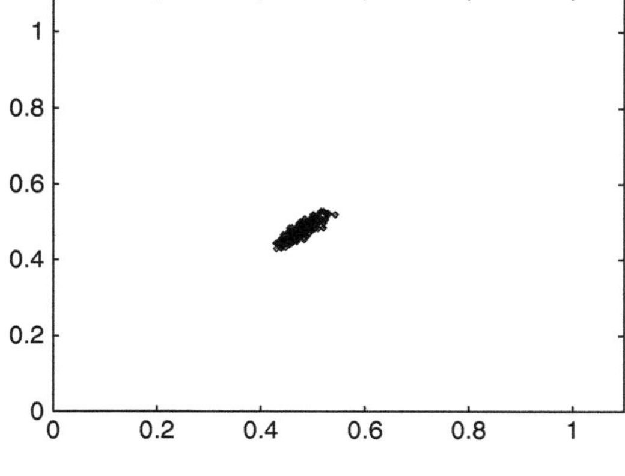

Fig. 9.12 Return map F_{t+1} vs. F_t with fixed $r = 3$ and $\varepsilon = 0.44$, for Model A, $\mathbb{G}(10^4, 25)$. The figure shows the uniqueness of the Gibbs state for $\varepsilon < 1/2$

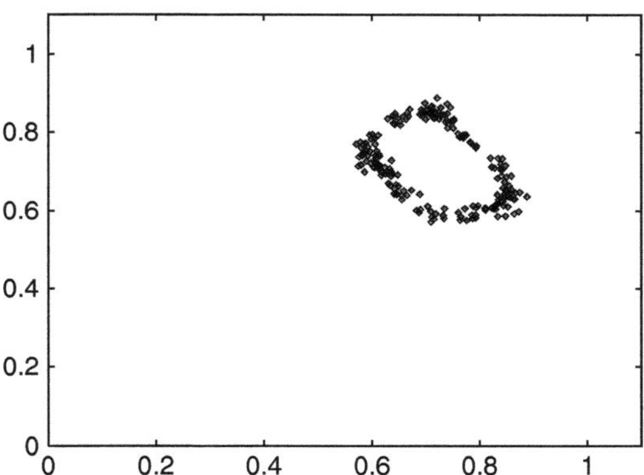

Fig. 9.13 Return map F_{t+1} vs F_t. Model A, $\mathbb{G}(10^4, 25)$; parameters are $r = 3$ and $\varepsilon = 0.54$. A 3-periodic collective motion

tend to the fixed point where the map Ψ is still polynomial, the consequent pitchfork bifurcations are still observable up to the very end of the turbulent window.

The return maps at different values of the coupling ε manifest the collective nontrivial behavior in the network. The return maps F_{t+1} vs. F_t for the network $\mathbb{G}(10^4, 25)$ show that before the onset of bifurcations, the sustained turbulent state in the system corresponds to a fixed point with normal statistical fluctuations, as seen in Fig. 9.12. For $\varepsilon = 0.54$ in Fig. 9.13, the turbulent fraction shows a period three motion. Other nontrivial collective states can be observed at different parameter values and for random networks with different values of k. For example, Fig. 9.14 shows that the instantaneous turbulent fraction F_t displays a period-six collective behavior in a RCMN spanned by the random graph $\mathbb{G}(10^4, 30)$ at $\varepsilon = 0.56$ and $r = 3.0$.

For large enough connectivities k, a re-laminarization process is observed in the systems. That is, at some $\varepsilon'_c > \varepsilon_c$ the mean turbulent fraction again vanishes, establishing a well defined window of spatiotemporal intermittency. This phenomenon has also been observed in globally coupled Chaté-Manneville maps (Cosenza and

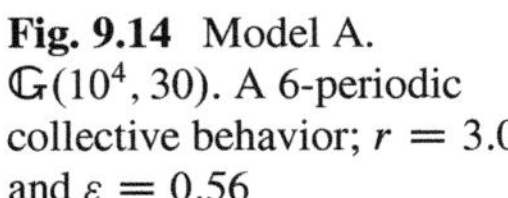

Fig. 9.14 Model A.
$\mathbb{G}(10^4, 30)$. A 6-periodic
collective behavior; $r = 3.0$
and $\varepsilon = 0.56$

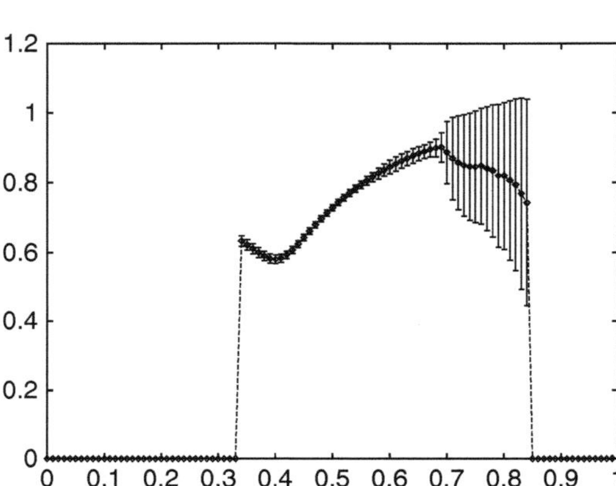

Fig. 9.15 Model A.
$\mathbb{G}(10^4, 10)$. The average
turbulent fraction $< F >$ vs ε

Parravano 1996). This suggests that the collective properties of randomly coupled
map networks and globally coupled maps are similar.

Figure 9.15 shows $\langle F \rangle$ vs. ε for the RCMN induced by the random graph
$\mathbb{G}(10^4, 10)$. The turbulent window is established within the interval $\varepsilon \in [0.33, 0.85]$.
Both the forward and backward transitions to the turbulent state appear as discon-
tinuous jumps in the mean turbulent fraction, similar to the windows of turbulence
in globally coupled maps (Cosenza and Parravano 1996). However, for $k > 10$, $\langle F \rangle$
decreases gradually, as shown in Figs. 9.16, 9.17. For connectivities $15 \leq k \leq 40$,
the mean turbulent fraction scales as

$$\langle F \rangle \propto (\varepsilon_c' - \varepsilon)^{-\gamma}$$

close to the second critical value ε_c'. The second critical exponent is approximately
the same for different k and was estimated at $\gamma = 0.117 \pm 0.003$, for fixed $r = 3$.

As the connectivity k is increased, the windows of turbulence shrink and
eventually disappear, as it can be seen from Figs. 9.15, 9.16, 9.17. We have plotted
the location and the width of the turbulent windows on the coupling parameter axis

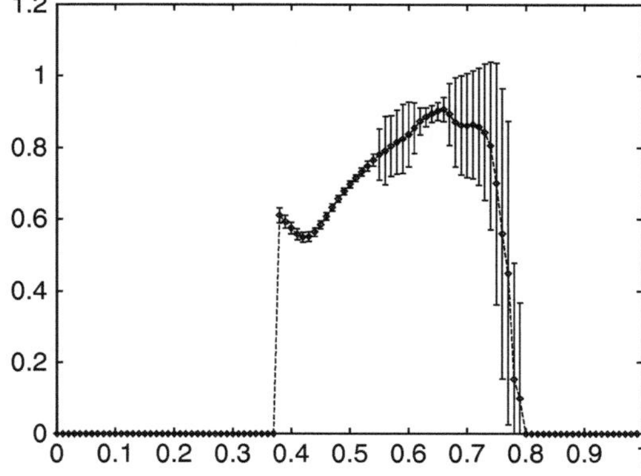

Fig. 9.16 Model A.
$\mathbb{G}(10^4, 15)$. The average
turbulent fraction $< F >$ vs ε

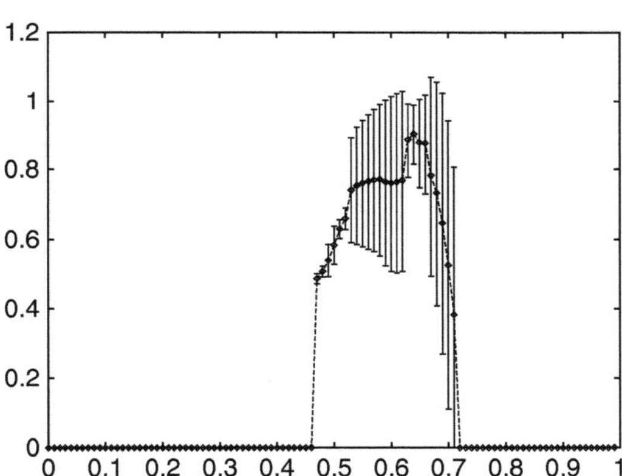

Fig. 9.17 Model A.
$\mathbb{G}(10^4, 35)$. The average
turbulent fraction $< F >$ vs ε

as a function of the connectivity k for both model A and model B in Fig. 9.18a and
b, respectively. In model A, with frozen connectivity, the turbulent window persists
for larger values of k.

9.3.3   The Evolution of $\mathbb{G}(N, k)$ with k

In order to explain the threshold phenomena occurring in random networks of
coupled maps, we consider the evolution of $\mathbb{G}(N, k)$ with k and count the number of
their small and Hamiltonian cyclic components. The previous observations reported
in Chaté and Manneville (1992a) and Manrubia and Mikhailov (1999) indicate that
the detailed evolution of the dynamical clustering depends very much on the entire
architecture of the particular network.

If the connectivity k is fixed, the incoming degree $\mathfrak{I}(\omega_i)$ of the vertex ω_i in
a random graph is a random variable distributed in accordance with the Poisson
distribution

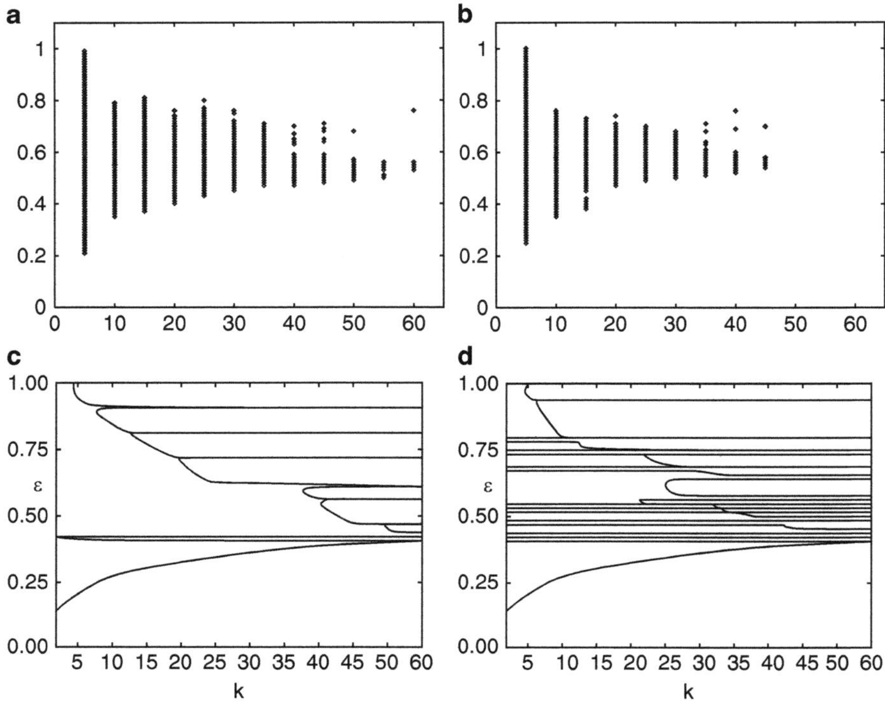

Fig. 9.18 The contraction of the turbulent window on the coupling parameter axis as a function of k; $r = 3$. (**a**) Model A. (**b**) Model B. Examples of phase diagrams $\varepsilon - k$ bounding the region where a nontrivial solution for the system (9.29, 9.30) exists. (**c**) $\lambda\rho_F = 0.5$, $N = 15$. d) $\lambda\rho_F = 0.5$, $N = 100$

$$\mathrm{Po}(z) = \frac{z^n e^{-z}}{n!}$$

where

$$z = kN/(N - 1)$$

is the average number of incoming links (Janson et al. 2000; Newman et al. 2001).

For $k = 1$, all components of $\mathbb{G}(N, 1)$ are small trees or unicycles comprising of $m \ll N$ vertices. As the connectivity approaches $k = 2$, very quickly all the largest components merge into a giant one roughly of $O(N^{2/3})$ vertices. The size distribution of remaining small clusters behaves as

$$P_\mu \propto \mu^{-3/2} \exp(-\mu)$$

(see Newman et al. 2001 for details). The appearance of the giant component at $k = 2$ does not guarantee that there are no isolated vertices in the graph, and that each vertex can be reachable from a given one. In fact, at $k = 2$ the random graph still consists of small disjoint clusters of sizes $m \ll N$.

Let us denote the numbers of small cycles of length $m \leq N - 1$ appearing in $\mathbb{G}(N, k)$ as X_m, such that $X_1 = 0$ is the number of loops, X_2 is the number of double connected vertices, X_3 is the number of triangles, X_4 is the number of squares, etc. All X_m are random variables in a random graph, however, their distributions converge jointly in $\mathbb{R}^\infty$ to the Poisson distributions $\mathrm{Po}(\mu_m)$ where μ_m are the expectations for X_m as $N \to \infty$ (Janson et al. 2000). These expectations can be readily calculated as

$$\mu_m = \frac{1}{m} \prod_{i=1}^{m} \Im(\omega_i) \simeq_{N \to \infty} \frac{k^m}{m},$$

where the factor $1/m$ comes from all permutations of vertex indexes within a cycle.

Hamilton cycles $\mathbb{H}$ are the cycles of length N. The standard analysis in the spirit of Janson et al. (2000) gives

$$\mu_N(k) = (N - 1)! \frac{(kN - 2N - 1)!!}{(kN - 1)!!} \cdot k^N \prod_{i=1}^{N} \Im(\omega_i). \tag{9.33}$$

It is obvious that asymptotically almost surely there are no Hamilton cycles in the graph $\mathbb{G}(N, k)$ for $k \leq 2$ as $N \to \infty$.

When $k \geq 3$, the number of Hamilton cycles in $\mathbb{G}(N, k)$ exhibits a threshold:

$$\mu_N(k \geq 3) \sim_{N \to \infty} \sqrt{\frac{\pi}{2N}} \left[\frac{(k - 2)^{k/2 - 1}}{k^{k/2 - 2}} \right]^N, \tag{9.34}$$

the quantity within the square brackets in (9.34) is greater than 1 for any $k \geq 3$, therefore, $\mu_N(k \geq 3) \to \infty$ as $N \to \infty$, and graphs in $\mathbb{G}(N, k \geq 3)$ have lots of Hamilton cycles and no isolated vertices.

The difference between onsets of intermittency exhibited by the $\mathbb{G}(N, 2)-$ and $\mathbb{G}(N, 3)-$ networks relies upon the properties of sharp and coarse thresholds occurring in random graphs and allows for a qualitative insight into the intermittency mechanism itself. In accordance to Friedgut (1999), a graph property that depends on containing a large subgraph have sharp thresholds. Otherwise, a monotone graph property with a coarse threshold (usually characterized by a power law) may be approximated by the property of containing at least one of a certain (finite) family of small graphs as a subgraph.

The case of $k = 4$ is of particular interest since the number of edges $E = 2N$, and the probability measure defined on $\mathbb{G}(N, 4)$ is contiguous to that one defined on a simple sum of two independent Hamilton cycles $\mathbb{H}_1(N) \oplus \mathbb{H}_2(N)$ as $N \to \infty$ (Janson et al. 2000). The error bars shown in Figs. 9.8 and 9.9 correspond to a pitchfork bifurcation of the mean chaotic fraction into two distinct chaotic "states" characterized by the different values of $\langle F \rangle$ at the same couplings.

For $k > 4$, there are many of independent Hamilton cycles in the $\mathbb{G}(N, k)$ (Janson et al. 2000), and $\langle F \rangle$ exhibits multiple bifurcations at different values of ε.

9.4   Thermodynamics of Random Networks of Coupled Maps

In the literature related to physics, the transitions to spatio-temporal intermittency and back, to a uniformly inhibited state, deserve the name of a phase transition since the behavior of the mean chaotic fraction (which plays the role of an order parameter) close to a critical value of coupling ε_c resembles the behavior of thermodynamical quantities close to a critical point in critical phenomena theory.

For $k = 2$, the system of randomly coupled maps exhibits a scaling behavior

$$\rho_\phi \propto (\varepsilon - \varepsilon_c)^\gamma$$

as $\varepsilon \to \varepsilon_c+$ (similar to the CMLs defined on the regular Euclidean lattice, Chaté and Manneville 1992a) that is typical for the *second order* phase transitions. If the connection is more dense, this transition appears as a discontinuous jump of $\langle F \rangle$ as the coupling parameter exceeds a threshold value ε_c (as well as for the globally coupled maps, Cosenza and Parravano 1996) and resembles the *first order* phase transitions. For the backward transition from the intermittency to a uniformly synchronized state, one can see that for minimal connectivities (as well as for the regular Euclidean lattices, Chaté and Manneville 1992a) such a transition does not occur for any $\varepsilon < 1$. However, for large connectivities ($k \geq 10$) (as well as for the globally coupled maps, Cosenza and Parravano 1996) this transition appears as a discontinuous jump. Data show convincingly that the formal analogy with phase transitions occurring in ferromagnetic physics does not provide us an adequate classification for the critical phenomena observing in the random networks of coupled maps.

In the literature which is close to mathematics, the notion of a phase transition expresses a non-uniqueness of probability invariant measures defined on all symbolic configurations ξ representing the dynamics of one-dimensional coupled maps. The general idea of this approach is to study such representations via the Gibbs states for the two-dimensional systems and goes back to the studies (Sinai 1972; Ruelle 1978).

Provided the symbolic representation for the dynamics exists (i.e., there exists a semi-conjugacy T such that for any dynamical orbit $\Phi^t x$ a symbolic code

$$\xi = \xi_1 \xi_2 \ldots \xi_t, \quad t \in \mathbb{Z}_+,$$

can be assigned satisfying the relation

$$T(\sigma \xi) = \Phi T(\xi),$$

where σ is a sub-shift on the symbolic configuration ξ), the probability invariant measure,

$$\rho_F = \lim_{T \to \infty} \frac{1}{T} \int_0^T F_t(\Phi^t x)\, dt,$$

can be defined as a Gibbs state,

$$\rho_F(\xi) = \frac{1}{Z} \exp\left[-\beta H(\xi)\right] \tag{9.35}$$

on all symbolic configurations ξ such that the correspondent orbits $\Phi^t x$, $t \in \mathbb{Z}_+$, are characterized by the positive Lyapunov exponents (Gielis and MacKay 2000)

$$\lambda_n > 0, n = 1 \ldots N,$$

In (9.35), Z is a normalization constant, $\beta > 0$ is an "inverse temperature" parameter characterizing a width of distribution (usually taken as $\beta = 1$).

The Hamiltonian $H(\xi)$ in (9.35) differs essentially from those of physics, because even in the simplest uncoupled case ($\varepsilon = 0$) it has a nontrivial interaction in the time direction (is not local in time). An example of such a Hamiltonian for the extended coupled maps ($N \to \infty$) has been given in Gielis and MacKay (2000),

$$H(\xi) = \sum_{t>0} \log\left|\det\left[J_\Phi(T(\sigma^t \xi))\right]\right|, \tag{9.36}$$

in which J_Φ is the Jacobian matrix of the transformation (9.29). The expression (9.36) can be evaluated for the model (9.29, 9.30) explicitly.

Since $(\Phi x)_\omega$ is linear, the Jacobian J_Φ does not depend of x_ω. Consequently, the Hamiltonian (9.36) is independent of the certain configuration ξ and is just a function $H(\varepsilon, r, k)$. A nice property of random networks is that the configuration of $\mathbb{G}(N, k)$ depends neither of time T nor of the network size N in the thermodynamic limit $N \to \infty$ (Janson et al. 2000).

In order to estimate the Hamiltonian (9.36), following Bricmont and Kupiainen (1996) and Jiang and Pesin (1998), we write the Jacobian matrix in the form

$$J_\Phi = J_0(\mathbf{1} - J_0^{-1} J_1),$$

where J_0 is the diagonal part correspondent to the uncoupled maps, and J_1 comes from the coupling,

$$\log|\det J_\Phi| = \log|\det J_0| + \log\left|\det(\mathbf{1} - J_0^{-1} J_1)\right|. \tag{9.37}$$

Directly from definitions, for the first term in (9.37), we have

$$\log|\det J_0| = NF_t\lambda + N\log(1 - \varepsilon), \tag{9.38}$$

in which $\lambda = \log r$ is the Lyapunov exponent of the local map (9.30) being in the chaotic state and NF_t is a number of such maps at time t. In accordance to Gielis and MacKay (2000), Bricmont and Kupiainen (1996) and Jiang and Pesin (1998), we use the standard relation $\det = \exp \mathrm{Tr} \log$ to transform the second term in (9.37) and then expand logarithm into power series,

$$\log \left| \det(\mathbf{1} - J_0^{-1} J_1) \right| = \sum_{n>0} \frac{\mathrm{Tr}\left[(J_0^{-1} J_1)^n \right]}{n}. \tag{9.39}$$

Let us note that the matrix $J_0^{-1} J_1$ is just a "weighted" adjacency matrix of the graph $\mathbb{G}(N, k)$ spanning the random network. Namely, $[J_0^{-1} J_1]_{\omega\omega'} = 0$ if the vertices ω and ω' have no connections.

$$\left[J_0^{-1} J_1 \right]_{\omega\omega'} = -\frac{\varepsilon}{k(1-\varepsilon)}$$

if the vertices ω and ω' are connected and cells are synchronized in the same state (either chaotic or inhibited).

$$\left[J_0^{-1} J_1 \right]_{\omega\omega'} = -\frac{r}{\varepsilon/k(1-\varepsilon)}$$

if the vertices ω and ω' are connected and the $\omega-$cell is in the chaotic state, but the $\omega'-$cell is in the alternative state.

$$\left[J_0^{-1} J_1 \right]_{\omega\omega'} = -\frac{\varepsilon}{rk(1-\varepsilon)}$$

if the vertices ω and ω' are connected and the $\omega-$cell is in the inhibited state, but the $\omega'-$cell is in the chaotic one.

It is easy to check that, up to a factor, $\mathrm{Tr}\left[(J_0^{-1} J_1)^n \right]$ equals to the total number of cycles of length n, X_n, presented in the graph (Lind and Marcus 1995),

$$\mathrm{Tr}\left[(J_0^{-1} J_1)^n \right] = X_n \left(-v/k \right)^n, \quad v \equiv \frac{\varepsilon}{(1-\varepsilon)}. \tag{9.40}$$

and is independent of r. We recall that loops are not allowed for the model in question, so that $X_1 = 0$, and $X_{n>N} = 0$, since the Hamiltonian cycle is a cycle of maximal length.

In a random graph, the numbers X_n are the random variables distributed in accordance to the Poisson law $\mathrm{Po}(\mu_n)$ specified by the mean values $\mu_n = k^n/n$ independently of time as $N \to \infty$. While interesting in the thermodynamic limit $(N \to \infty, T \to \infty)$ one can replace X_n with its mean μ_n and use ρ_F instead of the sum $T^{-1} \sum_{t=1}^{T} F_t$,

$$H \simeq NT \left(\lambda \rho_F + \log(1 - \varepsilon) + \lim_{N \to \infty} \frac{1}{N} \sum_{n>1}^{N} \frac{(-v)^n \mu_n}{n^2 k^n} \right). \tag{9.41}$$

The first and second terms in (9.41) are irrelevant to the topology of couplings which is taken into account by the third term. It is contributed separately from small cycles and from Hamiltonian cycles traversing all nodes of the network as $k > 2$,

$$\lim_{N \to \infty} \sum_{n>1}^{N} \frac{(-v)^n \mu_n}{n^2 k^n} = \left[\lim_{N \to \infty} \sqrt{\frac{\pi}{2}} \frac{v^N}{N^{5/2}} \left[\frac{k-2}{k} \right]^{\left(\frac{k}{2}-1\right)N} \right]_{k>2} + \mathrm{Li}_2(-v) + v \tag{9.42}$$

where the dilogarithm function is

$$\mathrm{Li}_2(z) = \sum_{n=1}^{\infty} z^n / n^2,$$

and we have used the expression (9.34) instead of $\mu_N (k > 2)$.

The dilogarithm function is analytic if $|v| < 1$ and is defined by analytic continuation otherwise with the use of the integral representation (Lewin 1981; Abramovitz and Stegun 1986)

$$\mathrm{Li}_2(-z) = - \int_0^z \frac{\ln(1+t)}{t} \, dt. \tag{9.43}$$

The branches other than the principal branch (which is analytic at $z = 0$), the point $z = 0$ is a branch point (another branch point is $z = -1$, but $z > 0$ in the problem we consider), and the branch cut is taken to be the positive real axis. We use the principal branch of dilogarithm.

As a result, in the thermodynamic limit ($N \to \infty, T \to \infty$), one arrives at the following asymptotic expression for the Hamiltonian:

$$H(\varepsilon, k, r) \simeq T(\lambda \rho_F + v) + T \cdot \mathrm{Li}_2(-v) \tag{9.44}$$

$$+ T \lim_{N \to \infty} \left[\sqrt{\frac{\pi}{2}} \frac{v^N}{N^{5/2}} \left[\frac{k-2}{k} \right]^{\left(\frac{k}{2}-1\right)N} \Bigg|_{k>2} + N \log(1 - \varepsilon) \right],$$

where $v \equiv \varepsilon/(1 - \varepsilon)$, $\mathrm{Li}_2(z)$ is the dilogarithm function, and

$$\lambda = \log r$$

is the Lyapunov exponent of the uncoupled map, and we have used ρ_F, instead of the limit

$$\lim_{T \to \infty} T^{-1} \sum_{t=1}^{T} F_t.$$

In the important case of the uncoupled maps ($\varepsilon = 0$) as well as for the minimal connectivities ($k = 1, 2$) the Hamiltonian (9.44) is independent from the topology of couplings. The only k-dependent contribution to H, for $k > 2$, tends to zero as $N \to \infty$ if

$$\varepsilon \geq \frac{1}{1 + \left(\frac{k-2}{k}\right)^{\left(\frac{k}{2}-1\right)}} \tag{9.45}$$

and is unbounded otherwise. For any finite value of T and given values of $\lambda \rho_F$ and N, the requirement

$$|H(\varepsilon, k, r)| < \infty,$$

defines a phase diagram $\varepsilon - k$ bounding the region where a nontrivial solution for the system (9.29, 9.30) exists. Two examples of such diagrams for different values of $\lambda \rho_F$ and N are displayed in Figs. 9.18c and d.

Substituting the result (9.44) back, into (9.35), and then solving a transcendental equation for ρ_F, one obtains the following expression

$$\rho_F = \frac{1}{\beta \lambda N T} \mathrm{W}\left(\frac{\beta \lambda N T}{Z} e^{-\beta T \left[N \log(1-\varepsilon) + \mathrm{Li}_2(-\nu) + \nu + \frac{\nu^N}{N^{5/2}} \sqrt{\frac{\pi}{2} \left(\frac{k-2}{k}\right)^{N(k-2)}}\right]}\right) \tag{9.46}$$

where $\mathrm{W}(z)$ is the Lambert function satisfying the equation

$$\mathrm{W}(z) \exp\left[\mathrm{W}(z)\right] = z.$$

For each $z \neq 0$, this equation has an infinite number of solutions, consequently, W has an infinite number of branches $\mathrm{W}(n, z)$, one of which $\mathrm{W}(0, z)$ is analytic at $z = 0$ (the principal branch). All other branches have a branch point at $z = 0$, and the branch cut dividing them is the negative real axis (Corless et al. 1996).

The Lambert function is closely related to the tree generating function in the analysis of algorithms discipline (Corless et al. 1996), and allows us for a physical interpretation of the result (9.46). Namely, if $\mathcal{N}[n]$ counts the number of distinct oriented trees with n labeled vertices, then

$$\mathcal{N}[n] = (-1)^n \beta \lambda N T \frac{d^n \rho_F(z)}{dz^n} \tag{9.47}$$

where z is the argument of Lambert's function in (9.46).

The transition to turbulence via spatiotemporal intermittency has been investigated in Cosenza and Tucci (2002), in the context of coupled maps defined on small-world networks. The critical boundary separating laminar and turbulent regimes was calculated on the parameter space of the system, given by the coupling strength and the rewiring probability of the network. It was observed that windows of re-laminarization are present in some regions of the parameter space. New features arise in small-world networks; for instance, the character of the transition to turbulence changes from second-order to a first-order phase transition at some critical value of the rewiring probability. A linear relation characterizing the change in the order of the phase transition is found. The global quantity used as order parameter for the transition also exhibits nontrivial collective behavior for some values of the parameters. These models may describe several processes occurring in nonuniform media where the degree of disorder can be continuously varied through a parameter.

In a conclusion, we have studied the main features associated with the transitions to spatiotemporal intermittency and relaminarization and the transitions to collective behavior occurring in the randomly coupled Chaté-Manneville minimal maps. Numerical simulations as well as a theoretical framework for these systems have been presented. The thermodynamic formalism based on the symbolic representation for the dynamics of randomly coupled chaotic maps has been introduced.

We have found that for low connectivities the transition to turbulence via the spatiotemporal intermittency in the randomly coupled chaotic maps occurs as a power law close to a critical value of the coupling parameter. In contrast, a discontinuous jump in the mean turbulent fraction $\langle F \rangle$ takes place for medium and large connectivities. As the connectivity increases, a synchronization of the system toward the uniformly laminar state occurs at another critical value of the coupling $\varepsilon'_c > \varepsilon_c$. Similarly to the case of globally coupled Chaté-Manneville maps, windows of turbulence are established. The onset of relaminarization appears as a discontinuous jump to a uniformly laminar state if the connectivity is around $k = 10$, and as a power law decay of the mean turbulent fraction $\langle F \rangle$ close to ε'_c for $k > 10$. We have shown that the turbulent windows contracts with increasing of the connectivity k, until they vanish when k is about several tenths. Additionally, the periodic collective behavior arises within the windows of turbulence, as in the globally coupled maps. Although randomly coupled maps and globally coupled maps have different topological properties, our results show that these two classes of networks behave collectively in analogous ways. Discontinuous phase transitions, well defined turbulent windows and the nontrivial collective behavior are common and distinctive features emerging in both classes of networks.

The observed collective properties of the system have been analyzed through the thermodynamic formalism. We have used a symbolic representation for the dynamics of coupled maps and studied this representation via the Gibbs states. A nice feature of the linear maps is that the Hamiltonian of the coupled system is independent from the particular symbolic configuration and is a function of the coupling strength, the connectivity, and the parameter of the map. In the thermodynamic limit, the configuration of the random graph is independent of time and the lattice size that very much simplifies the calculations. The non-uniqueness of the Gibbs state stipulates the complex collective periodic behavior observed in numerical simulations.

9.5 Large Gene Expression Regulatory Networks

The process of gene expression controlled by other genes expressed at the same time by the external signals is called the *gene regulation* (Ptashne and Gann 2002). In fact, the gene regulation is the control of transcription driven by a *promoter* of a specific gene. Transcription begins downstream from the promoter at a particular sequence of DNA that is recognized by the polymerase as the start site of transcription. A chemical sequence of DNA known as the start *codon* codes for the region of the gene that is converted into amino acids, the protein building blocks (Jacob and Monod 1961; Dickson et al. 1975). Being a single biochemical reaction, transcription is actually a complex sequence of reactions (von Hippel 1998) accomplished by a number of regulatory proteins called transcription factors and can occur in a positive or negative sense. Positive regulation, or *activation*, occurs when a protein increases transcription through biochemical reactions that enhance polymerase binding at the promoter region. Negative regulation, or *repression*, involves the blocking of polymerase binding at the promoter region.

The protein-DNA *feedback circuits* arise when the translated protein is capable of interacting with the promoter that drives its own production or promoters of other genes (Glass and Kauffman 1973; McAdams and Shapiro 1995) and are responsible for much of the nonlinearity that arises in genetic networks (Keller 1995; Smolen et al. 1998; Wolf and Eeckman 1998) such as the multi-stability and oscillations in the steady-state protein concentrations.

A central goal of post-genomic biology is the elucidation of the regulatory relationships among all constituents that together comprise the *genetic network* of a cell. This is a tough problem at present due to insufficient availability of micro-array data and due to the fact that post-transcriptional regulatory interactions are reflected only indirectly in mRNA expression measurements (Farkas et al. 2003). The cooperative functions between genes can be visualized through a graph Γ where nodes denote genes and links do activating or repressing effects on transcription (Rho et al. 2006).

In the present section, we discuss a toy discrete time deterministic model of a large regulatory network comprising of dynamically coupled synchronously updated units defined on both the homogeneous and scalable initial graphs studied in Volchenkov and Lima (2005). For a small regulatory network comprising of just a few elements, a direct logical analysis of such a model is possible (de Jong and Lima 2005; Coutinho et al. 2006). Herewith, the positive and negative feedback circuits drive its dynamics (Snoussi and Thomas 1993; Thomas and Kaufman 2001). However, for the real gene expression regulatory network that can consist of many thousands of interacting genes, such a direct logical analysis becomes a difficult problem because of its enormous complexity. In particular, in the large regulatory networks, the proteins often regulate their own production in a complex web of asynchronous interactions (Klemm and Bornholdt 2005a). Being interested in the highly reproducible dynamical patterns of regulatory processes in the toy model, we consider a statistical ensemble of such networks in which any layout of switching parameters featuring step-like interactions between the network units can be possible with some probability. It is important to note that the biological systems rely on the robust internal information processing; however, the biological information processing elements (genetic switches, neurons, etc.) are intrinsically noisy (Klemm and Bornholdt 2005b). Such a noise poses the severe stability problems to the entire system behavior, as it tends to desynchronize system dynamics (Hasty et al. 2000; Bialek 2001; Bornholdt 2005).

The effect of noisy switches onto a continuous time model of gene regulation has been discussed in Hasty et al. (2000), Bialek (2001) and Walczaka et al. (2005). The random shuffling of switching parameters would also model a source of fluctuations in the gene expressions (Bilke 2000; Bilke et al. 2003). Another important issue we concern is the effect of network topology on the dynamics and vice versa. This effect can be observed in many extended dynamical systems, for instance, in the problem of epidemic spreading over the graphs of different topology (Murray 1993), and has also been intensively discussed in the various post-genomic studies (see Maslov and Sneppen 2002; Farkas et al. 2003; Klemm and Bornholdt 2005a; Shaw 2003; Groenlund 2004).

The dynamical behavior of the toy model proposed in Volchenkov and Lima (2005) is characterized with the multi-stationarity and the multi-periodicity of oscillations of transcription factors that is governed by the circular sequences of interactions (*the feedback circuits*). The transcription degrees of gene regulations, the number of cycles in the interaction graph, and their lengths control the statistics of their appearance in the toy model.

9.5.1   A Model of a Large Gene Expression Regulatory Networks

It is usual in the modeling of gene transcription process (Wagner 1994) to assume that the expression of genes in the network is regulated exclusively on the transcription level, that each gene of the network produces only one species of an active transcription regulator, and that the enhancer elements mediating the regulator's effect on the expression of a target gene act independently from enhancer elements for other regulators of the same gene. In addition to the above assumptions, we suppose that the transcription activation (repression) effect on a target gene can alternate neither upon the concentration of the target gene nor upon the concentrations of other transcription regulators. It is believed that strong cooperative effects of transcription activation by individual transcription factors are mainly responsible for the expression of a target gene (Wagner 1994).

We consider the *maximal* directed graph Γ where $N \gg 1$ nodes denote genes and links account for *all* possible activating ($s_{ik} = +1$) or repressing ($s_{ik} = -1$) effects on expression. One can define an instantaneous *active subgraph* $\Gamma^t \subseteq \Gamma$ including all interactions *efficient* at time $t \geq 0$.

Following Volchenkov and Lima (2005), de Jong and Lima (2005) and Coutinho et al. (2006), we consider a discrete time dynamical system which state at time $t + 1$

is a function of its state at time t, in the form of a coupled map lattice (Kaneko 1993). For each gene located at the node i of the graph Γ, we define two dynamical variables: $x_i^t \in [0, 1]$, the normalized admissible concentration of the correspondent regulator within a cell at time $t \geq 0$, and $y_i^t \in [0, 1]$, the cumulative effect of other genes of network onto the expression of i at time $t \geq 0$. In the *irreducible* graphs (wherein each node has incoming edges), the exertion y_i^t can be defined as the mean,

$$y_i^t = \frac{\sum\limits_{k} e_{ik} A_{ik}^t}{\sum\limits_{k} e_{ik} A_{ik}^0}, \tag{9.48}$$

in which A_{ik}^0 and A_{ik}^t are the adjacency matrices of the graphs Γ and Γ^t respectively, $e_{ik} \in [0, 1]$ is the *effectiveness* expressing the *strength* of interaction of the product of gene k with gene i, the weight of the link. The effect of regulations inside the network can be obscured by the inhomogeneous e_{ik}, and we set is to $e_{ik} = 1$. The adjacency matrix A_{ik}^t is updated synchronously at each time step $t > 0$, in accordance to the current values of $\mathbf{x}^t \in [0, 1]^N$,

$$A_{ik}^t = A_{ik}^0 \Theta \left(s_{ik} \left(x_i^t - T_{ik} \right) \right),$$

in which the step function $\Theta(z) = 1$, for $z > 0$, and $\Theta(z) = 0$, for $z \leq 0$, and $T_{ik} \in [0, 1]$ is the threshold value for the action of i onto k, and $s_{ik} = \pm 1$ is the sign of interaction (the transcription degree). Therefore the link belongs to the active subgraph at time $t \geq 0$ as soon as $x_i^t > T_{ik}$ for the transcription activation $s_{ik} = +1$ and as $x_i^t < T_{ik}$ for the transcription repression. The discrete time synchronous coupling,

$$\mathbf{x}^{t+1} = \alpha \mathbf{x}^t + (1 - \alpha) \mathbf{y}^t \tag{9.49}$$

generates the flow φ^t in the phase space $[0, 1]^N \times [0, 1]^N$ transforming the initial point $(\mathbf{x}^0, \mathbf{y}^0)$ into $(\mathbf{x}^t, \mathbf{y}^t)$ for some $t > 0$ due to cross-regulation and auto-regulation of the expression of genes. The parameter $0 \leq \alpha < 1$ is a positive constant expressing the decay rate of concentrations, the second term in (9.49) describes the rates of protein synthesis. The protein decay rate α determines a time scale of system,

$$t \rightarrow (1 - \alpha)t,$$

which depends on biochemical parameters, such as the rate of transcription or the time necessary to export mRNA into the cytoplasm for translation (Wagner 1994) that does not affect the asymptotic behavior of system. For $\alpha \approx 1$, the system (9.49) is decoupled, and the concentrations decay exponentially fast with time. The initial state $(\mathbf{x}^0, \mathbf{y}^0)$ may be a response to an extracellular signal imposed onto the network by the *upstream* genes that are not regulated by other members of the network (Wagner 1994). In their turn, the genes of network would regulate the transcription of *downstream genes*.

Although the map (9.48, 9.49) is similar to those discussed in the pioneer work (Wagner 1994) and a number of papers on the threshold networks in the context of gene regulation published since then, it is conceptually different from them in that the model (9.48, 9.49) constitutes a synchronously updated array of elements evolving discretely in time that is typical for the coupled map lattices (Kaneko 1993). Several low dimensional examples of (9.48, 9.49) have also been considered in de Jong and Lima (2005) and Coutinho et al. (2006).

Being interested in the reproducible stable expression patterns attained through the regulation process governed by the model (9.48, 9.49), in the current lack of full data on the large gene networks, we treat the switching parameters T_{ik} and s_{ik} as the discrete random variables: we suppose that *any* layout of them can be possible with some nonzero probability, and study the statistical behavior of the regulation flow φ^t over the statistical ensemble of such networks. Namely, we suppose that the threshold value assigned to each directed edge linking genes in the network can take one of $n \leq N^2$ possible values,

$$0 \leq T_1 < \ldots < T_n \leq 1,$$

uniformly distributed over the unit interval $[0, 1]$. We vary the number of distinct thresholds n from several tens to several hundreds tuning the coarse graining of phase space while $N = 10^3$ nodes. We also suppose that the transcription degree s_{ik} takes the value $+1$ with some probability $0 \leq \eta \leq 1$, and $s_{ik} = -1$ with the probability $(1 - \eta)$. In the numerical simulations reported below, we have checked out 500 different *random* arrays of initial conditions $\mathbf{x}^0$ for each of 500 *random* layouts of T_{ik} and s_{ik}. Any observed stable pattern would induce expression of specific downstream genes affecting the phenotype of the organism (Wagner 1994).

Dynamical behavior observed in the discrete time model (9.48, 9.49) with the random layouts of switching parameters depends very much upon the structure chosen for the maximal graph Γ. Indeed, the formation and topology of the functional feedback circuits driving the dynamical behavior in the regulatory networks is closely related to the occurrence of cycles in Γ (Thieffry 2000; Gleiss et al. 2001; Conant and Wagner 2003). A highly inhomogeneous *scale-free* structure in the gene regulatory network topology has been reported recently from the statistical analysis of available genome data (Farkas et al. 2003; Wagner 2003). It is worth a mention that the vertex degree statistics alone does not fix the statistics of cycles of length ℓ that seems to define a preferential topology for the feedback circuits. The distributions of the numbers of cycles of length ℓ in random scale free graphs look nearly Gaussian, and the widths of these distributions crucially depend upon the used graph generation algorithm. In the present section, we use the graph generating algorithm proposed in Volchenkov et al. (2002a), which produces a random directed scale free graph with a sharp distribution of C_ℓ at $\ell = 2$ (the *bidirectional* edges).

Since the scale free graphs are relatively depleted in cycles, the effect the cycles take onto the dynamics of the network (9.48, 9.49) is obscured. To quantify it, we study the map (9.48, 9.49) on a family of inhomogeneous random scale free like graphs characterized with the different fractions $0 < \mu \leq 1$ of bidirectional edges constituting the cycles of length $\ell = 2$. We have obtained them by the random adding of complimentary edges to the initial scale free random graph.

We have also studied the model (9.48, 9.49) defined on the homogeneous graphs: the fully connected graph and the regular random graph in which each node has precisely $k \gg 1$ neighbors (for the fully connected graph, $k = N - 1$, and the probability that each node is connected to any other $p = 1$).

9.5.2   Numerical Analysis of Large Gene Expression Regulatory Networks

We commence the analysis with the networks defined on homogeneous graphs. After a short transient process, the stochastic system (9.48, 9.49) settles into a statistically stable dynamical regime which depends very much upon the fraction of negative regulations allowed between genes, $1 - \eta$. For any random initial string $\mathbf{x}^0$, it approaches to a fixed point exponentially fast as $\eta \to 1$. The transient decay rates can be tracked out with the velocities,

$$\mathbf{v}^t = \mathbf{x}^{t+1} - \mathbf{x}^t, \quad u^t = N^{-2} \sum_{ik} \left| A_{ik}^{t+1} - A_{ik}^t \right|. \tag{9.50}$$

It is important to note that the active subgraph gets stabilized first $\Gamma^t \to \Gamma_\star$ (see Fig. 9.19), and then the concentrations tends to a fixed point, $\mathbf{x}^t \to \mathbf{x}_\star$, which depends upon both the initial condition $\mathbf{x}^0$ and the certain layout of switching parameters that is an evidence of multi-stationarity in the model.

Because of the dynamical map (9.48, 9.49), the initial concentration of a protein x^0 changes in time and settles at a fixed point $x_\star$ located randomly within $[0, 1]$. The empirical probability density distributions $p(x_\star)$ have a broad bell shape, as the positive interactions prevail (see Fig. 9.20). To display the dependence of $p(x_\star)$ upon the choice of $\mathbf{x}^0$, we have collected the data over 500 random initial conditions and presented them as a box-plot in Fig. 9.20. Therein, a lower line of a box shows the first quartile, and an upper line of a box shows the third quartile. Half of the difference between the third quartile and the first quartile (the semi-interquartile range or the quartile deviation) is a measure of the dispersion of the data. Two lines extending from the central box of maximal length 3/2 the interquartile range.

The active subgraph $\Gamma_\star$ constitutes a random graph (*half-dense* in comparison with the fully connected graph $\Gamma(10^3)$) for any choice of $\mathbf{x}^0$ and any layout of switching parameters. In Fig. 9.21, we have shown the probability degree distributions (the circles are for the incoming degrees k_+, and the diamonds are for the outgoing degrees k_-) for the nodes of $\Gamma_\star$ formed at $\eta = 99\%$, the solid line on Fig. 9.21 displays the Gaussian degree statistics which is typical for the Erdös and Rényi random graphs (Erdös and Rényi 1960).

Increasing the fraction of negative regulations ($s_{ik} = -1$) in the model (9.48, 9.49) approximately up to $1 - \eta \approx 10\%$, one arrives at a complicated spatiotemporal

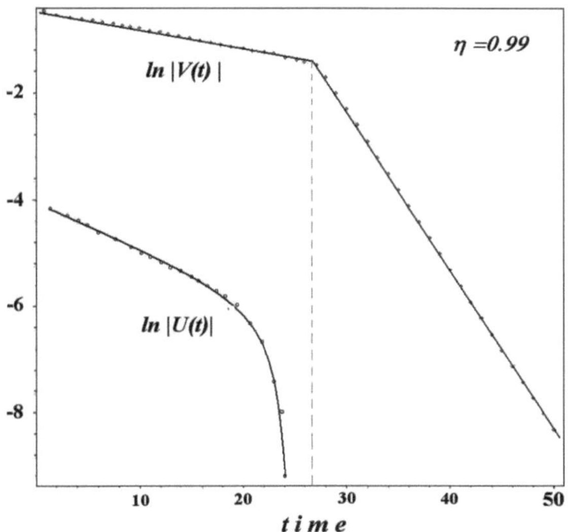

Fig. 9.19 The model (9.48, 9.49) defined on homogeneous initial graphs approaches a stable stationary state exponentially fast for any random initial condition $\mathbf{x}^0$ provided the positive regulations (activations) prevail in the system ($\eta \rightarrow 1$). The transient decay rate $|v^t|$ exhibits a crossover between two consequent phases of transient regime: before and after the structure of the active subgraph has been congealed. The exponential decay rate of transients is independent on neither the choice of initial conditions nor the certain layout of switching parameters, but the stationary asymptotic configuration $\mathbf{x}_*$ depends upon both

behavior, in which the nodes characterized by the stationary concentrations of regulators coexist and interleave with those of oscillating concentrations. In contrast to many synchronously updated discrete time dynamical systems defined on the regular lattices (Chate and Manneville 1988a; Chaté and Manneville 1992a) and on the regular random graphs (Sequeira et al. 2002), in the *threshold network* (9.48, 9.49), the dynamical state (oscillations) is restricted to a number of domains and does not propagate across the rest of network. Domains of oscillating concentrations are bounded by the nodes with the amplitudes of oscillations insufficient to cross the next thresholds. The number of nodes with the oscillating concentrations of regulators N_{osc} grows up to a half of network size as the fraction of negative regulations $1 - \eta$ increases; it plays the role of an *order parameter* describing the phase transition in the system,

$$\delta = \frac{N_{osc}}{N}. \tag{9.51}$$

Numerical analysis shows that being defined on the homogeneous graphs Γ the stochastic map (9.48, 9.49) exhibits $\delta > 0$ as $\eta < \eta_{cr} \approx 0.92$ and $\delta \rightarrow$ as $\eta \rightarrow 0$. If the gene interactions with the negative regulations prevail in the network, it exhibits the anomalously long transient processes (following power law decay in time as $\eta \rightarrow 0$) which eventually bring in oscillations. The formation and growth of domains

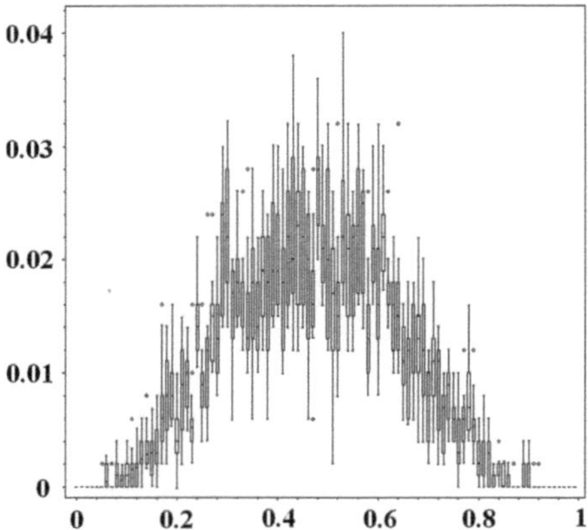

Fig. 9.20 The regulation flow of the model (9.48, 9.49) with $\alpha = 0.6$ defined on the homogeneous graphs with the randomly shuffled switching parameters transforms the initial point x^0 into some stationary configuration $x_\star$ located randomly within $[0, 1]$ provided the positive regulations (activations) prevail in the system ($\eta \to 1$). The empirical probability density distributions $p(x_\star)$ have a broad bell shape. The data is collected over 500 different layouts of 100 distinct threshold values uniformly distributed over $[0, 1]$. The probability distribution is computed for the fully connected graph $\Gamma(10^3)$. The box-plot displays the dependence of $p(x_\star)$ upon the choice of x^0 over 500 random initial conditions. A lower line of box shows the first quartile, and an upper line of a box shows the third quartile. Half of the difference between the third quartile and the first quartile (the semi-interquartile range or the quartile deviation) is a measure of the dispersion of the data. Two lines extending from the central box of maximal length 3/2 the interquartile range

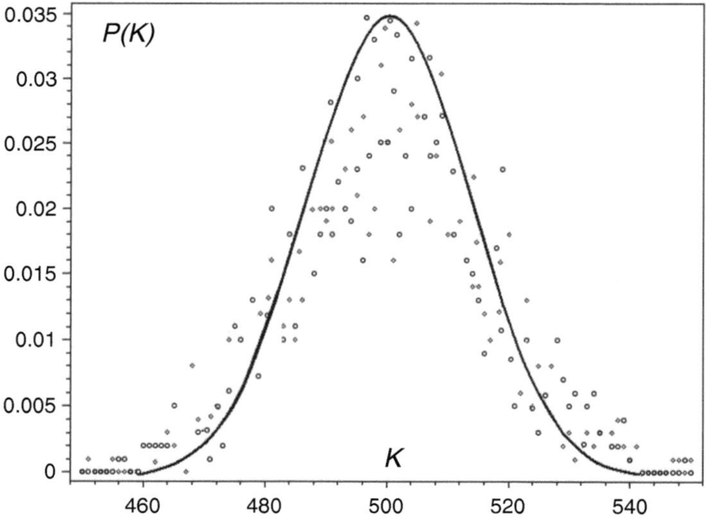

Fig. 9.21 In a stable stationary state, the active subgraph $\Gamma_\star$ constitutes a random graph (a *half-dense* one in comparison with the initial fully connected graph $\Gamma(10^3)$), for any choice of x^0 and any layout of switching parameters. The probability degree distributions for the nodes of $\Gamma_\star$ formed at $\eta = 0.99$. The circles are for the incoming degrees k_+, and the diamonds are for the outgoing degrees k_-, the solid line displays the Gaussian degree statistics that is typical for the Erdös and Rényi random graphs

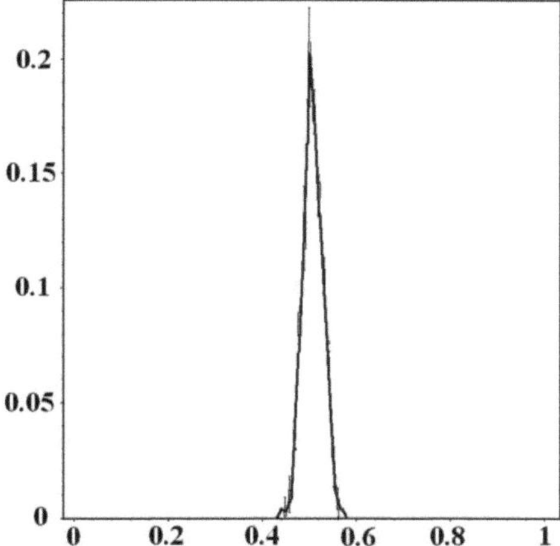

Fig. 9.22   The empirical probability density distributions $p(x_\star)$ form a sharp peak at $x = 1/2$ as the negative regulations prevail in the network ($\eta = 0.1$), with $\alpha = 0.6$ defined on the fully connected graph $\Gamma(10^3)$. The data is collected over 500 different layouts of 100 distinct threshold values uniformly distributed over $[0, 1]$. The box-plot displays the dependence of $p(x_\star)$ upon the choice of $\mathbf{x}^0$ over 500 random initial conditions

characterized with the periodically changed concentrations of regulators come along with the synchronization of the rest of system at $x = 1/2$. In the dynamically stable regime, the amplitudes of oscillations remain small.

The empirical probability density distributions $p(x_\star)$ form a sharp peak at $x = 1/2$ as the negative regulations prevail in the network (see Fig. 9.22). The synchronization of the system near $x = 1/2$ is sensitive to neither the initial conditions nor the certain layouts of T_{ik} and s_{ik}.

In Fig. 9.23, we have displayed the decreasing and vanishing of the order parameter δ as η grows up, in the model defined on the homogeneous graphs. The solid line is for the Gaussian curve,

$$\delta \simeq \frac{\exp\left(-\eta^2/2\sigma\right)}{\sqrt{2\pi\sigma^2}}, \tag{9.52}$$

in which the variance $\sigma \approx 0.55$ fits the data well for small η. Boxes in Fig. 9.23 show the fluctuations of δ over the ensemble of networks (9.48, 9.49) with the randomized switching parameters, and the bold points present the means of collected data.

Oscillations of many different periods can arise in the model (9.48, 9.49) defined on the homogeneous graphs as the negative regulations prevail. The returns maps $\log|\mathbf{x}^{t+1}|$ vs. $\log|\mathbf{x}^t|$ and $\log|u^{t+1}|$ vs. $\log|u^t|$ shown in Fig. 9.24a display the nice periodicity of oscillations observed in the network at parity of activation and repression regulations. In Fig. 9.24b, we have presented the same maps for $\eta = 0.1$ (90% of negative regulations) that reveals the *multi-periodicity* of oscillations.

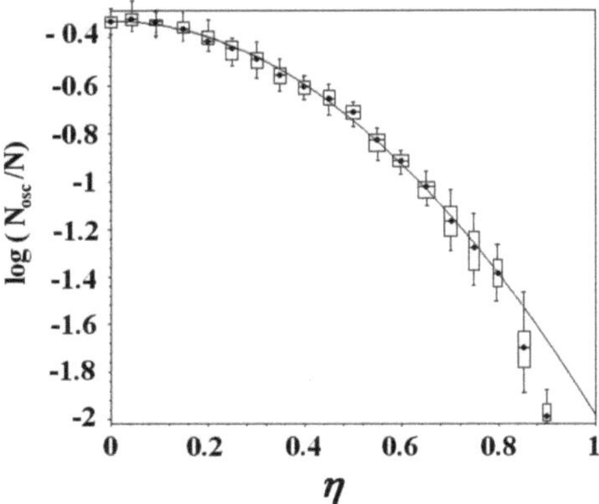

Fig. 9.23 The number of nodes with the oscillating concentrations of regulators N_{osc} decreases as the fraction η of positive regulations (activations) increases; N_{osc} vanishes at $\eta \approx 0.92$. The negative feedback circuits responsible for the appearance of persistent oscillations take effect, as there are enough negative regulations between the genes. Oscillations are bounded strictly to the elements engaged into the negative feedback circuits and do not propagate throughout the network. The model (9.48, 9.49) defined on the homogeneous graphs predicts a nearly Gaussian probability distribution for the appearance of a "functional" negative feedback circuit when the negative regulations prevail. The solid line is for the Gaussian curve, in which the variance $\sigma \approx 0.555$ fits the data well for small η. Boxes show the fluctuations of N_{osc} over the ensemble of networks with the randomized switching parameters (500 different layouts of 100 distinct threshold values uniformly distributed over [0,1]), and the bold points present the means of collected data. The data has been collected for the fully connected graph $\Gamma(10^3)$

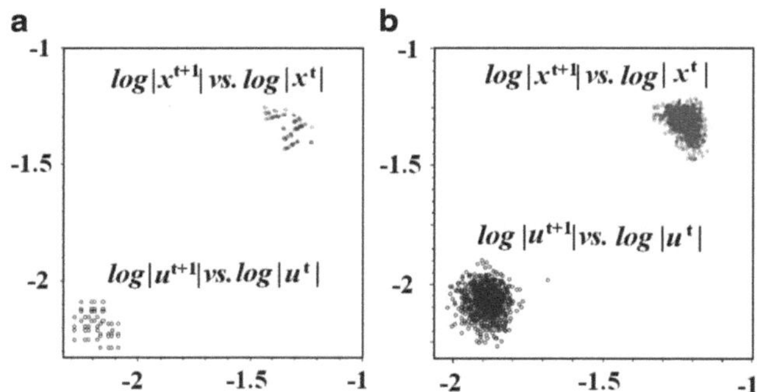

Fig. 9.24 Oscillations of many different periods arise in the model (9.48, 9.49) defined on the homogeneous graphs as the negative regulations prevail. The return maps for the model (9.48, 9.49), with $\alpha = 0.6$ defined on $\Gamma(10^3)$ for the fixed layout of 100 distinct thresholds uniformly distributed over [0, 1] (**a**) for $\eta = 0.5$; (**b**) for $\eta = 0.1$

The cooperative effects of transcription activation by the individual transcription factors observed in the model (9.48, 9.49) defined on the inhomogeneous graphs and on the scale free graphs in particular crucially depend upon the topology details determined by the used graph generating algorithm.

We have also studied the regulatory network defined on the directed scale free graph, in which both the statistics of incoming and outgoing edges follow the power law $P(k) \propto k^{-2.2}$ which has been generated in accordance to the random process introduced in Volchenkov et al. (2002). Formation of feedback circuits, in such a directed scale free graph, relays primarily upon the bidirectional edges $\overline{E}$ connecting the pairs of mutually interacting nodes, so that their fraction,

$$\mu = \frac{|\overline{E}|}{|E|} \tag{9.53}$$

can be used as a convenient parameter featuring the network behavior. Furthermore, to tune the value of μ in the interval $[0, 1]$, we modified the initial directed scale free graph Γ by the random adding (removal) of complimentary edges. Indeed, this made changes to the scale free structure of the graph, nevertheless it remained highly inhomogeneous.

In Fig. 9.25, we have presented a phase diagram displaying the ranges of η and μ for which the persistent oscillations would arise in the system in a conclusion of the transient processes. The inter-phase boundary looks indented deeply and could contain the small islands of the alternative phase. The fine structure of the boundary is sensitive to the initial conditions $\mathbf{x}^0$ and to the certain layout of switching parameters.

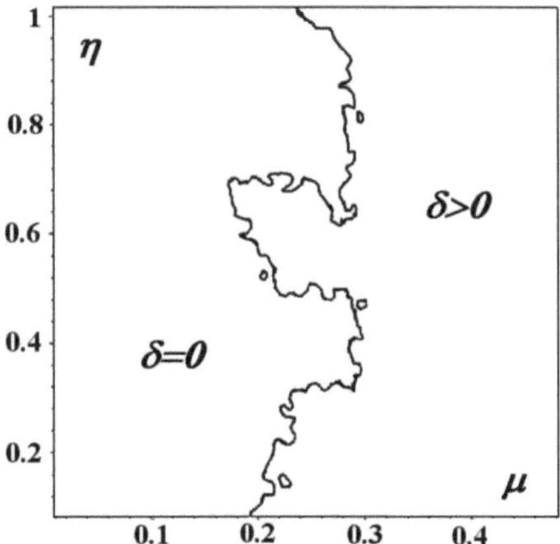

Fig. 9.25   The phase diagram (η, μ) displays the ranges of parameters for which the model (9.48, 9.49), with $\alpha = 0.3$, defined on the scalable graph exhibits the persistent oscillations of concentrations. The inter-phase boundary looks indented deeply and could contain the small islands of the alternative phase. The fine structure of the boundary is sensitive to the initial conditions $\mathbf{x}^0$ and to the certain layout of switching parameters. The graph is obtained from the directed scale free graph $\Gamma(10^3, 2.2)$ by the random removal (adding) of complimentary edges. The feedback circuits have been formed primarily by the bidirectional edges (the simplest 2-cycles). Oscillations fade out if the fraction of bidirectional edges is small $\mu \ll 1$. However, they arise and persist at whatever $\eta > 0$ for $\mu > 0.3$

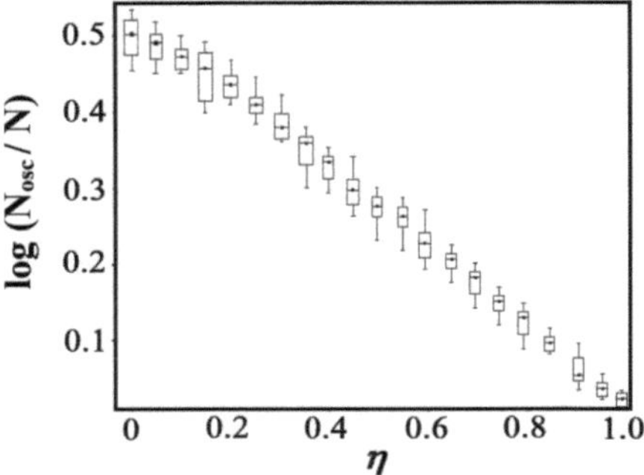

Fig. 9.26 Oscillations persist in the model (9.48, 9.49) at $\alpha = 0.74$ defined on the undirected ($\mu = 1$) scale free graph $\Gamma(10^3, 2.2)$ at whatever $\eta > 0$. The fraction of oscillating nodes δ decreases with η. Boxes present the fluctuations of data collected over the ensemble of 500 different random layouts of switching parameters and initial conditions. The bold points stand for the means

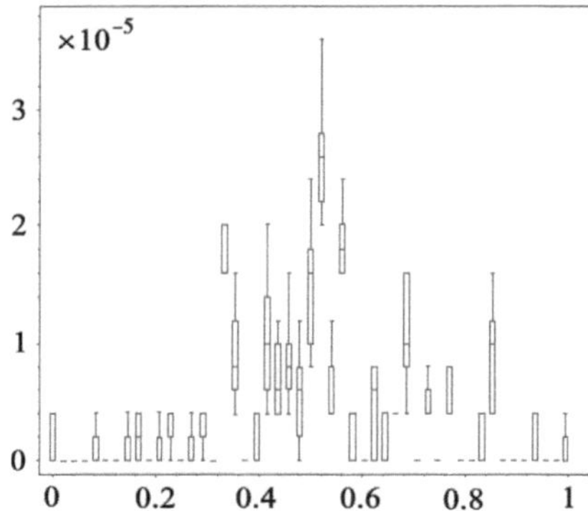

Fig. 9.27 The empirical probability density distributions $p(x_\star)$ in the model defined on the undirected scale free graph $\Gamma(10^3, 2.2)$ with a given configuration of 100 distinct thresholds uniformly distributed over $[0, 1]$ at $\alpha = 0.7$. Fluctuations shown by the boxes reveal the dependence of $p(x_\star)$ upon the certain choice of initial conditions and the layouts of switching parameters (500 different configurations had been used to collect the data) for $\eta = 0.99$ (i.e., 1% of negative regulations)

In what follows, we consider $\mu = 1$ that corresponds to the undirected scale free graphs. We have studied the decay of the fraction of nodes with the oscillating concentrations of regulators δ as the fraction of negative regulations η grows up (see Fig. 9.26) and the empirical probability density distributions $p(x_\star)$ for two marginal cases, at $\eta = 1$ (Fig. 9.27) and $\eta = 0$ (Fig. 9.28). As usual, the fluctuations shown by the boxes in Figs. 9.26–9.28 reveal the dependence of the measured quantities upon the certain choice of initial conditions and switching parameters. In contrast to the model (9.48, 9.49) defined on the homogeneous networks, the oscillations of concentrations of regulators would persist at whatever value of control parameter

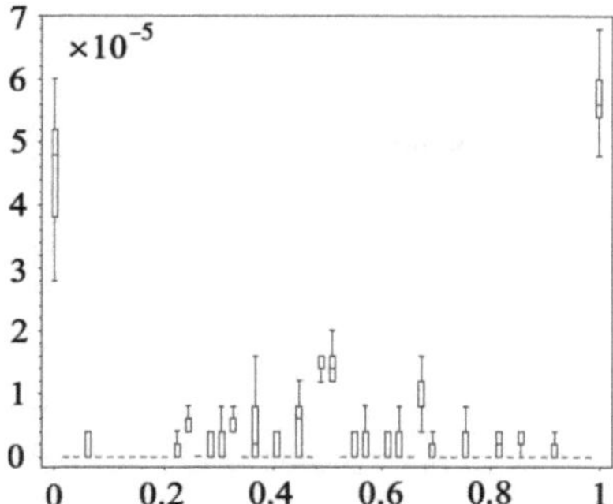

Fig. 9.28   The empirical probability density distributions $p(x_\star)$ in the model defined on the undirected scale free graph $\Gamma(10^3, 2.2)$ with a given configuration of 100 distinct thresholds uniformly distributed over $[0, 1]$ at $\alpha = 0.7$. Fluctuations shown by the boxes reveal the dependence of $p(x_\star)$ upon the certain choice of initial conditions and the layouts of switching parameters (500 different configurations had been used to collect the data) for $\eta = 0.01$ (i.e., 99% of negative interactions)

$\eta > 0$. In the empirical probability density distributions $p(x_\star)$ for the model (9.48, 9.49) defined on the scalable graphs, there exist "gaps", the intervals of concentrations of regulators that can not be observed starting from any initial condition at any layout of switching parameters. When the negative regulations present in abundance ($\eta = 0$), a valuable fraction of nodes, in the scalable regulatory network, is synchronized either at $x_\star = 0$ or at $x_\star = 1$.

9.6   Mean Field Approach to the Large Transcription Regulatory Networks

In accordance to (9.48, 9.49), the expression of a target gene j is due to the cooperative effects of transcription activation by individual transcription factors of other genes acting on it and proportional to the fraction of active arcs incident to the node j at time t. While shuffling the switching parameters in the regulatory network, we suppose that the positive transcription degree $s = +1$ is selected with the probability $0 < \eta \le 1$, and that the value T_k of activation threshold is chosen randomly among the set $\{T_1, \ldots T_n\}$ with some probability $0 < P_{ik} \le 1$, such that $\sum_k P_{ik} = 1$ for each i regulating j.

The distribution of values T_k over $[0, 1]$ can be defined by the set of integrable functions $g_k : [0, 1] \to \mathbb{R}^+$ satisfying the natural normalization condition and such that the probability that the threshold value T_k chosen for the regulation does not exceed x reads as

$$P(T_k \le x) = \int_0^x g_k(z)\, dz.$$

Then the probability that an *outgoing* arc is *active* at time t equals to

$$\Lambda_\eta\left(x_j^t\right) = \sum_{k=1}^{n} P_{ik}\left(\eta \int_0^{x_j^t} g_k(z)dz + (1-\eta)\int_{x_j^t}^1 g_k(z)\,dz\right), \qquad (9.54)$$

so that the expected exertion of the gene j at time t is

$$\langle y_j^t\rangle = \frac{\sum_i e_{ij} A_{ij}^0 \Lambda_\eta(x_j^t)}{\sum_i e_{ij} A_{ij}^0}. \qquad (9.55)$$

The main idea of the mean field approach to the large regulatory networks with shuffled switching parameters is to replace the exact values y_i^t in (9.49) with their expectations (9.55),

$$\langle \mathbf{x}^{t+1}\rangle = \alpha\langle \mathbf{x}^t\rangle + (1-\alpha)\langle \mathbf{y}^t\rangle. \qquad (9.56)$$

Generally speaking, the problem (9.56) is as complicated as the original one. However, in some cases it can reveal the entire dynamical mechanisms driving the system. In the simplest case of the uniform probabilities $P_{ik} = 1/n$, the system of N coupled equations (9.55) is linearized for the uniform distributions, $g_k = 1$. The transient processes extinguish exponentially fast, and independently upon the initial conditions the system (9.56) with constant P_{ik} and g_k settles into the stationary configuration (we also suppose that all $e_{ij} = 1$),

$$x_\star = \frac{n(1-\eta)}{(N-1)(n+1-2\eta)}.$$

For the "micro-canonical" distributions of thresholds,

$$g_k(x) = \delta(x - T_k),$$

with the randomly chosen threshold values T_k uniformly distributed over $[0,1]$ and $P_{ik} = 1/n$, the dynamics of expectations over the ensemble of large networks ($N \gg 1$) with the randomly shuffled switching parameters is given by

$$\langle x_j^{t+1}\rangle = \alpha\langle x_j^t\rangle + (1-\alpha)\left[(1-\eta)f^t(T_m) + \eta\left(1 - f^t(T_m)\right)\right] \qquad (9.57)$$

where the "mean field"

$$f^t(T_m) = \frac{N(x^t > T_m)}{N}$$

is the instantaneous fraction of nodes with concentrations $x^t > T_m$. The "mean" threshold value T_m is determined upon the certain set of randomly chosen thresholds $\{T_k\}$ by the (9.54) in the limit $N \gg 1$. In this case, the equation (9.57) represents a system with a feedback loop. Herewith, depending upon the value of η and T_m it can be either positive or negative. The system governed by (9.57) comes to a dynamically stable state very fast, in particular, it exhibits the oscillations of different periods if $\eta < T_m < 1 - \eta$, for $\eta < 1/2$, and $T_m < 1 - \eta$, $T_m > \eta$, for $\eta > 1/2$. In the case of inhomogeneous sets P_{ik}, for $N >> 1$, the (9.57) can be generalized to

$$\left\langle x_j^{t+1} \right\rangle = \alpha \left\langle x_j^t \right\rangle + (1 - \alpha) \left[(1 - \eta) f^t (T_{mi}) + \eta \left(1 - f^t (T_{mi}) \right) \right] \qquad (9.58)$$

where T_{mi} is the local "mean" threshold related to gene i. The next step can be done if one assumes some symmetry properties for the sets of P_{ik} that would dramatically reduce the number of independent local "mean" thresholds T_{mi} to just a few. Then the entire system can be considered as a set of coupled positive and negative feedback loops whose dynamical behavior can be a subject of detailed analysis. In particular, in the system with several coupled negative feedback circuits characterized by the distinct thresholds T_{mi}, the synchronous oscillations may persist for $\eta > 0$.

Large-scale gene expression studies have shown that even seemingly simple physiological changes entail expression changes in vast numbers of genes (Wagner 2000). Despite an impressive progress that has been made recently, the behavior of large gene expression regulatory networks is still far from being understood. There is still no a common opinion on whether they are organized hierarchically or contains a plenty of cross interactions, might be of quite irregular structure, organized in a form of connected sets of small sub-networks. The relations between the global structure of networks and their local dynamical properties are also still unclear.

A useful approach to the regulatory networks comprising of just a few elements consists of modeling their interactions using deterministic differential equations or Boolean models (Gardner et al. 2003; Kauffman et al. 2003). In their context, the circular sequences of interactions (the *feedback circuits*) have been shown to play the key dynamical roles: whereas positive circuits are able to generate multi-stationarity, negative circuits may generate oscillatory behavior (Thomas and Kaufman 2001). The fully connected Boolean networks represent genetic networks where each element interacts with all elements including it. A feedback circuit can be formally defined as a combination of terms of the Jacobian matrix of the system, with indexes forming a circular permutation. Flexibility in the network design is introduced by the use of Boolean parameters, one associated with each interaction of group of interactions affecting a given element. Within this formalism, a feedback circuit will generate its typical dynamical behavior (either stationary or oscillating) only for appropriate values of some of its logical parameters (Thieffry and Romero 1999). Each element of a circuit exerts an indirect effect on itself which has the same sign for all elements of the circuit, leading to the definition of the "circuit sign." It depends on the parity of the number of negative regulations involved in the circuit: if it is even, then the circuit is positive, otherwise, it is negative (Thieffry 2000).

What makes these concepts important is that specific biological and dynamical properties can be associated with each of theses two classes of feedback circuits. The relation between the presence of positive feedback loops and the occurrence of multiple states of the gene expression has been at a focus of many investigations (see Thieffry et al. 1995 and references therein). In particular, the positive loop(s) are necessary for multi-stationarity, and the negative circuits are required for the stable periodicity of behavior (Thomas 1981). Biologically, this means that positive circuits are required for the differentiated decisions and negative circuits are responsible for the homeostasis (Thieffry et al. 1995; Thomas 1994).

However, in the large regulatory networks, each element can participate in many different feedback structures "functional" and "passive" at once. Simulations convinced us that the "functional" circuits and the rest "passive" elements are tightly related, so that a dislocation made to an element of inactive circuits would result a change to some "functional" circuits and even cause they dissolve. Therefore, a statistical description of feedback circuits seems plausible for the large regulatory networks, in the framework of proposed approach.

Numerical simulations discussed in the previous section show that an abundance of positive regulations between the elements of homogeneous regulatory networks drive the system to the asymptotically stationary states (the fixed points) brought about by the positive feedback circuits and characterized by the broad empirical probability distributions of concentrations over $[0, 1]$. Thereat, the strain of negative feedback circuits related to just a few negative regulations allowed in the network remains statistically negligible. However, the negative feedback circuits take effect, as there are enough negative regulations between the genes, resulting in the oscillations of regulator concentrations. Oscillations are bounded strictly to the elements engaged into the negative feedback circuits and do not propagate throughout the network.

The model (9.48, 9.49) defined on the homogeneous graphs predicts a nearly Gaussian probability distribution for the appearance of a functional negative feedback circuit when the negative regulations prevail. At the end of anomalously long transient processes, in the stable dynamical regime, the concentrations of regulators for the genes engaged into the functional negative feedback circuits oscillate with many different periods, while the rest of system is synchronized at $x_\star = 1/2$.

The formation of feedback circuits in the model (9.48, 9.49) defined on the inhomogeneous scalable graphs crucially depends upon their detailed topological properties. In the graphs we have used in the simulation, it relays primarily upon the bidirectional edges connecting the pairs of mutually regulating genes (the simplest 2-cycles). Furthermore, the cooperative effects of transcription activation in the highly inhomogeneous networks are controlled by the individual transcription factors of genes located at the hubs (nodes with high degrees). Therefore, the sole link with a negative transcription degree inherent to such a hub would form many negative feedback circuits at once provided there are enough cycles traversing the hub. As a result, the persistent oscillations of concentrations are observed in the model (9.48, 9.49) at whatever value of $\eta > 0$. It is noteworthy that the absence of critical thresholds ($\eta_c = 1$) seems to be a special feature of the extended dynamical systems defined on the scale free graphs (for instance, let us recall the problem of epidemic spreading Murray 1993, Pastor-Satorras and Vespignani 2001, 2002; Volchenkov et al. 2002). Another special feature of scale free networks is their high error tolerance: only selected concentrations of proteins are allowed.

9.7 Summary

We have discussed some strongly nonlinear dynamical process on complex networks for which a feedback mechanism plays the important role. Topologically, feedback circuits in complex networks relay upon cycles which are not controlled by the degree statistics of graphs.

To be certain, we have analyzed the spread of epidemics on a variety of graphs, including the scale free graphs, in particular.

We have investigated the process of synchronization of an array of coupled Chaté-Manneville maps in random regular graphs. Discontinuous phase transitions, well defined turbulent windows and the nontrivial collective behavior are common and distinctive features emerging in the model. The observed collective properties of the system have been analyzed through the thermodynamic formalism.

Finally, we have considered the simplest model of a self-regulation network motivated by large gene interaction networks. The network in our model might adjust its structure in accordance to the current regime of dynamical behavior. We have shown that the performance of such a large self-regulation networks are governed by a complex nexus of interactions between the functional positive and negative feedback circuits, from one hand, and the rest "passive" elements which stabilize the dynamical behavior in entire network.

Chapter 10
Critical Phenomena on Large Graphs
with Regular Subgraphs

Geometry and topology of underlying graphs strongly influence on the physical properties of dynamical processes in complex networks. On the one hand, it is interesting to discuss the properties of graphs which might affect the dynamical behavior of a model. On the other hand, the study of large complex networks calls for statistical methods to give an effective description of observed collective dynamical behavior. Here, following Volchenkov and Blanchard (2007a) we discuss the nonlinear transport phenomena on generalized trees which contain the simple nodes and *supernodes* formed by either well-structured regular subgraphs, or those with many triangles. In particular, we show that the nonlinear transport process is a superdiffusion, for the highly connected nodes, while it is Brownian, for the rest of the nodes. Transport within a supernode is affected by the finite size effects vanishing as $N \to \infty$. For the even dimensions of space, $d = 2, 4, 6, \ldots$, the finite size effects break down the perturbation theory at small scales and can be regularized by using the heat-kernel expansion.

Diffusion processes defined on the small world networks in $d = 1$ has been considered in Monason (1999). It appears that the mean field theory breaks down for the small world models in dimensions $d < 2$ (Kozma et al. 2004) due to the emergence of strong site-to-site fluctuations of the Green's functions. In Hastings (2004), the ϵ-expansion for small worlds has been developed where

$$\epsilon = 2 - d$$

quantifies the departure of the space dimension from 2. The small word network has been constructed by adding random shortcuts to a regular lattices in d-dimensions while the density of links p has been taken small,

$$pa^d \ll 1$$

where a is the lattice scale. The average properties of Green's functions have been computed for $d = 2$, but breaks down for $\epsilon = 1$ due to traps in the system appearing at $d = 1$ if p is taken small.

P. Blanchard and D. Volchenkov, *Random Walks and Diffusions on Graphs and Databases*, Springer Series in Synergetics 10, DOI 10.1007/978-3-642-19592-1_10, © Springer-Verlag Berlin Heidelberg 2011

In contrast to the above mentioned works, we consider the diffusion processes on the graphs which contain regular subgraphs, but, as a whole, being far from the regular lattices approaching the networks known in the literature as "froths". The space tiled by the froth can be curved. In this case, the intrinsic dimension of the cellular system does not coincide with the dimension of the embedding space. Transport phenomena had been extensively studied for the disordered media (Limoge and Bocquet 1990; Zanette and Alemany 1995; Vera and Durian 1996) and fractals (Boettcher and Moshe 1995; Bender et al. 1995; Cassi and Regina 1996). They revealed the dependence of transport properties upon the intrinsic dimension d_f of disordered systems, in particular, the random walk probability

$$P \propto t^{-\frac{d_f}{2}} f\left(\frac{x}{\sqrt{t}}\right),$$

where f is a scaling function. Similar results have been reported in the nonlinear diffusion processes (Goldenfeld et al. 1990; Bricmont and Kupiainen 1992; Bricmont et al. 1996a; Antonov and Honkonen 2002).

In our model (Volchenkov and Blanchard 2007a), we consider the complex networks as the generalized trees with two types of nodes: the simple nodes (probably of low connectivity) and the super nodes which are either the subgraphs ample with polygons or the k-regular subgraphs. In a continuous setting, the supernodes can be treated as the complete Riemann curved surfaces characterized by the finite areas and subjected to compactification. The compact islands of the curved Riemann surfaces are bridged with the tree components.

The nonlinear term included into the diffusion equation models the effect of varying space dimension, the possible fluctuations of transport coefficient, the diffusion-reaction processes, and the possible queuing due to a bounded transport capacity of edges bridging between supernodes. The process of diffusion is treated as a generalized Brownian motion with arbitrary boundary conditions described by a functional integral. Here, we study the long time large scale asymptotic behavior of the Green function for the nonlinear diffusion equation defined on the large generalized trees. The transport through such a complex network is affected by both the varying effective space dimension between super nodes and the space curvature within a super node.

Strictly speaking, the possibility to replace a finite graph by a compact continuous manifold for a nonlinear diffusion process is a challenging question. For the linear differential operators acting on the periodic functions, it requires that the spectra associated to the discrete and continuous problems are similar (de Verdiére 1998). In particular, the multiplicities of the eigenvalues should be equal for both problems.

However, by this time, it is still not much known even about the spectral properties of Schrödinger operators defined on curved compact manifolds. Some estimations on the multiplicities of eigenvalues and the Euler characteristic of a surface can be found for the 2D-sphere, for the Klein's bottle, and for the 2D-torus in Cheng (1976), Besson (1987), Nadirashvili (1973), Sévennec (1994) and de Verdiére (1987). Therefore, even in this case, the justification of the validity for the approximation of processes defined on the graphs by that ones defined on the compact continuous manifolds is a difficult problem. Concerning the nonlinear diffusion process, we should confess that such an approximation is always an assumption.

10.1   Description of the Model and the Results

We are interested in the large scale asymptotic behavior of Green function for the diffusion equation for the density $u(x,t)$ defined on the curved Riemannian manifold with metric tensor g_{ij}, with the nonlinear term $\propto u^\alpha$ ($\alpha > 1$) included. We assume the such a model can be considered as an approximation of nonlinear diffusion process defined on the graphs containing the large regular subgraphs. This approximation, however, is rather intuitive and cannot be proved nowadays rigorously. Let us note that such a model can be considered as a generalization of the well-known Lévy flight random walks (Shlesinger et al. 1995) in which the increments are distributed according to a "heavy-tailed" probability distribution. The exponential scaling of the step lengths gives Lévy flights a scale invariant property, and they are used to model data that exhibits clustering.

If our assumption is true, then $\log_2 \deg(x)$ plays the role of effective local space dimension at node x on the graph, $\deg(x)$ being the degree of x. The nonlinear diffusion term is relevant to the large scale asymptotic behavior if

$$\alpha \leq 1 + \frac{2}{\log_2 \deg(x)}$$

and is irrelevant otherwise.

In the critical phenomena theory, the physical degrees of freedom are replaced by the scaling ones related to each other by the RG transformations. The large scale asymptotic behavior of physical system is then determined by the properties of scaling invariant system at a certain stable fixed point of the renormalization group differential equation. The derived results are irrelevant to the order of space and time (infinite) limits.

Our main result is that for the large scale asymptote of Green function,

$$G(t) \sim \begin{cases} t^{-\log_2 \deg(x)/2}, & \log_2 \deg(x) > \dfrac{2}{\alpha - 1}, & x \in \Gamma, \\[2mm] t^{-1/\alpha - 1}, & \log_2 \deg(x) \leq \dfrac{2}{\alpha - 1}, & x \in \Gamma, \end{cases}$$

The main technical difficulty of computations in the curved space metrics is the lack of the global frequency-momentum representation. In order to regularize the effective action at small scales in the one-loop order, we have used the heat-kernel expansion which does not depend upon the space-time topology. In the regular graphs endowed with the standard orientation of edges (the cycling ordering of edges to be the same for all nodes), the effect of curvature within the supernodes reveals itself by the finite size corrections to the scaling behavior.

10.2 The Regular Subgraphs Viewed as Riemann Surfaces

We consider the flat Euclidean space $\mathbb{R}^d$ as the limit $a \to 0$ of a regular lattice $\mathcal{L}_a = a\mathbb{Z}^d$ with the lattice scale length a. While simulating the diffusion equation $u_t = \Delta u$ for the scalar function u defined on the lattice, one uses its discrete representation,

$$u^{t+1}(x) = \frac{1}{2^d a^2} \left[\sum_{y \in U_x} u^t(y) - 2^d u^t(x) \right], \tag{10.1}$$

where U_x is the lattice neighborhood of $x \in \mathcal{L}_a$. The cardinal number 2^d is uniform for the given $\mathcal{L}_a$, and d is interpreted as the dimension of Euclidean space.

Being defined on an arbitrary connected graph $\Gamma = (V_\Gamma, E_\Gamma)$, in which V_Γ is the set of its nodes and E_Γ is the set of edges linking them, the discrete Laplace operator has actually the same form as in (10.1) excepting for the cardinality number changed to 2^{δ_x} where

$$\delta_x = \log_2 \deg(x)$$

and $\deg(x)$ is the degree of node $x \in \Gamma$. By counting the number of nodes N_Γ, the number of edges E_Γ, and the number of faces F_Γ, one obtains the Euler characteristic of a planar graph Γ,

$$\chi(\Gamma) = N_\Gamma - E_\Gamma + F_\Gamma.$$

We are interested in the subgraphs of Γ ample with polygons, in particular, the triangles which can be used as a base for a local mesh rendering a Riemann surface (the Delaunay triangulation).

An intuition on the deep relation between the triangle structures in the complex networks and their curvature has been expressed in Eckman and Moses (2002): "Triangles capture transitivity, which we measure by the associated notion of curvature". In Collet and Eckmann (2002), the upper and lower bounds on the number of graphs of fixed degree which have a positive density of triangles had been estimated. In particular, it has been shown that the triangles seem to cluster even at low density. Then, in Sergi (2005) it has been proved that the probability for a randomly selected node to participate in T triangles decays with T following a power-law, in a family of scale free random graphs with the degree exponent satisfying $2 < \beta < 2.5$. Moreover, a finite density of triangles appears in such graphs as $\beta = 2 + 1/3$.

The Laplace-Beltrami operator on a triangle mesh has been defined in concern with the various graphics applications such as mesh fairing, smoothing, surface editing in $3D$-space (Kim et al. 2002; Salisburym et al. 1995; Schneider and Kobbelt

2001; Kim and Rossignac 2004). The stiffness matrix correspondent to it can be computed for each triangle specified by its nodes $\mathbf{x} = (x_1, x_2, x_3)$, $\mathbf{y} = (y_1, y_2, y_3)$, $\mathbf{z} = (z_1, z_2, z_3)$ as

$$\mathbf{K} = S_\Delta \mathbf{B}\mathbf{B}^T, \quad \mathbf{B} = \frac{1}{2S_\Delta}\begin{pmatrix} x_2 - x_3 & y_2 - y_3 & z_2 - z_3 \\ x_3 - x_1 & y_3 - y_1 & z_3 - z_1 \\ x_1 - x_2 & y_1 - y_2 & z_1 - z_2 \end{pmatrix}$$

where S_Δ is the area of triangle.

However, from the various models of complex networks, it seems rather easier to count the number of edges which connect a node to others than to check out if its neighbors are really connected forming a triangle. Instead of hunting for triangles, while analyzing the graphs of real world networks, one can search for their 3-regular subgraphs $_{3R}\Gamma$ for which there are precisely three edges incident at each node (loops are counted twice, multiple edges are allowed). These graphs form a honeycomb. The idea of using 3-regular subgraphs to study the topological properties of rendered Riemann surfaces is originated in Buser et al. (1984). In Brooks and Makover (2001) and Mangoubi (2001), it has been shown that for each 3-regular graph $_{3R}\Gamma$ of $2N$ nodes with an orientation $\mathcal{O}$ (which assigns to each node of $_{3R}\Gamma$ a cyclic ordering of edges incident at it) one can construct a complete Riemann surface $S(_{3R}\Gamma, \mathcal{O})$ by associating the ideal hyperbolic triangles to each node of $_{3R}\Gamma$ and gluing sides together according to the edges of the graph $_{3R}\Gamma$ and the orientation $\mathcal{O}$. The resulting surface is endowed with a metric excepting for the finitely many points (cusps) where the metric could be undefined. The surface area is finite and equals to $2\pi N$. The conformal compactification of $S(_{3R}\Gamma, \mathcal{O})$ are dense in the space of Riemann surfaces. The standard orientation on the 3-regular graph which is the 1-skeleton of a cube contains six left-hand-turn paths on which a traveler always turns left, giving that the associated surface is a sphere with six punctures. The choice of different orientations can give the surfaces of genus 0, 1, and 2 (Brooks and Makover 2001).

The closed paths of length k on $_{3R}\Gamma$ are homotopic to the closed geodesic lines on $S(\Gamma, \mathcal{O})$. Their cardinality is defined by the spectral density of $_{3R}\Gamma$-subgraph that is the density of eigenvalues of its adjacency matrix (Farkas et al. 2001),

$$\varrho(\lambda) = \frac{1}{N}\sum_{j=1}^{N}\delta(\lambda - \lambda_j),$$

which converges to a continuous function with $N \to \infty$ (λ_j is the j-th largest eigenvalue of adjacency matrix $\mathbf{A}$). The k-th moment M_k of $\varrho(\lambda)$ scales with number of cycles C_k of length k,

$$M_k = \frac{1}{N}\sum_{j=1}^{N}\lambda_j^k = \frac{1}{N}\mathrm{Tr}(\mathbf{A}^k) = \frac{k}{N}C_k,$$

in which C_k can be computed with the help of the zeta-function,

$$\zeta(z) = \frac{1}{\det(\mathbf{1} - z\mathbf{A})}.$$

All roots of the polynomial $\det(\mathbf{1} - z\mathbf{A})$ lay on the unit circle and coincide with the inverted eigenvalues λ_j^{-1} of adjacency matrix excepting for the zeros, $\{\lambda = 0\}$. The generating property of $\zeta(z)$ is

$$C_k = \frac{1}{(k-1)!} \frac{d^k}{dz^k} \zeta(z)\Big|_{z=0}.$$

It is worth to mention that $S(_{3R}\Gamma, \mathcal{O})$ allows for a conformal compactification (Brooks and Makover 2001). The latter property is of importance for the renormalization group application.

10.3 Nonlinear Diffusions Through Complex Networks

In the previous subsection, we have supposed that a regular subgraph in a complex network can be treated, in a continuous setting, as a Riemann space of finite area characterized by some metric tensor. For a particular time slice, the line element ds between each pair of neighboring points on the spatial surface is given by

$$ds^2 = \sum_{ij} g_{ij}(x)dx^i dx^j$$

where dx^k denote the coordinated differences between neighboring points, and g_{ij} denotes the metric tensor. Then, the complex network as a whole can be considered as a disordered media in which the compact islands S of curved Riemann space (with the "effective" space dimensions δ_S) are bridged by the tree like graph components (see Fig. 10.1) in which the local space dimension δ_x can vary from point to point.

The transport properties through such a disordered media is essentially of nonlinear nature. In the previous studies of nonlinear diffusion (Goldenfeld et al. 1990; Bricmont and Kupiainen 1992; Bricmont et al. 1994; Bricmont and Kupiainen 1996; Teodorovich 1999; Antonov and Honkonen 2002), the authors had introduced various nonlinear terms into the diffusion equation modelling the possible fluctuations of transport coefficient, the diffusion-reaction processes, and the queuing due to a bounded transport capacity of edges. We also introduce it for accounting the effect of varying dimension of space in the complex network (see the discussion below).

To be certain, let us consider the equation for the scalar density field $u(\mathbf{x}, t)$ defined on a Riemann surface,

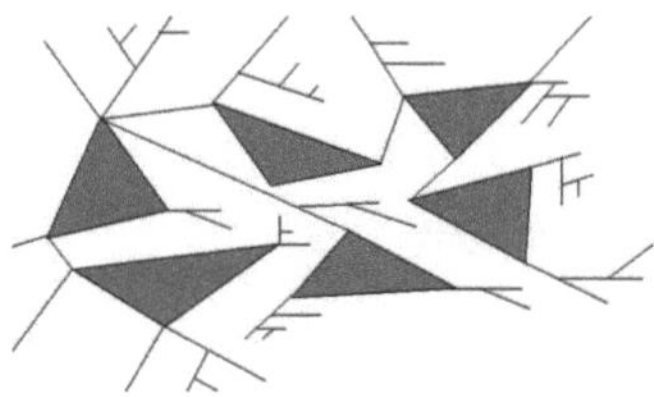

Fig. 10.1 In the model, we consider the complex networks as the generalized trees with two types of nodes: the simple nodes (perhaps of low connectivity) and the supernodes which are either the subgraphs with many triangles or the k-regular subgraphs

$$\nabla_t u = g_0^{ij}\, \nu_0 \frac{\partial^2 u}{\partial x_i\, \partial x_j} - \xi_0 \nu_0 R u - \eta_0 \nu_0 u^\alpha, \quad \nabla_t \equiv \partial_t + b_0^i \frac{\partial}{\partial x_i}. \tag{10.2}$$

All 0-subscripted variables denote their bare values before the application of the renormalization group transformation. In (10.2), the Riemann metric tensor g_0^{ij} depends upon the chosen conformal parameterization of regular subgraphs, R is the scalar curvature. We prefer to keep the entries of g_0^{ij} dimensionless therefore we have introduced the parameter ν_0 having the dimension of a viscosity. We examine metric rescaling that are space-time constants (we suppose that the subgraph of Γ is regular and the edges do not rewire with time). However, it is possible to consider the effect of a rescaling given by a space-time dependent function. The covariant derivative ∇_t contains the curvature drift term proportional to

$$b_0^i = g_0^{jk}\, \Gamma_{jk}^i$$

(the curvature drift velocity) which expresses the local anisotropy of space because of its curvature. Γ_{jk}^i are the Christoffel symbols calculated out from the metric tensor g^{ij} in the standard way,

$$\Gamma_{jk}^i = g^{il}\frac{(g_{kl,j} + g_{jl,k} - g_{jk,l})}{2}.$$

While being interested in the long time large scale ranges, one usually keeps only the first order derivative term ∂_i if it presents in the linear part of equation since its contribution $O(k)$ should dominate over the diffusion $O(k^2)$ for small k. Nevertheless, we keep both terms to ensure the convergence of integrals in time, in the limit of flat metric. However, in the general curved space-time, the homogeneity required for the existence of a global momentum space representation is lacking. The real valued parameters ξ_0 and η_0 are the bare coupling constants governing the coupling of configurations u to the scalar curvature of space R and to the varying effective space dimensionality δ_x respectively. We assume them to be small. The nonlinearity exponent is $\alpha > 1$ (α is not necessary integer) (Goldenfeld et al. 1990;

Bricmont and Kupiainen 1992; Bricmont et al. 1994; Bricmont and Kupiainen 1996; Teodorovich 1999; Antonov and Honkonen 2002).

Let us explain the role played by the nonlinear term in (10.2) in more details. In the critical phenomena theory, the physical degrees of freedom are replaced by the scaling degrees of freedom. In particular, one considers the canonical dimension d_F instead of time and space physical dimensions of the quantity F. The analysis of canonical dimensions allows for the selection of relevant interactions among all possible interaction which could arise in the model. In the spirit of critical phenomena theory, the nonlinear term that could affects substantially the large scale asymptotic behavior typical for the diffusion process should have the same canonical dimension as the normal diffusion. If the canonical dimension of the nonlinear term added to the equation is less than of the ordinary diffusion, it has to be neglected. The opposite is also true: if the nonlinear term provides the leading contribution to the asymptotic behavior, then the diffusion term has to be dropped.

Therefore, it is interesting to consider the model in which the nonlinear term would play an important role. Below, we demonstrate that the exponent α is related to the dimension of space d and the dimension $[\eta_0]$ of coupling constant η_0. If we assume for a moment that in the vicinity of some point the dimension of space d is changed to some other value δ, then, strictly speaking the (10.2) could have no sense therein: either the diffusion term or the nonlinearity should be neglected. For given α and $[\eta_0]$, there is only one value d at which (10.2) is relevant with respect to the large scale asymptotic behavior of diffusion process.

If we consider the plane of parameters $[\eta_0]$ and α, then the relevant space dimensions d is a line on it. Therefore, by tuning the values of α and $[\eta_0]$, one can "modify" the space dimension d in the model of nonlinear diffusion. It is indeed unphysical to change the nonlinearity exponent α (we suppose that α is a property of a certain physical process), however, one can tune the value of $[\eta_0]$ and use it as the small expansion parameter of perturbation theory (like the parameter $\varepsilon = 4 - d$ in the Wilson's theory of critical phenomena).

A similar idea is used in the usual dimensional regularization of Feynman diagrams. In the continuous Euclidean space, the dimensional regularization scheme does not look natural and therefore is usually treated as a formal trick which helps to reformulate the singularities arisen in the Feynman graphs in the form of poles in ε. However, if the dimension of physical space could vary, than the dimensional regularization would acquire the natural meaning provided the nonlinearity exponent α is fixed and the correspondent nonlinear diffusion process is relevant with respect to the large scale asymptotic behavior.

As we have mentioned above, such a relevance can be justified by means of dimensional analysis (Bricmont and Kupiainen 1992; Antonov and Honkonen 2002). Dynamical models have two scales: the length scale L and the time scale T. The physical dimension of viscosity is

$$[\nu_0] = L^2 T^{-1},$$

of the scalar curvature is

$$[R] = L^{-2},$$

and of the drift velocity is

$$[b_0^i] = LT^{-1}.$$

Let us choose the physical dimension for the coupling constant to be $[\eta_0] = L^{-2\varepsilon}$ (assuming that $\varepsilon = 0$ in the logarithmic theory when the nonlinear interaction is marginal) and note that the dimension of a scalar quantity is $[u] = L^{-d}$. In the linear part of (10.2), $\partial_t \propto \partial^2$, thus $L^2 \sim T$. All terms in (10.2) should be of the same canonical dimension, in particular

$$[\partial_t u] = [\eta_0 \nu_0 u^\alpha],$$

and therefore

$$-2 + \log_L[u] = \alpha \log_L[u] - 2\varepsilon$$

that leads to

$$[u] = L^{\frac{2(1-\varepsilon)}{(1-\alpha)}},$$

from which it follows that

$$d = 2 \cdot \frac{1-\varepsilon}{1-\alpha}. \tag{10.3}$$

The above relation gives us a hint of that the space dimension in the model of nonlinear diffusion can be effectively tuned by the parameters α and ε. We choose the parameter ε to quantify the local irregularity of the graph by measuring the relative deviation of the node degree 2^{δ_x} from the cardinality number 2^d in the regular lattice,

$$\varepsilon = 1 - \frac{d}{\delta_x}. \tag{10.4}$$

The nodes with $\deg(x) < 2^d$ correspond to $\varepsilon < 0$, while the nodes for which $\deg(x) \geq 2^d$ are described by $\varepsilon \geq 0$.

We supply (10.2) with the locally integrable initial condition $u(\mathbf{x}, 0)$ and study the standard Cauchy problem being interested in the large scale asymptotic Green's functions $G(\mathbf{x}, \mathbf{x_0}; t, t_0)$. For a curved space, the natural way to proceed is to examine the change to the Green's functions as the metric is scaled. This can be achieved by moving the points along the geodesics (cycles in the graph Γ) connecting them or alternatively by scaling the geodesic distance function (the metric).

The Green's functions of nonlinear problem (10.2) supplied with the integrable initial conditions can be formally calculated by the perturbation series with respect to the nonlinearity (as the coupling parameter η_0 is small) followed by the integrations over the initial condition $u(\mathbf{x}, 0)$. Some integrals estimating corrections to the linearized diffusion problem diverge logarithmically since the integration domain is not compact. If we introduce the ε-parameter in accordance to (10.4), the divergences reveal themselves by the poles in

$$\varepsilon = 1 + d \cdot \frac{1 - \alpha}{2}.$$

Therefore, the nonlinear interaction is irrelevant (in the sense of Wilson) for $\varepsilon < 0$ ($d > 2/(\alpha - 1)$), but is essential as $\varepsilon \geq 0$ when the ordinary perturbation expansion (in the form of series in η_0) fails to give the correct large scale asymptotic behavior and the whole series has to be summed up. For instance, it happens at $d = 2$ for $\alpha = 2$.

In other words, the logarithmic (marginal) value of α is determined by comparison of the nonlinear contribution with that of the linear dissipative term,

$$\alpha_{\log} = 1 + 2/d.$$

While introducing the parameter ε accounting for the local change of connectivity in the graph, we effectively pass from $\alpha_{\log}$ to its new value

$$\alpha'_{\log} = 1 + \frac{2}{\delta_x}.$$

Then, it turns that the nonlinear contribution to the long range asymptotic transport through the rims ($\varepsilon < 0$) is irrelevant in comparison with the linear diffusion and therefore can be neglected. In contrast to it, the contribution coming from hubs ($\varepsilon \geq 0$) is more essential than the linear one and has to be taken into account in all orders of perturbation theory since the relevant fluctuations dominate the diffusion at large scales.

We calculate the asymptotic Green's functions for the model (10.2) in the logarithmic theory (on the regular subgraphs with the cardinality number 2^d) in curved space metric and develop the ε−expansion accounting for the corrections in the long time large scale region due to the irregularity of graph. The relevant contributions to the nonlinear transport coming from hubs are summed by the field-theoretic renormalization group method. Herewith, the real values of parameter ε has to be taken as

$$\varepsilon_x = 1 - \frac{d}{\delta_x},$$

the excess of hub's connectivity over the regular cardinality number 2^d.

We also remark that the problem of renormalization in a curved space-time has been discussed extensively in the literature (de Witt 1975; Nelson and Panangaden 1982; Toms 1982 and by other authors). However, it has never been studied in connection with the critical phenomena theory. The analysis of transport through the graphs would provide us with such a model. It is worth to mention that in contrast to the case of the gravitational field, we are not restricted on graphs by the equivalence principle, so that the curved space has not to be flattened in any sufficiently small region.

10.4 Diffusion as a Generalized Brownian Motion

It is well known that many problems of stochastic dynamics (and of the transport through a disordered media, in particular) can be treated as a generalized Brownian motion,

$$P(u) = \langle \delta \left(u - u\left(\mathbf{x}, t\right)\right)\rangle,$$

in which the average is taken over all configurations of field $u(\mathbf{x}, t)$ satisfying the dynamical equation

$$\nabla_t u - v_0 \Delta_{LB} u + v_0 \eta_0 u^\alpha + \xi_0 v_0 R u = \frac{1}{\sqrt{g}} \delta(t - t_0)\delta(\mathbf{x} - \mathbf{x}_0), \tag{10.5}$$

for the integrable initial condition $u(\mathbf{x}, 0)$ and *arbitrary* conditions on the boundaries of the graph. The Laplace-Beltrami operator Δ_{LB} is given by (10.2), and $g = |\det g_0^{ij}|$.

We use the functional representation of the $\delta-$function for expressing the probability,

$$P(u) = \int \mathcal{D}u \int \mathcal{D}u' \exp \left(u' \left(u - u(\mathbf{x}, t)\right)\right), \tag{10.6}$$

in which u marks the position of a "particle", and the auxiliary field u' (of the same nature as u) is not inherent to the original model, but appears since we treat the dynamics as a Brownian motion. The formal convergence requires the field u to be real and the field u' to be purely imaginary. Should a unique solution of dynamic equation exists, we perform the natural change of variables in (10.6),

$$\begin{aligned}
(u - u(\mathbf{x}, t)) \rightarrow &-\nabla_t u + v_0 \Delta_{LB} u \\
&- v_0 \eta_0 u^\alpha - \xi_0 v_0 R u \\
&- \frac{1}{\sqrt{g}} \delta(t - t_0)\delta(\mathbf{x} - \mathbf{x}_0) \\
= &\ 0,
\end{aligned}$$

from which it follows that

$$P(u) = \int \mathcal{D}u \int \mathcal{D}u' \exp \mathcal{S}(u, u') \det \mathbf{M}$$

where $\det \mathbf{M}$ is the Jacobian associated to the change of variable, and $\mathcal{S}(u, u')$, the action functional,

$$S = \mathcal{S}_0 - \eta_0 v_0 \mathrm{Tr}_g \left(u' u^\alpha\right) - \frac{1}{\sqrt{g}} u'(\mathbf{x}_0, t_0), \tag{10.7}$$

$$\mathcal{S}_0 = \mathrm{Tr}_g \left(-u' \nabla_t u + v_0 u' \Delta_{LB} u - \xi_0 v_0 R u' u\right).$$

The trace Tr_g means the summation over the discrete indexes and the integration $\int dv_x \int dt$ over the invariant volume element on the d-dimensional manifold, $dv_x = \sqrt{g(x)}d^d x$.

The Jacobian $\det \mathbf{M}$ deserves a thorough consideration. The linear part of the variable transformation can be factorized from it,

$$\det \mathbf{M} = \det \mathbf{M}_0 \det(1 - \Delta_{uu'}\mathbf{M}_1) \tag{10.8}$$

where

$$\mathbf{M}_0 = -\nabla_t + \nu_0 \Delta_{LB} - \xi_0 \nu_0 R,$$

and the interaction part

$$\mathbf{M}_1 = \alpha \nu_0 \eta_0 u^{\alpha-1}\delta(t - t'),$$

and $\Delta_{uu'}$ is the Feynman propagator in curved Riemann space (Bunch and Parker 1979; Balakrishnan 2000), defined as the solution of linearized problem

$$(-\nabla_t + \nu_0 \Delta_{LB} - \xi_0 \nu_0 R)\Delta_{uu'}(\mathbf{x} - \mathbf{x}', t - t') = \frac{1}{\sqrt{g}}\delta(t - t')\delta(\mathbf{x} - \mathbf{x}'). \tag{10.9}$$

We are interested in the explicit solution of (10.9), in the d-dimensional curved space metric, for $G(\mathbf{x}, \mathbf{x}', t)$ satisfying

$$\lim_{\mathbf{x}' \to 0} G(\mathbf{x}, \mathbf{x}', t) = G(\mathbf{x}, t)$$

and

$$\lim_{t \to 0} G(\mathbf{x}, \mathbf{x}', t) = \delta(\mathbf{x} - \mathbf{x}').$$

The details can be found in Bunch and Parker (1979) for field theories and in Balakrishnan (2000) (see also references therein) for models of classical statistical physics.

The general solution is

$$\Delta_{uu'}(\mathbf{x}, \mathbf{x}', t) = \theta(t)\frac{e^{(-\xi_0 \eta_0 R t)}}{(4\pi t)^{d/2}}e^{\left(-\frac{\sigma(\mathbf{x},\mathbf{x}')}{2t}\right)}\Delta^{1/2}(\mathbf{x}, \mathbf{x}')\Omega(\mathbf{x}, \mathbf{x}', t) \tag{10.10}$$

where $\theta(t)$ is the Heaviside function, $\sigma(\mathbf{x}, \mathbf{x}')$ equals to half the square of the geodesic distance between $\mathbf{x}$ and $\mathbf{x}'$, and $\Delta(\mathbf{x}, \mathbf{x}')$ is the van Vleck determinant,

$$\Delta(\mathbf{x}, \mathbf{x}') = -\frac{\det\left(\partial_i \partial_j \sigma(\mathbf{x}, \mathbf{x}')\right)}{\sqrt{g(\mathbf{x})g(\mathbf{x}')}},$$

which reduces to unity in flat space, R is the scalar curvature. The function $\Omega(\mathbf{x}, \mathbf{x}', t)$ allows for the following series expansion in the limit $\mathbf{x}' \to \mathbf{x}$:

$$\lim_{\mathbf{x}' \to \mathbf{x}} \Omega(\mathbf{x}, \mathbf{x}', t) = 1 + \sum_{l=1}^{\infty} t^l E_l(\mathbf{x})$$

valid in the limit $t \to 0$ where $E_l(\mathbf{x})$ are known in the literature as Gilkey coefficients (Gilkey 1975; de Witt 1965; Parker and Toms 1985a,b; Parker 1979).

For the (10.2), in a flat metric, the only coefficient which contributes is E_0 and we recover the well known standard diffusion kernel.

The planar 3-regular graph of order $2N$ with the standard orientation (the cyclic ordering of edges is taken the same for each node), a $2N$-honeycomb, corresponds to a sphere of radius $\rho = \sqrt{N/2}$. The Ricci scalar curvature is

$$R = \frac{2}{\rho^2} = \frac{4}{N},$$

and the Gaussian curvature equals to $\kappa = 2/N$. The Gilkey coefficients (Gilkey 1975) reduces to

$$E_0 = 1, \quad E_1 = \frac{2}{3N},$$

$$E_2 = \frac{4}{15}\frac{1}{N^2}, \quad E_3 = \frac{32}{315}\frac{1}{N^3}. \tag{10.11}$$

Then, in the limit $\mathbf{x}' \to \mathbf{x}$ and $t \to 0$, the Feynman propagator exhibits the following dependence on the size of 3-regular subgraph:

$$\Delta_{uu'}(\mathbf{x}, t) = \theta(t)\frac{e^{-\frac{x^2}{4v_0 t}}}{4\pi v_0 t}\left(1 + \frac{1}{3}\left(\frac{2v_0 t}{N}\right) + \frac{1}{15}\left(\frac{2v_0 t}{N}\right)^2 + \frac{4}{315}\left(\frac{2v_0 t}{N}\right)^3 + \ldots\right). \tag{10.12}$$

It is important for us that the propagator $\Delta_{uu'}$ (10.10) is proportional to the Heaviside function $\theta(t - t')$ as a consequence of causality principle.

The first factor in (10.8) does not depend upon fields and therefore can be scaled out of the functional integration. The second factor in (10.8) can be expanded into the "diagram" series,

$$\log \det(1 - \Delta_{uu'}\mathbf{M}_1) = -\mathrm{Tr}_g\left(\Delta_{uu'}\mathbf{M}_1 + \frac{1}{2}\Delta_{uu'}\mathbf{M}_1\Delta_{uu'}\mathbf{M}_1 + \ldots\right)$$

comprising of cycles of retarded Feynman propagators proportional to the Heaviside functions and therefore being trivial, excepting for the very first term, $\mathrm{Tr}_g(\Delta_{uu'}\mathbf{M}_1)$, in which the operator $\mathbf{M}_1$ contains $\delta(t - t')$. The first term in the expansion is proportional to the undefined quantity $\theta(0)$ which value is usually taken as $1/2$. However, in the critical dynamics, another convention is used (Adzhemyan et al. 1998), $\theta(0) = 0$, under which the Jacobian $\det \mathbf{M}$ turns to be just a constant and therefore may be scaled out away at the irrelevant cost of changing only the normalization.

The action functional of type (10.7) in the problem of nonlinear diffusion in the flat metric has been introduced in Antonov and Honkonen (2002). In Volchenkov and Lima (2008a) and Volchenkov (2009), the functional with an ultra-local interaction term like in (10.7) has been derived as a limiting one in the framework of MSR formalism (stochastic quantization, Martin et al. 1973; de Dominicis 1976). The renormalization of field theoretic models with ultra-local terms, located on surfaces, had been studied in Symanzik (1981) in details.

Further insight into the field theory representation of Brownian motion and the properties of auxiliary field u' can be obtained from the equations for the saddle-point configurations. The first such an equation,

$$\frac{\delta \mathcal{S}}{\delta u'} = 0,$$

recovers the original Cauchy problem. Another one,

$$\frac{\delta \mathcal{S}}{\delta u} = 0,$$

reads as following

$$\nabla_t u' + \nu_0 \Delta_{LB} u' - \xi_0 \nu_0 R = \alpha \eta_0 \nu_0 u' u^\alpha$$

and is characterized by a negative viscosity. One can conclude from it that the auxiliary field should be trivial for positive time, $u'(t > 0) = 0$, and decays as $t \to -\infty$.

In the framework of field theory approach, the Green function $G(x, t; x_0, t_0)$ for the Cauchy problem (10.5) can be computed as the functional average,

$$G(x, t; x_0, t_0) = \langle u \rangle = \frac{\int \mathcal{D}u \int \mathcal{D}u' u(x, t) \exp S(u, u')}{\int \mathcal{D}u \int \mathcal{D}u' \exp (S_0)}, \tag{10.13}$$

in which $\mathcal{S}$ is the action functional (10.7). The Green function (10.13) and all higher moments of fields u and u' allow for the standard Feynman diagram series expansions. The diagram technique with the ultra-local interaction terms has been discussed in Antonov and Honkonen (2002). A special feature of such diagrams is that the final point of any diagram corresponds to (x_0, t_0). It is worth to mention that diagrams could formally contain a non-integer number of lines (since the nonlinearity exponent could deviate from an integer number). Diagrams are drawn of three elements: (i) the final point (x_0, t_0) with an arbitrary number of attached u'-legs (we mark them by a slash, see Fig. 10.2); (ii) the interaction node with one u'-leg and α u-legs attached to it (we put the letter α inside the loop to stress that it is not necessarily integer); (iii) the propagator $\Delta_{uu'}$ is only available. The first three diagrams for the Green function (10.13) are sketched in Fig. 10.2.

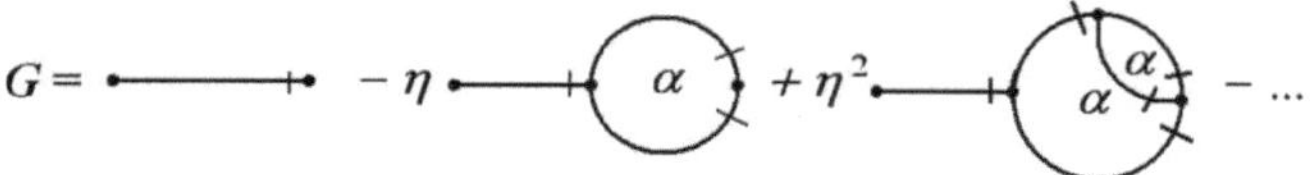

Fig. 10.2 Three diagrams of the Feynman diagram expansion for the Green function of the nonlinear diffusion equation

10.5   Scaling of a Scalar Field Coupled to a Complex Network

Power-counting arguments (Volchenkov and Blanchard 2007a; Volchenkov 2009) show that the divergent Feynman diagrams are those which involve any number of external u'-legs (lines with slashes in Fig. 10.2). For all these functions, the formal index of divergence equals to zero, therefore all divergent contributions are logarithmic (the correspondent counterterms are constants). It is worth to mention that the model is renormalizable despite the fact that it requires an infinite number of counterterms. It is sufficient to renormalize the only "one-particle-irreducible" Green function (the only diagram block which can be drawn using the elements (i)–(iii) mentioned above) to render the model finite.

Moreover, the counter-term corresponding to this block is sufficient to regularize all higher moments of fields u and u', since any diagram of perturbation series is expressed as a convolution of equivalent blocks. Diagrams contain no additional superficial divergences (Volchenkov and Blanchard 2007a; Volchenkov 2009). We recall that all loops which could arise in the diagram expansions are created by the single local vortex with any number of u'−legs incident at it. The counter-term corresponding to the elementary divergent block is constant and local in configuration space, i.e., $\propto \delta(t - t_0)\delta(\mathbf{x} - \mathbf{x}_0)$. In the action functional (10.7), the same local term appears, so that the model is renormalized multiplicatively, and only the renormalization constant Z is required. To keep the interaction coupling constant η dimensionless, one must introduce a regularization mass parameter μ. The bare parameters are related to the renormalized parameters by

$$\nu_0 = \nu, \quad \xi = \xi_0,$$

$$\eta_0 = \eta \mu^{2\varepsilon} Z^{\alpha-1}. \tag{10.14}$$

The auxiliary fields and the Green function are related to their renormalized analogs by:

$$u' = u'_R Z,$$

$$G(\nu_0, \xi_0, \eta_0) = Z^{-1} G_R(\nu, \xi, \eta, \mu),$$

$$Z = 1 + \sum_{l=1}^{\infty} \frac{c_l(\xi, \eta, \mu)}{\varepsilon^l} \tag{10.15}$$

where the amplitudes c_l are defined to be precisely those needed to subtract the poles in the corresponding Feynman integrals.

The derivation and analysis of the renormalization group equation for the model in question was carried in Volchenkov and Blanchard (2007a) and Volchenkov (2009) in details. To quantify the dilatation of the background metric g_{ij} in curved space, the new mass parameter ϖ had been introduced. In particular, it was shown that the renormalization group transformation has a unique fixed point stable in the large scale long time limit.

The canonical scale invariance of the renormalized Green function $G_R(\xi, \mu, v, \eta)$ with respect to dilatations of all variables requires that is has the following form,

$$G_R(t,r) = (vt)^{-d/2} \chi\left(\frac{r^2}{vt}, \varpi vt, \frac{1}{t\mu^2 v}\right), \quad r = |\mathbf{x} - \mathbf{x}'|, \tag{10.16}$$

where χ is a scaling function of dimensionless variables. The dependence on η in χ is not displayed explicitly, because the derivatives with respect to this parameter do not enter into the scaling equation (see Volchenkov and Blanchard 2007a; Volchenkov 2009). Furthermore, the scaling function can be expressed as

$$\chi(s, y, z) = s^{\frac{\gamma_*}{2}} \psi(s, y) \tag{10.17}$$

where ψ is an arbitrary function of the first and second arguments, and the value of anomalous dimension at the fixed point of the renormalization group transformation is

$$\gamma_* = -\frac{2}{\alpha - 1}\left(1 - \frac{\delta_x}{d}\right)\bigg|_{\alpha = \alpha_{\log}} = \delta_x - d. \tag{10.18}$$

Finally, for the Green function (10.13), we obtain

$$G(t,r) \sim t^{-\frac{d}{2} - \frac{\gamma_*}{2}} \psi\left(\frac{r^2}{vt}, \varpi vt\right) = t^{-\frac{\delta_x}{2}} \psi\left(\frac{r^2}{vt}, \varpi vt\right), \quad \delta_x > d. \tag{10.19}$$

We conclude the section with a remark on the arguments of scaling function ψ for the planar 3-regular graphs of order $2N$ with the standard orientation (i.e., the $2N$-honeycomb). We have already mentioned that they are equivalent to the sphere of radius ρ. The surface area of the sphere equals to $2\pi N = 4\pi\rho^2$, and therefore $\rho = \sqrt{N/2}$. One can see that the corrections to the standard diffusion kernel risen by the space curvature can be naturally interpreted as the corrections due to the finite size of the 3-regular subgraph. Then, the scaling function ψ is

$$\psi = \psi\left(\frac{r^2}{vt}, \frac{2vt}{N}\right),$$

and it can be calculated by means of diagram expansion. In the thermodynamic limit, $N \to \infty$, (when the graph Γ is large and regular) these corrections vanish and the scaling function depends only on one argument r^2/vt relevant to the flat space. The result on the critical scaling (10.19) is still valid in the thermodynamic limit.

10.6 Summary

We have studied the transport through large complex networks containing regular subgraphs. We considered such networks as generalized trees in which two types of nodes are allowed: simple nodes and supernodes. We supposed that the supernodes are either the subgraphs ample with polygons or k-regular subgraphs. In particular, we discussed the case of 3-regular subgraphs which can be treated as the complete Riemann curved surfaces characterized by finite areas and subjected to the compactification.

The diffusion process taking place on such a complex network is considered as a generalized Brownian motion with arbitrary boundary conditions. Its random dynamics is then described by the functional integral. We studied the long time large scale asymptotic behavior of the Green function for the nonlinear diffusion equation defined on the complex network supplied with an integrable initial condition. The transport trough the complex network is of a strongly nonlinear nature being affected by both the varying effective space dimension between supernodes and the space curvature within a supernode.

We modeled the effect of varying space dimension by a nonlinear term included into the diffusion equation. It takes into account the possible fluctuations of transport coefficient, the diffusion-reaction processes, and the possible queuing due to a bounded transport capacity of edges bridging the supernodes. The fluctuations have been treated in the framework the field theory and summed up by means of the renormalization group method. The ε-expansion has been developed where the parameter $\varepsilon = (\delta_x - d)/\delta_x$ quantifies the deviation of local space dimension $\delta_x = \log_2 \deg(x)$ at $x \in \Gamma$ from the space dimension d intrinsic to the regular subgraphs. The renormalization group method predicts for the nodes with $\delta_x > d$ the long time asymptotic of Green function in the form of

$$ G(t) \sim t^{-\delta_x/2}, \quad \delta_x > d, $$

while it is $G(t) \sim t^{-d/2}$ for the nodes characterized by $\delta_x \leq d$.

In the regular graphs endowed with the standard orientation of edges (the cycling ordering of edges to be the same for all nodes), the effect of curvature within the supernodes reveals itself by the finite size corrections to the scaling behavior. In the even space of dimensions, the curvature results in the additional singularities of Green function at small scales and calls for the special regularization. We have to stress that in contrast to the case of gravitational field, we are not restricted by the equivalence principle while considering models of complex networks. The main technical difficulty of computations in the curved space metrics is the lack of the global frequency- momentum representation. We have used the heat-kernel expansion which does not depend upon the space time topology to regularize the effective action in the small scales in the one-loop order.

References

A.G. Ableton, *Live 8* [computer software] (Berlin, 2009)

M. Abramovitz, I.A. Stegun, The Editors of the *Handbook of Mathematical Functions* (Dover, New York, 1986)

L.Ts. Adzhemyan, N.V. Antonov, A.N. Vasiliev, *Field Theoretic Renormalization Group in Fully Developed Turbulence* (Gordon and Breach, London, 1998)

R.P. Agaev, P.Yu. Chebotarev, On determining the eigenprojection and components of a matrix. Automat. Rem. Contr. **63**(10), 1537 (2002)

R.K. Ahuja, T.L. Magnanti, J.B. Orlin, T.L. Magnanti, *Network Flows: Theory, Algorithms, and Applications* (Prentice Hall, United States, 1993)

R. Albert, H. Jeong, A.-L. Barabási, The diameter of the World Wide Web. Nature **401**, 130–131 (1999)

R. Albert, A.-L. Barabási, Statistical mechanics of complex networks. Rev. Mod. Phys. **74**(1), 47 (2002)

D.J. Aldous, J. Pitman, Brownian bridge asymptotic for random mappings. Random. Struct. Algor. **5**, 487 (1994)

D.J. Aldous, J.A. Fill, *Reversible Markov Chains and Random Walks on Graphs* (University of California, Berkeley, 2002)

A. Anderson, Y. Sinoto, New radiocarbon ages for colonization sites in East Polynesia. Asian Perspect. **41**, 242 (2002)

N.V. Antonov, J. Honkonen, Field theoretic renormalization group for a nonlinear diffusion equation. Phys. Rev. E **66**, 046105 (2002)

A. Arenas, A. Diaz-Guilera, J. Kurths, Y. Moreno, C. Zhou, Synchronization in complex networks. Phys. Rep. **469**, 93 (2008)

N. Aubry, On the hidden beauty of the proper orthogonal decomposition. Theor. Comp. Fluid Dyn. **2**, 339 (1991)

N. Aubry, R. Guyonnet, R. Lima, J. Stat. Phys. **64**, 683 (1991)

N. Aubry, L. Lima, *Spatio-temporal Symmetries* (Centre de Physique Theorique, Luminy, Marseille, France, 1993), Preprint CPT-93/P.2923

F.R. Bach, M.I. Jordan, Learning spectral clustering. Technical report, UC Berkeley, http://www.cs.berkeley.edu/fbach. Accessed 2003; Tutorial given at ICML 2004 international conference on machine learning, Banff, Alberta, Canada, 2004

J. Balakrishnan, Spatial curvature effects on molecular transport by diffusion. Phys.Rev. E **61**, 4648 (2000)

R.B. Bapat, T.E.S. Raghavan, in *Nonnegative Matrices and Applications*, Encyclopedia of Mathematics and its Applications, ed. by R. Doran, P. Flajolet, M. Ismail, T.-Y. Lam, E. Lutwak, (Cambridge University Press, New York, 1997)

P. Blanchard and D. Volchenkov, *Random Walks and Diffusions on Graphs and Databases*, Springer Series in Synergetics 10, DOI 10.1007/978-3-642-19592-1,
© Springer-Verlag Berlin Heidelberg 2011

R.B. Bapat, I. Gutman, W. Xiao, A simple method for computing resistance distance. Z. Naturforsch. **58a**, 494 (2003)

N.T.J. Bailey, *The Mathematical Theory of Infectious Diseases*, 2nd edn. (Griffin, London, 1975)

Ph. Baldi, in *The Foundations of Latin*, Trends in Linguistics: Studies and Monographs, vol. 117, ed. by Gast, Volker, (Mouton de Gruyter, Berlin, New York, 2002)

A.-L. Barabási, R. Albert, Emergence of scaling in random networks. Science **286**, 509 (1999)

G.P. Basharin, A.N. Langville, V.A. Naumov, The life and work of A.A. Markov. Linear Algebra Appl. **386**, 3–26 (2004)

V. Batagelj, T. Pisanski, D. Keržic, Automatic clustering of languages. Comput. Linguist. **18**(3), 339 (1992)

M. Batty, *A New Theory of Space Syntax*, CASA Working Paper, vol. 75 (UCL Centre For Advanced Spatial Analysis Publications, London, UK, 2004)

P. Bellwood, P. Koon, Lapita colonists leave boats unburned! Antiquity **63**(240), 613 (1989)

C.M. Bender, S. Boettcher, P.N. Meisinger, Universality in random walk models with birth and death. Phys. Rev. Lett. **75**, 3210 (1995)

M. Ben Hamed, F. Wang, Stuck in the forest: Trees, networks and Chinese dialects. Diachronica **23**(1), 29–60 (2006)

A. Ben-Israel, A. Charnes, Contributions to the theory of generalized inverses. J. Soc. Ind. Appl. Math. **11**(3), 667 (1963)

A. Ben-Israel, TH.N.E. Greville, *Generalized inverses: Theory and applications*, 2nd edn. (Springer, New York, 2003)

S.N. Bernstein, Sur lextension du théoremé limite du calcul des probabilies. Math. Ann. Bd. **97**, 1–59 (1926)

G. Besson, Sur la multiplicité des valeurs propres du laplacien. Séminaire de théorie spectrale et géométrie (Grenoble) **5**, 107–132 (1986–1987)

W. Bialek, Stability and noise in biochemical switches, in *Advances in Neural Information Processing Systems*, vol. 13, ed. by T.K. Leen, V. Tresp. Papers from Neural Information Processing Systems (NIPS) 2000 (MIT, Denver, CO, USA, 2001), pp. 75–81

G. Bianconi, N. Gulbahce, A.E. Motter, Local structure of directed networks. Phys. Rev. Lett. **100**, 118701 (2008)

N. Biggs, *Algebraic Graph Theory*, 2nd edn. (Cambridge Mathematical Library, Cambridge, 1993)

N.L. Biggs, E.K. Lloyd, R.J. Wilson, *Handbook of Combinatorics*, vol. 2 (MIT, Cambridge, MA, USA, 1996)

S. Bilke, *Shuffling Yeast Gene Expression Data (2000)*, Lund University - preprint TP 00-18; arXiv:physics/0006050 (2000)

S. Bilke, C. Peterson, Topological properties of citation and metabolic networks. Phys. Rev. E **64**, 036106 (2001)

S. Bilke, T. Breslin, M. Sigvardsson, Probabilistic estimation of microarray data reliability and underlying gene expression. BMC Bioinformatics **4**, 40 (2003)

T. Biyikoğlu, W. Hordijk, J. Leydold, T. Pisanski, P.F. Stadler, Graph Laplacians, nodal domains, and hyperplane arrangements. Linear Algebra Appl. **390**, 155 (2004)

T. Biyikoglu, J. Leydold, P.F. Stadler, Laplacian eigenvectors of graphs – Perron-Frobenius and Faber-Krahn type theorems, in *Springer Lecture Notes in Mathematics*, vol. 1915, ed. by J.-M. Morel, B. Teissier (Springer-verlag, Berlin, Heidelberg, 2007)

A. Björner, L. Lovász, P. Shor, Chip-firing games on graphs. Eur. J. Combin. **12**, 283 (1991)

A. Björner, L. Lovász, Chip-firing games on directed graphs. J. Algebr. Comb. **1**, 305 (1992)

Ph. Blanchard, D. Volchenkov, Intelligibility and first passage times in complex urban networks. Proc. R. Soc. A **464**, 2153 (2008). doi:10.1098/rspa.2007.0329

Ph. Blanchard, D. Volchenkov, in *Mathematical Analysis of Urban Spatial Networks*, Springer series: Understanding Complex Systems, vol. XIV, ed. by Kelso, J.A. Scott, (Springer-verlag, Berlin, Heidelberg, 2009)

Ph. Blanchard, D. Volchenkov, Probabilistic embedding of discrete sets as continuous metric spaces. Stochastics: Int. J. Prob. Stoch. Proc. (formerly: Stochastics and Stochastic Reports) **81**(3), 259 (2009)

Ph. Blanchard, F. Petroni, M. Serva, D. Volchenkov, Geometric representations of language taxonomies. Comput. Speech Lang. (2010). doi: 10.1016/j.csl.2010.05.003

S. Boccaletti, V. Latora, Y. Moreno, M. Chavez, D.-U. Hwang, Complex networks: Structure and dynamics. Phys. Rep. **424**, 175–308 (2006)

S. Boettcher, M. Moshe, Statistical models on spherical geometries. Phys. Rev. Lett. **74**, 2410 (1995)

M. Boguñá, F. Papadopoulos, D. Krioukov, Sustaining the internet with hyperbolic mapping. Nat. Commun. **1**, 62 (2010)

B. Bollobas, *Graph Theory* (Springer-Verlag, New York, 1979)

B. Bollobas, *Modern Graph Theory* (Springer-Verlag, New York, 1998)

B. Bollobas, A. Thomason, *Combinatorics, Geometry and Probability* (Cambridge University Press, Cambridge 2004)

R.P. Bolton, *Building For Profit* (Reginald Pelham Bolton, New York, 1922)

M. Bona, *Combinatorics of Permutations* (Chapman Hall-CRC, Boca Raton, FL, 2004). ISBN 1-58488-434-7

S. Bornholdt, Less is more in modeling large genetic networks. *Science* **310**(5747), 449 (2005)

G. Brightwell, I. Leader, A. Scott, A. Thomason, *Combinatorics and Probability* (Cambridge University Press, Cambridge, 2007)

J. Bricmont, A. Kupiainen, Renormalization group and the Ginzbug-Landau equation. Commun. Math. Phys. **150**, 193 (1992)

J. Bricmont, A. Kupiainen, G. Lin, Renormalization group and asymptotics of solutions of nonlinear parabolic equations. Commun. Pur. Appl. Math. **47**, 893 (1994)

J. Bricmont, A. Kupiainen, High temperature expansions and dynamical systems. Commun. Math. Phys. **178**, 703 (1996)

J. Bricmont, A. Kupiainen, J. Xin, Global large time self-similarityof a thermal-diffusive combustion system with critical nonlinearity. J. Differ. Equ. **130**(1), 9 (1996)

R. Brooks, E. Makover, *Random Construction of Riemann Surfaces*, Preprint (Department of Mathematics, Technion, 2001)

C. Brown, *A Portrait of Mendelssohn* (Yale University Press, New Haven and London, 2003)

E. Bryant, *The Quest for the Origins of Vedic Culture: The Indo-Aryan Migration Debate* (Oxford University Press, New York, 2001)

T.S. Bunch, L. Parker, Feynman propagator in curved spacetime: A momentum-space representation. Phys. Rev. D **20**(10), 2499 (1979)

E.M. Burns, Intervals, scales, and tuning, in *The Psychology of Music*, 2nd edn. ed. by D. Deutsch (Academic, San Diego, CA, 1999)

P. Buser, On the bipartition of graphs. Discrete Appl. Math. **9**, 105 (1984)

T. Brylawski, J.G. Oxley, The tutte polynomial and its applications, in *Matroid Applications*, ed. by N. White (Cambridge University Press, Cambridge, 1992), pp. 123–225

S. Butler, Interlacing for weighted graphs using the normalized Laplacian. Electron. J. Linear. Al. **16**, 90 (2007)

S.L. Campbell, C.D. Meyer, N.J. Rose, Applications of the drazin inverse to linear systems of differential equations with singular constant coefficients. SIAM J. Appl. Math. **31**(3), 411 (1976)

S.L. Campbell, C.D. Meyer Jr., *Generalized Inverses of Linear transformations* (Dover Publications, New York, 1979)

J.W. Cannon, W.J. Floyd, R. Kenyon, W.R. Parry, Hyperbolic geometry, in *Flavors of Geometry*, vol. 31, ed. by S. Levy (Cambridge Univ. Press, Cambridge, 1997)

D. Cassi, S. Regina, Random walks on bundled structures. Phys. Rev. Lett. **76**, 2914 (1996)

L.L. Cavalli-Sforza, *Genes, Peoples, and Languages* (North Point Press, New York, 2000)

P. Chan, M. Schlag, J. Zien, IEEE Trans. CAD-Integrated Circuits and Syst. **13**, 1088 (1994)

A. Chan, C. Godsil, Symmetry and eigenvectors, in *Graph Symmetry, Algebraic Methods and Applications*, ed. by G. Hahn, G. Sabidussi (Kluwer, Dordrecht, The Netherlands, 1997), p. 75

A.K. Chandra, P. Raghavan, W.L. Ruzzo, R. Smolensky, P. Tiwari, The electrical resistance of a graph captures its commute and cover times. Comput. Complex. **6**(4), 312 (1996)

G. Chartrand, *Introductory Graph Theory* (Dover, New York, 1985). ISBN 0-486-24775-9

H. Chaté, P. Manneville, Spatiotemporal intermittency in coupled map lattices. Physica D **32**, 409–422 (1988)

H. Chaté, P. Manneville, Transition to turbulence via spatio-temporal intermittency. Europhys. Lett. **6**, 591–595 (1988)

H. Chaté, P. Manneville, Collective behaviors in coupled map lattices with local and nonlocal connections. Chaos **2**, 307–313 (1992)

H. Chaté, P. Manneville, Collective behaviors in spatially extended systems: Interactions and synchronous updating. Prog. Theor. Phys. **87**, 1 (1992)

J. Cheeger, A lower bound for the smallest eigenvalue of the Laplacian, in *Problems in Analysis, Papers dedicated to Salomon Bochner*, ed. by Gunning, Robert C., (Princeton University Press, Princeton, 1969), p. 195

H. Chen, F. Zhang, Resistance distance and the normalized Laplacian spectrum. Discrete Appl. Math. **155**, 654 (2007)

S.Y. Cheng, Eigenfunctions and nodal sets. Commun. Math. Helv. **51**, 43 (1976)

F.R.K. Chung, *Lecture Notes on Spectral Graph Theory* (AMS Publications Providence, 1997)

F. Chung, L. Lu, V. Vu, Spectra of random graphs with given expected degrees. Proc. Natl. Acad. Sci. U.S.A. **100**(11), 6313 (2003)

F. Chung, Laplacians and the cheeger inequality for directed graphs. Ann. Comb. **9**, 1 (2005)

P. Collet, J.-P. Eckmann, The number of large graphs with a positive density of triangles. J. Stat. Phys. **109**(5–6), 923 (2002)

G.C. Conant, A. Wagner, Convergent evolution in gene circuits. Nat. Genet. **34**(3), 264 (2003)

J.H. Conway, R.K. Guy, Arrangement numbers, in *The Book of Numbers*, ed. by J.H. Conway, (Springer-Verlag, New York, 1996)

D. Coppersmith, P. Tetali, P. Winkler. Collisions among random walks on a graph. SIAM J. Discrete Math. **6**(3), 363 (1993)

R.M. Corless, G.H. Gonnet, D.E.G. Hare, D.J. Jeffrey, D.E. Knuth, On the Lambert W function. Adv. Comput. Math. **5**, 329 (1996)

T.H. Cormen, C.E. Leiserson, R.L. Rivest, C. Stein, *Introduction to Algorithms*, 2nd edn. Chapter 21: Data Structures for Disjoint Sets (The MIT Press, Cambridge, Massachusetts, 2001), pp. 498–524. ISBN 0-262-03293-7

M.G. Cosenza, R. Kapral, Coupled maps on fractal lattices. Phys. Rev. A **46**(4), 1850 (1992)

M.G. Cosenza, R. Kapral, Spatiotemporal intermittency on fractal lattices. Chaos **4**, 99 (1994)

M.G. Cosenza, A. Parravano, Turbulence in globally coupled maps. Phys. Rev. E **53**, 6032 (1996)

M.G. Cosenza, K. Tucci, Transition to turbulence in coupled maps on hierarchical lattices. Chaos Soliton. Fract. **11**, 2039–2044 (2000)

M.G. Cosenza, K. Tucci, Turbulence in small-world networks. Phys. Rev. E **65**, 036223 (2002)

L.D.F. Costa, F.A. Rodrigues, G. Travieso, P.R.V. Boas, Characterization of complex networks: A survey of measurements. Adv. Phys. **56**, 167 (2007)

L.D.F. Costa, G. Travieso, Exploring complex networks through random walks. Phys. Rev. E **75**, 016102 (2007)

R. Coutinho, B. Fernandez, R. Lima, A. Meyroneinc, Discrete time piecewise affine models of genetic regulatory networks. J. Math. Biol. **52**(4), 0303–6812 (Print), 1432–1416 (Online) (2006)

T.M. Cover, J.A. Thomas, *Elements of Information Theory*, (Wiley, New York, 1991)

P. Crucitti, V. Latora, S. Porta, Centrality in networks of urban streets. Chaos **16**, 015113 (2006)

D.M. Cvetkovic, P. Rowlinson, S. Simic, *Eigenspaces of Graphs*, in series Encyclopedia of Mathematics and Its Applications, vol. 66, ed. by R. Doran, P. Flajolet, M. Ismail, T.-Y. Lam, E. Lutwak, (Cambridge University Press, Cambridge, UK, 1997)

D.M. Cvetkovic, M. Doob, H. Sachs, *Spectra of Graphs*, 3rd Rev. edn. (Academic, New York, 1980)

D. d'Urville, Sur les îles du Grand Océan. Bulletin de la Société de Géographie **17**, 1 (1832)

O.C. Dahl, *Avhandlinger utgitt av Egede-Instituttet*, vol. 3 (Arne Gimnes Forlag, Oslo, 1951)

C. Dahlhaus, Harmony, in *Grove Music Online*, ed. by L. Macy (2007), http://www. grovemusic. com. Accessed 24 Feb 2007

G.B. Dantzig, R. Fulkerson, S.M. Johnson, Solution of a large-scale traveling salesman problem. Oper. Res. **2**, 393 (1954)

Database. The database of 200 words most resistant to changes is available at http://univaq.it/~ serva/languages/languages.html

E.B. Davis, G.M.L. Gladwell, J. Leydold, P.F. Stadler, Discrete nodal domain theorems. Linear Algebra Appl. **336**, 51 (2001)

C. de Dominicis, Techniques de renormalisation de la théorie des champs et dynamique des phénomène critiques. J. Phys. (Paris) **37**, Colloq. C1, C1-247 (1976)

H. de Jong, R. Lima, Modeling the dynamics of genetic regulatory networks: Continuous and discrete approaches, in *Dynamics of Coupled Map Lattices and of Related Spatially Extended Systems*, vol. 671, ed. by J.-R. Chazottes, B. Fernandez. Lecture Notes in Physics (Springer, Berlin, Heidelberg, 2005), p. 307

Y.C. de Verdiére, Construction de laplaciens dont une partie finie du spectre est donnée. Ann. Sci. Ecole. Norm. S. **20**, 599 (1987)

Y.C. de Verdiére, *Spectres de Graphes*, Cours Spécialisés 4, (Société Mathématique de France (1998) (in French)

B.S. De Witt, *Dynamical Theory of Groups and Fields* (Gordon and Breach, New York, 1965)

B.S. De Witt, Quantum-field theory in curved space–time. Phys. Rep. **19**, 295 (1975)

M. Dyer, A. Frieze, R.Kannan, *A random polynomial-time algorithm for approximating the volume of convex bodies*, Journal of the ACM **38** (1), 1 (1991)

I. Dhillon, Y. Guan, B. Kulis, *A Unified View of Kernel k-means, Spectral clustering and Graph Cuts*, Technical Report TR-04-25, University of Texas at Austin, 2004

I. Dhillon, Y. Guan, B. Kulis, Kernel k-means: spectral clustering and normalized cuts, in *Proceedings of the 10th ACM SIGKDD international conference on Knowledge discovery and data mining*, Seattle, WA, USA, 2004

P. Diaconis, *Group Representations in Probability and Statistics* (Institute of Mathematical Statistics, Hayward, CA, 1988)

J.M. Diamond, Express train to polynesia. Nature **336**, 307 (1988)

R. Diestel, *Graph Theory* (Springer, Berlin, 2005)

R. Dickson, J. Ableson, W. Barnes, W. Reznikoff, Genetic regulation: the Lac control region. Science **187**, 27 (1975)

C. Ding, X. He, K-means clustering via principal component analysis, in *Proceedings of International Conference Machine Learning (ICML'2004)*, ACM, New York, July 2004, pp. 225–232

S. Dorogovtsev, J.F.F. Mendes, Evolution of networks. Adv. Phys. **51**, 1079 (2002)

S.N. Dorogovtsev, A.V. Goltsev, J.F.F. Mendes, Critical phenomena in complex networks. Rev. Mod. Phys. **80**, 1275 (2008)

S.N. Dorogovtsev, *Lectures on Complex Networks* (Oxford University Press, Oxford, 2010)

P.G. Doyle, J.L. Snell, *Random Walks and Electrical Networks* (Mathematical Association of America, Washington, DC, 1984); freely redistributable under the terms of the GNU General Public License (2000)

M.P. Drazin, Pseudo-inverses in associative rings and semigroups. Am. Math. Mon. **65**, 506 (1958)

I. Dyen, J.B. Kruskal, P. Black An Indo-european classification: A lexicostatistical experiment. Trans. Am. Philos. Soc. **82**(5), 1 (1992)

I. Dyen, J. Kruskal, P. Black, *Comparative Indo-European Database*, collected by Isidore Dyen. (Available at http://www.wordgumbo.com/ie/cmp/iedata.txt) Copyright (C) 1997 by Isidore Dyen, Joseph Kruskal, and Paul Black. The file was last modified on Feb 5 (1997). Redistributable for academic, non-commercial purposes

H. Dym, *Linear Algebra in Action*, in Series Graduate Studies in Mathematics, vol. 78 (AMS, 2007)

J.-P. Eckman, E. Moses, Curvature of co-links uncovers hidden thematic layers in the World Wide Web. Proc. Natl. Acad. Sci. U.S.A. **99**, 5825 (2002)

T.M. Ellison, S. Kirby, Measuring language divergence by intra-lexical comparison, in *Proceedings of the 21st International Conference on Computational Linguistics & 44th Annual Meeting of the Association for Computational Linguistics*, Sydney, Australia, 2006

S.M. Embelton, *Statistics in Historical Linguistics* (Brockmeyer, Bochum, 1986)

D. Eppstein, Spanning trees and spanners, in *Handbook of Computational Geometry*, ed. by J.R. Sack, J. Urrutia (Elsevier, Amsterdam, 1999), pp. 425–461

I. Erdélyi, On the matrix equation $Ax = \lambda Bx$. J. Math. Anal. Appl. **17**, 119 (1967)

P. Erdös, A. Rényi, On the evolution of random graphs. Publ. Math. Inst. Hungar. Acad. Sci. **5**, 17 (1960)

P. Erdös and A. Rényi, Asymmetric graphs. Acta Math. Acad. Sci. Hungar. **14** 295 (1963)

E. Estrada, J.A. Rodríguez-Velázquez, Subgraph centrality in complex networks. Phys. Rev. E **71**, 056103 (2005)

E. Estrada, J.A. Rodríguez-Velázquez, Spectral measures of bipartivity. Phys. Rev. E **72**, 046105 (2005)

European Environment Agency report *Urban sprawl in Europe. The ignored challenge* (2006). ISBN 92-9167-887-2

K. Fan, On a theorem of weyl concerning eigenvalues of linear transformations. Proc. Natl. Acad. Sci. U.S.A. **35**, 652 (1949)

I.J. Farkas, I. Derenyi, A.-L. Barabasi, T. Vicsek, Spectra of "real-world" graphs: Beyond the semicircle law. Phys. Rev. E **64**, 026704:1 (2001)

I. Farkas, I. Derényi, H. Jeong, Z. Neda, Z.N. Oltvai, E. Ravasz, A. Schubert, A.-L. Barabási, T. Vicsek, Networks in life: Scaling properties and eigenvalue spectra. Physica A **314**, 25 (2002)

I.J. Farkas, H. Jeong, T. Vicsek, A.-L. Barabasi, Z.N. Oltvai, The topology of the transcription regulatory network in the yeast, *S. cerevisiae*. Physica A **318**, 601 (2003)

M. Fiedler, Algebraic connectivity of graphs. Czech. Math. J. **23**(98), 298–305 (1973); **25**(146), (1975)

P. Forster, A. Toth, Toward a phylogenetic chronology of ancient Gaulish, Celtic, and Indo-European. Proc. Natl. Acad. Sci. U.S.A. **100**(15), 9079 (2003)

P. Fouracre, *The New Cambridge Medieval History* (Cambridge University Press, Cambridge, UK, 1995–2007)

D.M. Franz, Markov Chains as Tools for Jazz Improvisation Aanalysis, Master's Thesis, Industrial and Systems Engineering Department, Virginia Tech, 1998

L.C. Freeman, Centrality in social networks: Conceptual clarification. Soc. Networks **1**, 215 (1979)

E. Friedgut, Sharp thresholds of graph properties, and the k-sat problem. J. Am. Math. Soc. **12**, 1017 (1999)

J.S. Friedländer et al., Genetic structure of Pacific Islanders. PLoS Genet. **4**(1), e19 (2008)

M. Friendly, Corrgrams: Exploratory displays for correlation matrices. Am. Stat. **56**(4), 316 (2002)

P.M. Gade, H.A. Cerdeira, R. Ramaswamy, Coupled maps on trees. Phys. Rev. E **52**, 2478 (1995)

P. Gade, Synchronization of oscillators with random nonlocal connectivity. Phys. Rev. E **54**, 64 (1996)

Th.V. Gamkrelidze, V.V. Ivanov, The early history of Indo-European languages. Sci. Am. **262**(3), 110 (1990)

Th.V. Gamkrelidze, V.V. Ivanov, *Indo-European and the Indo-Europeans: A Reconstruction and Historical Analysis of a Proto-Language and a Proto-Culture*, in series Trends in Linguistics: Studies and Monographs, vol. 80, ed. by Gast, Volker, (Mouton de Gruyter, Berlin, New York, 1995)

F.R. Gantmacher, *The Theory of Matrices*. Trans. from the Russian by K. A. Hirsch, vols. I and II, (Chelsea, New York, 1959)

T.S. Gardner, D. di Bernardo, D. Lorenz, J.J. Collins, Inferring genetic networks and identifying compound mode of action via expression profiling. Science **301**, 102 (2003)

G. Gielis, R.S. MacKay, Coupled map lattices with phase transition. Nonlinearity **13**, 867 (2000)

P. Gilkey, The spectral geometry of a Riemannian manifold. J. Differ. Geom. **10**, 601 (1975)

M. Gimbutas, Old Europe in the fifth millenium B.C.: The European situation on the arrival of Indo-Europeans, in *The Indo-Europeans in the Fourth and Third Millennia*, ed. by E.C. Polomé (Karoma Publishers, Ann Arbor, 1982), pp. 1–60

M. Girvan, M.E.J. Newman, Community structure in social and biological networks. Proc. Natl. Acad. Sci. U.S.A. **99** (12), 7821 (2002)

E.L. Glaeser, J. Gyourko, *Why is Manhattan So Expensive?*, vol. 39, Manhattan Institute for Policy Research, Civic Report (2003)

L. Glass, S.A. Kauffman, The logical analysis of continuous non–linear biochemical control networks. J. Theor. Biol. **39**, 103 (1973)

P.M. Gleiss, P.F. Stadler, A. Wagner, D.A. Fell, Relevant cycles in chemical reaction network. Adv. Complex Syst. **4**, 207 (2001)

Ch. Godsil, G. Royle, *Algebraic Graph Theory*, in Springer Series: Graduate Texts in Mathematics, vol. 207 (Springer-Verlag, New York, 2001)

N. Goldenfeld, O. Martin, Y. Oono, F. Liu, Anomalous dimensions and the renormalization group in a nonlinear diffusion process. Phys. Rev. Lett. **64**(12), 1361 (1990)

G.H. Golub, Ch.F. Van Loan, *Matrix Computations*, 3rd edn. Johns Hopkins Studies in Mathematical Sciences (The Johns Hopkins University Press, Baltimore, MD, 1996)

J. Gomez-Gardenes, V. Latora, Entropy Rate of diffusion processes on complex networks. Phys. Rev. E **78**, 065102(R) (2008)

R.G. Gordon Jr. (ed.), *Ethnologue: Languages of the World*, 15th edn. (SIL International, Dallas, TX, 2005). Online version: http://www.ethnologue.com/

R. Gould, *Graph Theory* (Benjamin/Cummings, Menlo Park, CA, 1988)

A. Graham, *Nonnegative Matrices and Applicable Topics in Linear Algebra* (John Wiley & Sons, New York, 1987)

R.L. Graham, M. Grötschel and L. Lovász (eds.), *Handbook of Combinatorics* (Elsevier Science B.V., MIT, Amsterdam, Cambridge, MA 1995)

R.D. Gray, F.M. Jordan, Language trees support the express-train sequence of Austronesian expansion. Nature **405**, 1052 (2000)

R.D. Gray, Q.D. Atkinson, Language-tree divergence times support the Anatolian theory of Indo-European origin. Nature **426**, 435 (2003)

R.D. Gray, S.J. Greenhill, R.M. Ross, The pleasures and perils of darwinizing culture (with phylogenies). Biol. Theor. **2**(4), 360 (2007)

R.D. Gray, A.J. Drummond, S.J. Greenhill, Language phylogenies reveal expansion pulses and pauses in pacific settlement. *Science* **323**, 479 (2009)

P. Green, *The Greco-Persian Wars* (University of California Press, Berkeley, Los Angeles, London, 1996)

S.J. Greenhill, R. Blust, R.D. Gray, The austronesian basic vocabulary database: From bioinformatics to lexomics. Evol. Bioinform. **4**, 271 (2008). The Austronesian Basic Vocabulary Database is available at http://language.psy.auckland.ac.nz/austronesian

W.H. Greub, *Linear Algebra*, 4th edn. Graduate Texts in Mathematics (Springer, New York, 1981)

J. Gross, *Graph Theory*, Textbooks and Resources at http://www.graphtheory.com/

A. Groenlund, Networking genetic regulations and neural computation: Directed network topology and its effect on the dynamics. Phys. Rev. E **70**, 061908 (2004)

M.J. Hall Jr., *Combinatorial Theory*, 2nd edn. (Wiley, New York, 1986; 1998)

W.G. Hansen, How accessibility shapes land use. J. Am. Inst. Planners **25**, 73 (1959)

F. Harary, *Graph Theory* (Addison-Wesley, Reading, MA, 1969)

J.M. Harris, J.L. Hirst, M.J. Mossinghoff, *Combinatorics and Graph Theory* (Springer, New York, 2005)

R.E. Hartwig, More on the souriau-frame algorithm and the drazin inverse. SIAM J. Appl. Math. **31**(1), 42 (1976)

M.B. Hastings, An ε-expansion for small-world networks. Euro Phys. J. B. **42**, 297 (2004)

J. Hasty, J. Pradines, M. Dolnik, J.J. Collins, Noise-based switches and amplifiers for gene expression. Proc. Natl. Acad. Sci. U.S.A. **97**, 2075–2080 (2000)

P. Heggarty, Interdisciplinary indiscipline? Can phylogenetic methods meaningfully be applied to language data and to dating language? in *Phylogenetic Methods and the Prehistory of Languages*, ed. by P. Forster, C. Renfrew (McDonald Institute for Archaeological Research, Cambridge, 2006), p. 183

P. Heggarty, Splits or waves? Trees or webs? Network analysis of language divergence. AHRC Conference on Cultural and Linguistic Diversity, Great Missenden, 9–13 December 2008

H.W. Hethcote, The mathematics of infectious diseases. Soc. Indus. Appl. Math. **42**, 599 (2000)

L.A. Hiller, L.M. Isaacson, *Experimental Music-Composition with an Electronic Computer* (McGraw–Hill, New York, 1959)

B. Hillier, J. Hanson, *The Social Logic of Space* (Cambridge University Press, Cambridge, 1984). ISBN 0-521-36784-0

B. Hillier, *Space is the Machine: A Configurational Theory of Architecture* (Cambridge University Press, Cambridge, 1999). ISBN 0-521-64528-X

B. Hillier, *The Common Language of Space: A Way of Looking at the Social, Economic and Environmental Functioning of Cities on a Common Basis* (Bartlett School of Graduate Studies, London, 2004)

R.A. Horn, C.R. Johnson, *Matrix Analysis* (Cambridge University Press, Cambridge, 1990)

J.M. Houlrik, I. Webman, M.H. Jensen, Mean-field theory and critical behavior of coupled map lattices. Phys. Rev. A **41**, 4210 (1990)

B.A. Huberman, L.A. Adamic, Growth dynamics of the world wide web. Nature **401**, 131 (1999)

B.D. Hughes, *Random Walks and Random Environments* (Oxford University Press, New York, 1996)

M.E. Hurles, E. Matisoo-Smith, R.D. Gray, D. Penny, Untangling pacific settlement: The edge of the knowable. Trends Ecol. Evol. **18**, 531 (2003)

M.E. Hurles, B.C. Sykes, M.A. Jobling, P. Forster, The dual origins of the Malagasy in Island Southeast Asia and East Africa: Evidence from maternal and paternal lineages. Am. J. Hum. Gen. **76**, 894 (2005)

A. Hyvärinen, J. Karhunen, E. Oja, *Independent Component Analysis* (Wiley, New York, 2001)

F. Jacob, J. Monod, Genetic regulatory mechanisms in th synthesis of proteins. J. Mol. Biol. **3**, 318 (1961)

S. Janson, T. Łuszak, A. Rucinski, *Random Graphs* (Wiley, New York 2000), 333 p

M.R. Jerrum, A. Sinclair, Approximating the permanent. SIAM J. Comput. **18**(6), 1149 (1989)

M. Jiang, Ya.B. Pesin, Equilibrium measures for coupled map lattices. Commun. Math. Phys. **193**, 677 (1998)

B. Jiang, C. Claramunt, Topological analysis of urban street networks. Environ. Plann. B **31**, 151 (2004)

I.T. Jolliffe, *Principal Component Analysis*, vol. XXIX, 2nd edn. Springer Series in Statistics (Springer, New York, 2002)

K. Jones, Compositional applications of stochastic processes. Comput. Music J. **5**(2), (1981)

P.E.T. Jorgensen, E.P.J. Pearse, *Operator theory of electrical resistance networks*, arXiv:0806.3881; (2008)

P.E.T. Jorgensen, E.P.J. Pearse, A Hilbert space approach to effective resistance metric. Complex Anal. Oper. Th. **4**(4), 975–1030 (2009). doi: 10.1007/s11785-009-0041-1

M. Kac, On the notion of recurrence in discrete stochastic processes. Bull. Am. Math. Soc. **53**, 1002 (1947) [Reprinted in M. Kac *Probability, Number Theory, and Statistical Physics: Selected Papers*, K. Baclawski, M.D. Donsker (eds.), Cambridge, Mass: MIT Press, Series: *Mathematicians of our time* Vol. 14, 231 (1979)]

K. Kaneko, Period-doubling of kink-antikink patterns, quasi-periodicity in antiferro-like structures and spatial intermittency in coupled map lattices — toward a prelude to a "field theory of chaos". Prog. Theor. Phys. **72**, 480–486 (1984)

K. Kaneko, Spatiotemporal intermittency in coupled map lattices. Prog. Theor. Phys. **74**, 1033–1044 (1985)

K. Kaneko, Lyapunov analysis and information flow in coupled map lattices. Physica D **23**, 436–447 (1986)

K. Kaneko, Clustering, coding, switching, hierarchical ordering, and control in network of chaotic elements. Physica D **41**, 137–172 (1990)

K. Kaneko (ed.), *Theory and applications of coupled map lattices*, Nonlinear science: theory and applications (Wiley, New York, Chichester, 1993)

K. Karhunen, Zur spektraltheorie stochatischer prozesse. Ann. Acad. Sci. Fenn. **A:1**, 34 (1944)

S. Kauffman, C. Peterson, B. Samuelsson, C. Troein, Random boolean network models and the yeast transcriptional network. Proc. Natl. Acad. Sci. U.S.A. **100**, 14796 (2003)

M. Kayser et al., Melanesian and asian origins of polynesians: mtDNA and Y chromosome gradients across the pacific. Mol. Biol. Evol. **23**, 2234 (2006)

S.M. Keane, *Stock Market Efficiency* (Philip Allan Ltd, Oxford, 1983)

A. Keller, Model genetic circuits encoding autoregulatory transcription factors. J. Theor. Biol. **172**, 169 (1995)

F. Kelly, *Reversibility and stochastic networks* (Wiley, New York, 1979)

B.Kessler, Phonetic comparison algorithms. T. Philol. Soc. **103**(2), 243 (2005)

L. Kim, A. Kyrikou, M. Desbrun, G. Sukhatme, An implicit based haptic rendering technique. *Intelligent Robots and Systems*, **3**, 2943 (2002)

B.M. Kim, J. Rossignac, Localized bi-laplacian solver on a triangle mesh and its applications. GVU Technical Report Number: GIT-GVU-04-12, College of Computing, Georgia Tech. (2004).

P.V. Kirch, *The Lapita Peoples: Ancestors of the Oceanic World* (Blackwell, Cambridge, Mass, 1997)

P.V. Kirch, *On the road of the winds: An archaeological history of the Pacific Islands before European contact* (University of California Press, Berkley, CA, 2000)

D.J. Klein, M. Randić, Resistance distance. J. Math. Chem. **12**(4), 81 (1993)

K. Klemm, S. Bornholdt, Robust gene regulation: Deterministic dynamics from asynchronous networks with delay. Phys. Rev. E **72**, 055101 (2005)

K. Klemm, S. Bornholdt, Topology of biological networks and reliability of information processing. Proc. Natl. Acad. Sci. U.S.A. **102**, 18414 (2005)

V.F. Kolchin, *Random Mappings* (Optimization Software, New York, 1986)

B. Kolman, D.R. Hill, *Elementary Linear Algebra with Applications*, 9th edn. (Prentice Hall, New Jersey, USA, 2007)

B.O. Koopman, Hamiltonian systems and transformations in Hilbert space. Proc. Natl. Acad. Sci. U.S.A. **17**, 315 (1931)

B. Kozma, M.B. Hastings, G. Korniss, Roughness scaling for Edwards-Wilkinson relaxation in small-world networks. Phys. Rev. Lett. **98**(10), 108701 (2004)

K.S. Krell, Gimbutas kurgan-PIE homeland hypothesis: A linguistic critique, in *Archaeology and Language*, vol. II, ed. by R. Blench, M. Spriggs (Routledge, London 1998)

D. Krioukov, F. Papadopoulos, M. Kitsak, A. Vahdat, M. Boguñá, Hyperbolic geometry of complex networks. Phys. Rev. E **82**, 036106 (2010)

D. Krioukov, F. Papadopoulos, A. Vahdat, M. Boguñá, Curvature and temperature of complex networks. Phys. Rev. E **80**, 035101(R) (2009)

J. Lamping, R. Rao, P. Pirolli, A focus+context technique based on hyperbolic geometry for visualizing large hierarchies, *CHI '95 Proceedings of the SIGCHI conference on Human factors in computing systems*, ed. by I.R. Katz, R. Mack, L. Marks, M.B. Rosson, J. Nielsen, (ACM Press/Addison-Wesley Publishing Co. New York, NY, USA, 1995)

J.-L. Lagrange, *Œuvres*, vol. **1**, (Gauthier-Villars, Paris, 1867), pp. 72–79 (in French)

G. Larson et al., Phylogeny and ancient DNA of Sus provides insights into neolithic expansion in Island Southeast Asia and Oceania. Proc. Natl. Acad. Sci. U.S.A. **104**(12), 4834 (2007)

V. Latora, M. Marchiori, Efficient behavior of small-world networks. Phys. Rev. Lett. **87**, 198701 (2001)

V. Latora, M. Marchiori, Vulnerability and protection of infrastructure networks. Phys. Rev. E **71**, 015103(R) (2005)

E.A. Leicht, P. Holme, M.E.J. Newman, Vertex similarity in networks. Phys. Rev. E **73**, 026120 (2006)

A. Lemaître, H. Chaté, Nonperturbative renormalization group for chaotic coupled map lattices. Phys. Rev. Let. **80**, 5528 (1998)

V.I. Levenshtein, Binary codes capable of correcting deletions, insertions and reversals. Sov. Phys. Doklady **10**, 707 (1966)

D.A. Levitt, A representation for musical dialects, in *Machine Models of Music*, ed. by S. Schwanauer, D. Levitt (MIT, Cambridge, Massachusetts, 1993), pp. 456-469

L. Lewin, *Polylogarithms and Associated Functions*, (North Holland, New York, 1981)

P.J. Li, Types of lexical derivation of men's speech in Mayrinax. B. Inst. Hist. Philol. **54**(3), 1 (1983)

W. Li, On the relationship between complexity and entropy for markov chains and regular languages. Com. Sys. **5**, 381 (1991)

P.J. Li, The dispersal of the Formosan Aborigines in Taiwan. Lang. Linguist. **2**(1), 271 (2001)

L. Li, D. Alderson, W. Willinger, J. Doyle, A first principles approach to understanding the internet router-level topology, in *Proceedings of the ACM SIGCOMM'04*, Portland, August 2004

Y. Limoge, J.L. Bocquet, Temperature behavior of tracer diffusion in amorphous materials: A random-walk approach. Phys. Rev. Lett. **65**, 60 (1990)

N. Lin, *The Study of Human Communication* (The Bobbs-Merrill Company, Indianapolis, 1973)

D. Lind, B. Marcus, *An Introduction to Symbolic Dynamics and Coding* (Cambridge University Press, Cambridge 1995)

M. Loève, *Probability Theory* (van Nostrand, New York, 1955)

L. Lovász, Random walks on graphs: A survey, in *Bolyai Society Mathematical Studies*, vol. 2: *Combinatorics, Paul Erdös is Eighty*, (Keszthely, Hungary, 1993), p. 1

L. Lovász, P. Winkler, in *Mixing of Random Walks and Other Diffusions on a Graph*, Surveys in Combinatorics, Stirling. London Mathematical Society Lecture Note Series, vol. 218 (Cambridge University Press, 1995), p. 119

J.K. Lum, L.B. Jorde, W. Schiefenhovel, Affinities among melanesians, micronesians, and polynesians: A neutral, biparental genetic perspective. Hum. Biol. **74**, 413 (2002)

M.C. Mackey, *Time's Arrow: The Origins of Thermodynamic Behavior* (Springer, New York, 1991)

P. Mahadevan, D. Krioukov, K. Fall, A. Vahdat, A basis for systematic analysis of network topologies, in *Proceedings of the ACM SIGCOMM Conference*, Pisa, Italy, September 2006

J.P. Mallory, *In Search of the Indo-Europeans: Language, Archaeology, and Myth* (Thames & Hudson, London, 1991)

D. Mangoubi, *Riemann Surfaces and 3-regular Graphs*, Research MS Thesis, Technion-Israel Institute of Technology, Haifa, 2001

S.C. Manrubia, A.S. Mikhailov, Mutual synchronization and clustering in randomly coupled chaotic dynamical networks. Phys. Rev. E **60**, 1579 (1999)

A.A. Markov, Extension of the limit theorems of probability theory to a sum of variables connected in a chain, reprinted in Appendix B of: R. Howard. *Dynamic Probabilistic Systems*, vol. 1: *Markov Chains*, (Wiley, 1971)

Y. Marom, *Improvising Jazz With Markov Chains*, The report for the Honor Program of the Department of Computer Science, The University of Western Australia, 1997

P.C. Martin, E.D. Siggia, H.A. Rose, Statistical dynamics of classical systems. Phys. Rev. A **8**, 423 (1973)

S. Maslov, K. Sneppen, Specificity and stability in topology of protein networks. Science **296**(5569), 910 (2002)

E. Matisoo-Smith, J.H. Robins, Origins and dispersals of Pacific peoples: Evidence from mtDNA phylogenies of the pacific rat. Proc. Natl. Acad. Sci. U.S.A. **101**(24), 9167 (2004)

H.H. McAdams, L. Shapiro, Circuit simulation of genetic networks. Science **269**, 650 (1995)

K. McCann, A. Hastings, G.R. Huxel, Weak trophic interactions and the balance of nature. Nature **395**, 794 (1998)

J. McLeod, *The History of India* (Greenwood Publishing Group, Westport, CT, 2002)

A. McMahon, R. McMahon, *Language Classification by Numbers* (Oxford University Press, Oxford, UK, 2005)

A. McMahon, P. Heggarty, R. McMahon, N. Slaska, Swadesh sublists and the benefits of borrowing: An andean case study. T. Philol. Soc. **103**(2), 147 (2005)

C.D. Meyer, The role of the group generalized inverse in the theory of finite Markov chains. SIAM Rev. **17**, 443 (1975)

C.D. Meyer, Analysis of finite Markov chains by group inversion techniques. Recent Applications of Generalized Inverses, in *Research Notes in Mathematics*, vol. 66, ed. by S.L. Campbell (Pitman, Boston, 1982), pp. 50–81

C.D. Michener, R.R. Sokal, A quantitative approach to a problem in classification. Evolution **11**, 130 (1957)

H. Minc, *Nonnegative Matrices* (Wiley, New York, 1988). ISBN 0-471-83966-3

A. Möbius, *Der Barycentrische Calcul* (Johann Ambrosius Barth, Leipzig, 1827)

R. Monason, Diffusion, localization and dispersion relations on small-world lattices. Eur. Phys. J. B **12**, 555 (1999)

J.A. Moorer, Music and computer composition, in *Machine Models of Music*, ed. by S. Schwanauer, D. Levitt (MIT, Cambridge, Massachusetts, 1993)

T. Morris, *Computer Vision and Image Processing* (Palgrave Macmillan, Basingstoke, 2004). ISBN 0-333-99451-5

T. Muir, *Treatise on the Theory of Determinants* (revised and enlarged by W. H. Metzler), (Dover, New York, 1960)

J.D. Murray, *Mathematical Biology* (Springer-Verlag, Berlin, 1993)

Mutopia. All music in the Mutopia Project free to download, print out, perform and distribute is available at http://www.mutopiaproject.org, While collecting the data, we have also used the following free resources: http://windy.vis.ne.jp/art/englib/berg.htm (for Alban Berg), http://www.classicalmidi.co.uk/page7.htm, http://www.jacksirulnikoff.com/

N. Nadirashvili, Multiple eigenvalues of laplace operators. Math. USSR Sbornik **61**, 325 (1973)

B.S. Nelson, P. Panangaden, Scaling behavior of interacting quantum fields in curved spacetime. Phys. Rev. D **25**, 1019 (1982)

J. Nerbonne, W. Heeringa, P. Kleiweg, Edit distance and dialect proximity, in *Time Warps, String Edits and Macromolecules: The Theory and Practice of Sequence Comparison*, ed. by D. Sankoff, J. Kruskal (CSLI Press, Stanford, 1999), pp. 5–15

J. Nesetril, E. Milková, H. Nesetrilová, *Otakar Boruvka on Minimum Spanning Tree Problem (translation of the both 1926 papers, comments, history)*, CiteSeer, DMATH: Discrete Mathematics (2000)

M.E.J. Newman, S.H. Strogatz, D.J. Watts, Random graphs with arbitrary degree distribution and their applications. Phys. Rev. E **64**, 026118 (2001)

M.E.J. Newman, Assortative mixing in networks. Phys. Rev. Lett. **89**, 208702 (2002)

M.E.J. Newman, The structure and function of complex networks. SIAM Review **45**, 167 (2003)

M.E.J. Newman, Communities, clustering phase transitions, and hysteresis: Pitfalls in constructing network ensembles. Phys. Rev. E **68**, 026121 (2003)

M.E.J. Newman, Modularity and community structure in networks. Proc. Natl. Acad. Sci. U.S.A. **103**, 8577 (2006)

J. Nichols, T. Warnow, Tutorial on computational linguistic phylogeny. Lang. Linguist. Compass **2**(5), 760 (2008)

J. Nieminen, On centrality in a graph. Scand. J. Psychol. **15**, 322 (1974)

F. Ninio, A simple proof of the Perron-Frobenius theorem for positive symmetric matrices. J. Phys. A-Math. Gen. **9**(8), 1281 (1976)

H. Noguchi, Mozart: Musical game in C K.516f http://www.asahi-net.or.jp/~rb5h-ngc/e/k516f.htm. Accessed 1996

P. Novotná, V. Blažek, Glottochronolgy and its application to the Balto-Slavic languages. Baltistica **XLII**(2), 185 (2007)

E. Nummelin, *General irreducible Markov chains and non-negative operators* (Cambridge University Press, Cambridge, 2004). ISBN 0-521-60494-X

R.T. Paine, Food-web analysis through field measurement of per capita interaction strength. Nature **355**, 73 (1992)

L. Parker, in *Recent Developments in Gravitation: Cargese Lectures 1978*, ed. by M. Levy, S.Deser (Plenum Press, New York, 1979)

L.Parker, D.J. Toms, New form for the coincidence limit of the Feynman propagator, or heat kernel, in curved spacetime. Phys. Rev. D **31**, 953 (1985)

L.Parker, D.J. Toms, Explicit curvature dependence of coupling constants. Phys. Rev. D **31**, 2424 (1985)

R. Pastor-Satorras, A. Vespignani, Epidemic spreading in scale-free networks. Phys. Rev. Lett. **86**(14), 3200 (2001)

R. Pastor-Satorras, A. Vespignani, Epidemic dynamics in finite size scale-free networks. Phys. Rev. E **65**, 035108 (2002)

A. Penn, Space syntax and spatial cognition. or, why the axial line? in *Proceedings of the Space Syntax 3rd International Symposium*, ed. by J. Peponis, J. Wineman, S. Bafna (Georgia Institute of Technology, Atlanta, 2001), pp. 11.1–11.17

R. Penrose, A generalized inverse for matrices. Proc. Cambridge Philos. Soc. **51**, 406 (1955)

Perl MIDI. The software that allows to read, compose, modify, and write MIDI files is freely available at the web-page http://search.cpan.org/~sburke/MIDI-Perl-0.8

F. Petroni, M. Serva, Language distance and tree reconstruction. J. Stat. Mech. **2008**, P08012 (2008)

E. Prisner, *Graph Dynamics* (CRC Press, Boca Raton, FL, 1995)

M. Ptashne, A. Gann, *Genes and Signals* (Cold Spring Harbour Laboratory Press, Cold SpringHarbour, NY, 2002)

J.G. Ratcliffe, *Foundations of hyperbolic manifolds*, Springer series: Graduate Texts in Mathematics (Springer-Verlag, New York, London, 1994)

M. Ravallion, Urban poverty. Financ. Dev. **44**(3), 141–167 (2007)

Native Instruments Software Synthesis GmbH, *Reaktor 5.1* [computer software] (Berlin, 2005)

Propellerhead Software, *Reason 4* [computer software] (Stockholm, 2007)

C. Renfrew, *Archaeology and Language: The Puzzle of Indo-European Origins* (Cambridge University Press, New York, 1987)

C. Renfrew, Time depth, convergence theory, and innovation in Proto-Indo-European, in *Proceedings of the Conference Languages in Prehistoric Europe*, Eichstätt University, 4–6 October 1999, Heidelberg, published in 2003, p. 227

K. Rho, H. Jeong, B. Kahng, Identification of lethal cluster of genes in the yeast transcription network. Physica A **364**, 557 (2006)

P. Robert, On the group inverse of a linear transformation. J. Math. Anal. Appl. **22**, 658 (1968)

S. Roman, *Advanced Linear Algebra*, 2nd edn., in Springer series: Graduate Texts in Mathematics, (Springer, New York, 2005)

D. Ruelle, *Thermodynamic Formalism*, in 5 of Encyclopedia of Mathematics and its Applications (Addison-Wesley, Reading, Mass, 1978)

L.G. Sabidussi, The centrality index of a graph. Psychometrica **31**, 581 (1966)

V.N. Sachkov, *Combinatorial Methods in Discrete Mathematics*, Encyclopedia of Mathematics and its Applications (Cambridge University Press, Cambridge, 1996)

S. Saitoh, *Theory of Reproducing Kernels and its Applications* (Longman Scientific and Technical, Harlow, UK, 1988)

N. Saitou, N. Masatoshi, The neighborhood joining method: A new method of constructing phylogenetic trees. Mol. Biol. Evol. **4**, 406 (1987)

K. Salisburym D. Brock, T. Massie, N. Swarup, C. Zilles, Haptic rendering: Programming touch interaction with virtual objects, in *Proceedings of the 1995 Symposium on Interactive 3D Graphics* (ACM, New York, 1995), p. 123

L. Saloff-Coste, *Lectures on Finite Markov Chains*, Ecole d'Été, Saint-Flour, Lecture Notes in Mathematics, vol. 1664 (Springer, Berlin, 1997), pp. 301–413

R. Shaw, *The Dripping Faucet as a Model Chaotic System* (CA Aerial Press, Santa Cruz, 1984)

R. Schneider, L. Kobbelt, Generating fair meshes with G1 boundary conditions. Comput. Aided Geom. D. **4**(18), 159 (2001)

B. Schölkopf, A.J. Smola, K.-R. Müller, Nonlinear component analysis as a kernel eigenvalue problem. Neur. Comput. **10**, 1299 (1998)

A. Schrijver, *Combinatorial Optimization: Polyhedra and Efficiency* (Springer, Berlin, 2002)

T. Schürmann, P. Grassberger, Entropy estimation of symbol sequences. Chaos **6**(3), 414 (1996)

Ch. Seeger, Reflections upon a given topic: Music in universal perspective. Ethnomusicology **15**(3), 385 (1971)

D. Sergi, Random graph model with power-law distributed triangle subgraphs. Phys. Rev. E **72**, 025103 (2005)

M. Serva, F. Petroni, Indo-European languages tree by Levenshtein distance. Europhys. Lett. **81**, 68005 (2008)

M.F. Shlesinger, First encounters. Nature **450**(1), 40 (2007)

R. Sedgewick, Permutation generation methods. Comput. Surv. **9**, 137 (1977)

D. Volchenkov, S. Sequeira, Ph. Blanchard, M.G. Cosenza, Transitions to intermittency and collective behavior in randomly coupled map networks. Stoch. Dynam. **2**(2), 203 (2002)

B. Sévennec, *Multiplicité du spectre des surfaces: Une approche topologique* (Preprint ENS, Lyon, 1994)

Ya.G. Sinai, Gibbs measures in ergodic theory. Usp. Mat. Nauk **27**(4), 21-64 (1972) (in Russian); MR, **53**, # 3265; Russ. Math. Surv. **27**(4), 21-69 (1972) (in English)

Ya.G. Sinai, A.B. Soshnikov, A refinement of Wigner's semicircle law in a neighborhood of the spectrum edge for random symmetric matrices. Funct. Anal. Appl. **32**(2), 114 (1998)

S. Skiena, Implementing Discrete Mathematics: Combinatorics and Graph Theory with Mathematica. Reading, MA: Addison-Wesley, (1990)

C.E. Shannon, A mathematical theory of communication. Bell Syst. Tech. J. **27**, 379; 623 (1948)

C.E. Shannon, Prediction and entropy of printed english. Bell Syst. Tech. J. **30**, 50 (1951)

S. Shaw, Evidence of scale-free topology and dynamics in gene regulatory networks, in *Proceedings of the ISCA 12th International Conference on Intelligent and Adaptive Systems and Software Engineering*, ed. by A. Satyadas and S. Dascalu, (San Francisco, California, USA) 2003, p. 37

J. Shi, J. Malik, Normalized cuts and image segmentation. IEEE Trans. PAMI **22**(8), 888–905 (2000)

M. Shlesinger, G.M. Zaslavsky, U. Frisch (eds.), *Lévy Flights and Related Topics in Physics* (Springer-Verlag, New York, 1995)

T.S. Shores, *Applied Linear Algebra and Matrix Analysis*, in Springer series: Undergraduate Texts in Mathematics (Springer, 2006)

G.E. Shilov, B.L. Gurevich, *Integral, Measure, and Derivative: A Unified Approach*, Richard A. Silverman (trans. from Russian) (Dover Publications, New York, 1978)

A. Smola, R.I. Kondor, Kernels and regularization on graphs, in *Learning Theory and Kernel Machines*, ed. by B. Scholkopf, M.K. Warmuth (Springer, Berlin, New York, 2003), pp. 144–158

P. Smolen, D.A. Baxter, J.H. Byrne, Frequency selectivity, multistability and oscillations emerge from models of genetic regulatory systems. Am. J. Physiol.–Cell Ph. **43**, C531 (1998)

E.H. Snoussi, R. Thomas, Logical identification of all steady states: The concept of feedback loop caracteristic states. Bull. Math. Biol. **55**(5), 973 (1993)

D. Stassinopoulos, P. Alstrøm, Coupled maps: An approach to spatiotemporal chaos. Phys. Rev. A **45**, 675 (1992)

B. Su, L. Jin, P. Underhill, J. Martinson, N. Saha, S.T. McGarveyi, M.D. Shriver, J. Chu, P. Oefner, R. Chakraborty, R. Deka, Polynesian origins: Insights from the Y chromosome. Proc. Natl. Acad. Sci. U.S.A. **97**(15), 8225 (2000)

M. Swadesh, Lexico-statistic dating of prehistoric ethnic contacts. Proc. Natl. Acad. Sci. U.S.A. **96**, 452 (1952)

K. Symanzik, Schrödinger representation and casimir effect in renormalizable quantum field theory. Nucl. Phys. B **190**, 1 (1981)

B. Tadic, Exploring complex graphs by random walks. in *Modeling of complex systems: Seventh Granada Lectures*, ed. by P. Garrido, J. Marro, Granada, Spain, AIP Conference Proceedings, vol. 661 (American Institute of Physics, Melville, 2002), pp. 24–26

H. Tangmunarunkit, J. Doyle, R. Govindan, S. Jamin, W. Willinger, S. Shenker, Does AS size determine AS degree? in *ACM SIGCOMM Computer Communication Review* (ACM, New York, 2001)

E.V. Teodorovich, The renormalization group method in the problem of transport in the presence of nonlinear sources and sinks. J. Eksp. Theor. Phys. (Sov. JETP) **115**, 1497 (1999) [*JETP* **88**, 826 (1999)]

P. Tetali, Random walks and the effective resistance of networks. J. Theor. Probab. **4**(1), 101 (1991)

D. Thieffry, D. Romero, The modularity of biological regulatory networks. BioSystems **50**, 49 (1999)

D. Thieffry, E.H. Snoussi, J. Richelle, R. Thomas, Positive loops and differentiation. J. Biol. Syst. **3**(2), 457 (1995)

D. Thieffry, *Qualitative Analysis of Gene Networks* in the Memoire pour l'obtention d'Agrege de l'Eseignement Superieur, Universite Libre de Bruxelles (2000)

R. Thomas, On the relation between the logical structure of systems and their ability to generate multiple steady states or sustained oscillations. Springer Series Syne. **9**, 180 (1981)

R. Thomas, The role of feedback circuits] positive feedback circuits are a necessary condition for positive real eigenvalues in the jacobian matrix. Ber. Brunzen. Phys. Chem. **98**, 1148 (1994)

R. Thomas, M. Kaufman, Logical analysis of regulatory networks in terms of feedback circuits. Chaos **11**(1), 180 (2001)

W. Thomson, *Tonality in Music: A General Theory* (Everett Books, San Marino, CA, 1999)

D.J. Toms, Renormalization of interacting scalar field theories in curved space-time. Phys. Rev. D **26**, 2713 (1982)

W.T. Trotter, *Combinatorics and Partially Ordered Sets* (The Johns Hopkins University Press, Baltimore, MD, 2001)

W.T. Tutte, A contribution to the theory of chromatic polynomials. Canadian J. Math. **6**, 80 (1954)

W.T. Tutte, *Graph Theory* (Cambridge University Press, Cambridge, 2001)

Population Division of the Department of Economic and Social Affairs of the United Nations Secretariat, *World population prospects: The 2006 revision*, Dataset on CD-ROM. (United Nations, New York, 2007)

US Bureau of Census Data of Urbanized Areas, available at http://www.sprawlcity.org/

N. Utsurikawa, *A Genealogical and Classificatory Study of the Formosan Native Tribes* (Toko shoin, Tokyo, 1935)

L. Vaughan, The relationship between physical segregation and social marginalisation in the urban environment. World Archit. **185**, 88 (2005)

L. Vaughan, D. Chatford, O. Sahbaz, Space and exclusion: The relationship between physical segregation. Economic marginalization and povetry in the city. Paper presented to Fifth International Space Syntax Symposium, Delft, Holland, 2005

M.U. Vera, D.J. Durian, The angular distribution of diffusely transmitted light. Phys. Rev. E **53**, 3215 (1996)

D. Volchenkov, L. Volchenkova, Ph. Blanchard, Epidemic spreading in a variety of scale free networks. Phys. Rev. E **66**(4), 046137 (2002); Virtual J. Biol. Phys. Res. **4**(9), (2002)

D. Volchenkov, Ph. Blanchard, An algorithm generating scale free graphs. Physica A **315**, 677 (2002)

D.Volchenkov, R. Lima, Random shuffling of switching parameters in a model of gene expression regulatory network. Stoch. Dynam. **5**(1), 75–95 (2005)

D. Volchenkov, Ph. Blanchard, Nonlinear diffusion through large complex networks with regular subgraphs. J. Stat. Phys. **127**(4), 677 (2007)

D. Volchenkov, Ph. Blanchard, Random walks along the streets and channels in compact cities: Spectral analysis, dynamical modularity, information, and statistical mechanics. Phys. Rev. E **75**, 026104 (2007)

D. Volchenkov, Ph. Blanchard, Scaling and universality in city space syntax: Between Zipf and Matthew. Physica A **387**(10), 2353 (2008)

D. Volchenkov, R. Lima, Asymptotic series in dynamics of fluid flows: Diffusion versus bifurcations. Commun. Nonlin. Sci. Num. Simulat. **13**, 1329 (2008)

D. Volchenkov, Renormalization group and instantons in stochastic nonlinear dynamics. Eur. Phys. J. Spec. Top. **170**(1), 1–142 (2009)

D. Volchenkov, Random walks and flights over connected graphs and complex networks. Commun. Nonlin. Sci. Num. Simul. (2010), http://dx.doi.org/10.1016/j.cnsns.2010.02.016

P.H. von Hippel, Integrated model of the transcription complex in elongation, termination, and editing. Science **281**, 660 (1998)

U. von Luzburg, M. Belkin. O. Bousquet, Consistency of Spectral Clustering. Technical Report Number TR-134, Max-Planck-Institut fuer biologische Kybernetik, 2004

I. Vragovic, E. Louis, A. Diaz-Guilera, Efficiency of informational transfer in regular and complex networks. Phys. Rev. E **71**, 036122 (2005)

G. Wahba, *Spline Models for Observational Data*, CBMS-NSF Regional Conference Series in Applied Mathematics, Vol. 59 (SIAM, Philadelphia, 1990)

A. Wagner, Evolution of gene networks by gene duplications: A mathematical model and its implications. Proc. Natl. Acad. Sci. U.S.A. **91**, 4387–4391 (1994)

A. Wagner, Inferring lifestyle from gene expression patterns. Mol. Biol. Evol. **17**, 1985–1987 (2000)

A. Wagner, How large protein interaction networks evolve. Proc. R. Soc. Lond. B **270**, 457–466 (2003)

A.M. Walczaka, M. Sasaic, P.G. Wolynes, Self-consistent proteomic field theory of stochastic gene switches. Biophys. J. **88**, 828–850 (2005)

W.S.-Y. Wang, J.W. Minett, Vertical and horizontal transmission in language evolution. Trans. Philol. Soc. **103**(2), 121 (2005)

T. Warnow, S.N. Evans, D.A. Ringe Jr., L. Nakhleh, A stochastic model of language evolution that incorporates homoplasy and borrowing, in *Phylogenetic Methods and the Prehistory of Languages*, ed. by P. Forster, C. Renfrew (McDonald Institute for Archaeological Research, Cambridge, 2006), p. 75

S. Wasserman, K. Faust, *Social Network Analysis* (Cambridge University Press, Cambridge, England, 1994)

D.J. Watts, S.H. Strogatz, Collective dynamics of 'small-world' networks. Nature **393**(6684), 440 (1998)

S. Wichmann, A. Müller, V. Velupillai, Homelands of the world's language families. Diachronica **27**(2), 247 (2010)

H. Wiener, Structural determination of paraffin boiling points. J. Am. Chem. Soc. **69**, 17 (1947)

A.G. Wilson, *Entropy in Urban and Regional Modeling* (Pion Press, London, 1970)

H. Whitney, A logical expansion in mathematics. Bull. Am. Math. Soc. **38**, 572 (1932)

L. Wirth, *The Ghetto* (edition 1988), Studies in Ethnicity, (Transaction Publishers, New Brunswick, USA, London, UK, 1928)

D.M. Wolf, F.H. Eeckman, On the relationship between genomic regulatory element organization and gene regulatory dynamics. J. Theor. Biol. **195**, 167 (1998)

J. Wolfe, Speech and music, acoustics and coding, and what music might be 'for', in *Proceedings of the 7th International Conference on Music Perception and Cognition, Sydney*, ed. by C. Stevens, D. Burnham, G. McPherson, E. Schubert, J. Renwick (Causal Productions, Adelaide, 2002)

F.Y. Wu, Theory of resistor networks: The two-point resistance. J. Phys. A: Math. Gen. **37**, 6653 (2004)

B.Y. Wu, K.-M. Chao, *Spanning Trees and Optimization Problems* (CRC Press, Boca Raton, 2004)

I. Xenakis, *Formalized Music* (Indiana University Press, Bloomington, 1971)

W. Xiao, I. Gutman, Resistance distance and Laplacian spectrum. Theor. Chem. Acc. **110**, 284 (2003)

W. Xiao, I. Gutman, On resistance matrices. MATCH: Commun. Math. Co. **49**, 67 (2003)

W. Xiao, I. Gutman, Relations between resistance and Laplacian matrices and their applications. MATCH Commun. Math. Co. **51**, 119 (2004)

S.-J. Yang, Exploring complex networks by walking on them. Phys. Rev. E **71**, 016107, (2005)

S.X. Yu, J. Shi, Multiclass spectral clustering, in *Proceedings of International Conference on Computer Vision* (IEEE Computer Society, Washington, DC, USA, 2003), pp. 313–319

D.H. Zanette, P.A. Alemany, Thermodynamics of anomalous diffusion. Phys. Rev. Lett. **75**, 366 (1995)

H. Zha, C. Ding, M. Gu, X. He, H. Simon, *Neural Information Processing Systems* (NIPS 2001), vol. 14 (Vancouver, Canada, 2001)

H.-H. Zhang, W.-B. Yan, X.-S. Li Trace formulae of characteristic polynomial and Cayley-Hamilton's theorem, and applications to chiral perturbation theory and general relativity. Commun. Theor. Phys. **49**, 801 (2008)

D. Zicarelli, M and jam factory. Comp. Music J. **11**(4), 1329 (1987)

Glossary of Graph Theory

A

- *acyclic*: a directed graph that does not contain any directed cycle.
- *adjacent vertices*: two vertices joined by an edge or an arc.
- *anti-edge*: $u, v \in G$ forms an anti-edge whenever neither $u \nsim v$, nor $v \nsim u$.
- *arborescence*: is an oriented tree in which all vertices are reachable from a single vertex.
- *arc*: in a directed graph, a segment (link) which joins two consecutive vertices whether they are distinct or not.
- *automorphism* of a graph: a form of symmetry in which the graph is mapped onto itself while preserving the edge-vertex connectivity.

B

- *bipartite* graph: a graph whose vertices can be divided into two disjoint sets U and V such that every edge connects a vertex in U to one in V.
- *biregular* graph: one that has unequal maximum and minimum degrees and every vertex has one of those two degrees.
- *bridge*: an edge or arc that joins two disconnected parts of a graph.
- *branching*: see *arborescence*

C

- *center* of a graph: vertices of minimum eccentricity.
- *chain*: in an undirected graph, a series of successive edges forming a continuous curve passing from one vertex to another.
- *chromatic number* of a graph: the minimum number of different colors needed to color all of the vertices of a graph without any two adjacent vertices having the same color.
- *circuit*: in a directed graph, a path that begins and ends at the same vertex.
- *circulant* graph: a graph of N vertices in which the i-th vertex is adjacent to the $(i + j)$-th and $(i - j)$-th graph vertices for each j in the list of its nodes.
- *circumference*: the length of a longest (simple) cycle.
- *claw*: an induced star with 3 edges.
- *clique* in a graph: a set of pairwise adjacent vertices.
- *closed* walk: in which its first and last vertices are the same.

- *complement*: the graph that must be added to a graph to make a complete graph.
- *complete* graph: a graph in which every vertex is joined to every other vertex by exactly one edge.
- *connected* graph: a graph in which any one vertex can be linked directly or indirectly to any other vertex in the graph.
- *connected component*: a maximal connected subgraph of the graph.
- *connectivity* of a vertex: the number of edges connecting to a vertex.
- *cospectral* graphs: graphs that share the same graph spectrum.
- *critical path*: the longest path in a directed graph.
- *cubic* graph: a 3-regular graph.
- *cut* of a connected graph: is a set of vertices whose removal renders it disconnected.
- *cycle*: a path that begins and ends at the same vertex.

D

- *degree* of a vertex: the number of neighbors the vertex has in the graph.
- *degree sequence*: the list of degrees of a graph in non-increasing order.
- *dense* graph: a graph in which the number of edges is close to the maximal number of edges.
- *diameter* of a graph: the maximum *eccentricity* over all vertices in that.
- *dicircuit*: the pair of alternatively directed edges connecting two vertices.
- *directed* graph: a graph made up of a set of vertices and a set of directed edges (with arrows).
- *distance* between two (not necessary distinct) vertices: the length of a shortest path between them.
- *domination number*: the minimum size of a *dominating set*.
- *dominating set* of a graph: a vertex subset whose closed neighborhood includes all vertices of the graph.

E

- *eccentricity* of a vertex: the maximum distance from it to any other vertex.
- *edge*: a set of two basic elements of a graph (vertices); it is drawn as a line connecting two vertices.
- *edge-connectivity*: the size of a smallest edge cut.
- *edge cut*: a set of edges whose removal renders the graph disconnected.
- *edge-labeled* graph: one with labeled edges only.
- *empty graph*: a graph with zero or more vertices, but no edges.
- *Euler chain*: a chain that contains all edges of an undirected graph exactly once.
- *Euler circuit*: in a directed graph, a circuit that contains all the arcs of the graph.
- *Euler cycle*: a cycle that contains all the edges of an undirected graph exactly once.
- *Euler path*: a path that contains all the arcs of a directed graph exactly once.
- *even vertex*: vertex having an even degree.
- *even cycle*: a cycle that has even length.
- *extremal graph*: the largest graph of order N which does not contain a given graph of the same order as a subgraph.

F

- *factor*: see the *spanning subgraph*.
- *finite* graph: one that has finite number of vertices and edges.
- *forest*: a graph that does not contain any circuit or cycle.

G

- *girth* of a graph: the length of a shortest (simple) cycle in the graph.
- *graph*: a structure which consists of two types of elements – vertices and edges, such that edges form a subset of the Cartesian product of vertices.

H

- *Hamiltonian connected graph*: one that contains a Hamiltonian path for any given pair of (distinct) end vertices.
- *Hamilton chain*: a chain that passes through each vertex of an undirected graph exactly once.
- *Hamilton circuit*: in a directed graph, a circuit that passes through all the vertices of the graph.
- *Hamilton cycle*: a cycle that passes through each vertex of an undirected graph exactly once.
- *Hamilton path*: a path that passes through all the vertices of a directed graph exactly once.
- *head*: the terminal vertex of a directed edge.
- *homomorphic* graphs: such that if two vertices are adjacent in one of them then their corresponding vertices are also adjacent in another.

I

- *in-degree*: the number of edges entering a vertex in a directed graph.
- *infinite* graph: one that has infinitely many vertices or edges, or both.
- *incident edge*: an edge which is connected to a vertex.
- *induced subgraph*: a graph which has all the edges that appear in the host graph over the same vertex set.
- *independent paths*: paths that have any vertex in common, except the first and last ones.
- *internal vertex*: a non-leaf vertex.
- *isolated* vertex: a vertex of degree zero.
- *isomorphic* graphs: which allow a one-to-one correspondence between them.
- *isospectral* graphs: see the *cospectral graphs*.

K

- *k-ary tree*: a rooted tree in which every internal vertex has k children. A 1-ary tree is just a *path*. A 2-ary tree is also called a binary tree.
- *k-factor*: a k-regular spanning subgraph.
- *knot* in a directed graph: a collection of vertices and edges with the property that every vertex in the knot has outgoing edges, and all outgoing edges from vertices in the knot terminate at other vertices in the knot.
- *k-regular* graph: one, in which every vertex has degree k.

L

- *labeling*: the assignment of natural numbers to the edges and vertices of a graph.
- *leaf*: a vertex of degree 1.

- *leaf edge*: an edge incident to a leaf.
- *length* of a walk: the number of edges that it uses.
- *locally finite* graph: one where every vertex has finite degree.
- *loop*: an edge whose end vertices are the same vertex.
- *looped graph*: one that contains a loop at each vertex.

M

- *maximum (minimum) degree* of a graph: the largest (smallest) degree over all vertices of the graph.
- *minor* of a graph: one resulted from the given graph via repeated edge deletion and/or edge contraction.
- *mixed graph*: one that contains both directed and undirected edges.
- *multiplicity of an edge*: the number of multiple edges sharing the same end vertices.
- *multiplicity of a graph*: the maximum multiplicity of its edges.
- *multi-graph*: one that has multiple edges, but no loops

N

- *null graph*: one with no vertices and no edges.

O

- *odd vertex*: vertex having an odd degree.
- *odd cycle*: a cycle that has odd length.
- *open walk*: one that has its first and last vertices different.
- *order*: the number of vertices in a graph.
- *out-degree*: the number of edges leaving a vertex in a directed graph.

P

- *pancyclic graph*: one that contains cycles of every possible length (from 3 to the order of the graph).
- *path in a directed graph*: a route among vertices along one graphs edges such that no edge is used more than once.
- *path in an undirected graph*: see the *open walk*.
- *parallel edges*: more than one edge that connects two vertices.
- *peripheral vertices*: vertices with maximum eccentricity.
- *planar graph*: a graph which can be embedded in the plane.
- *pseudograph*: one that contains both multiple edges and loops.

R

- *random graph*: a graph that is generated by some random process.
- *radius of a graph*: the minimum eccentricity over all vertices in that.
- *regular graph*: one, in which every vertex has the same degree.
- *root*: a distinguished vertex of the tree.
- *rooted tree*: a tree with a root.

S

- *schlicht graph* : see *simple graph*
- *simple graph*: a finite, undirected graph without multiple edges or loops.
- *sink*: a vertex with 0 *out-degree*.
- *size*: the number of edges in a graph.

- *source*: a vertex with 0 *in-degree*
- *spanning subgraph*: a graph which has the same vertex set as the given one.
- *spanning tree*: a tree inside a connected graph that includes every vertex of the original graph.
- *sparse graph*: one with only a few edges.
- *spectrum of a graph*: the set of eigenvalues of either the adjacency matrix of the graph, or the Laplace operator matrix defined on that.
- *star*: a complete bipartite graph where the first vertex set consists of the only vertex (see *bipartite* graph).
- *strongly connected component* of a directed graph: a subgraph where all nodes in the subgraph are reachable by all other nodes in the subgraph.
- *strongly regular graph*: a regular graph such that any adjacent vertices have the same number of common neighbors as other adjacent pairs and that any nonadjacent vertices have the same number of common neighbors as other nonadjacent pairs.
- *subdivision* of a graph: results from inserting vertices into edges.
- *subgraph*: a graph whose vertex set is a subset of that of the host graph, and whose adjacency relation is a subset of that of the host graph restricted to this subset.
- *subtree*: a connected subgraph of a tree.
- *supergraph*: a graph of which the given graph is a subgraph.

T

- *tail*: the initial vertex of a directed edge.
- *total degree* of a graph: two times the number of edges, loops included.
- *tour*: see the *circuit*.
- *trail*: a walk in which all the edges are distinct.
- *traceable graph*: one that contains a Hamiltonian path.
- *traversable graph*: one that contains an Eulerian path.
- *tree*: a connected, undirected graph that does not contain any circuits.

U

- *unicyclic graph*: one that contains exactly one cycle.
- *universal graph* of a class $\mathcal{K}$: a simple graph in which every other graph in $\mathcal{K}$ can be embedded as a subgraph.

V

- *vertex*: a basic element of a graph.
- *vertex connectivity*: the size of a smallest *vertex cut*.
- *vertex cut*: see the *cut*.
- *vertex-labeled graph*: one with labeled vertices only.

W

- *walk*: an alternating sequence of vertices and edges, beginning and ending with a vertex, in which each vertex is incident to the two edges that precede and follow it in the sequence, and the vertices that precede and follow an edge are the end vertices of that edge.

- *weighted graph*: a graph in which each edge is assigned a real number that may be positive or zero.
- *Wiener index of a vertex*: the sum of distances between it and all other vertices.
- *Wiener index of a graph*: is the sum of distances over all pairs of vertices.

Z

- *zweieck* of an undirected edge: the pair of directed edges which form the simple directed dicircuit.

Index